The Earth Mother
and the Assault of Capitalism

ALSO BY JULIAN E. KUNNIE
AND FROM MCFARLAND

*The Cost of Globalization: Dangers
to the Earth and Its People* (2015)

The Earth Mother and the Assault of Capitalism

Living Sustainably with All Life

Julian E. Kunnie

McFarland & Company, Inc., Publishers
Jefferson, North Carolina

LIBRARY OF CONGRESS CATALOGING-IN-PUBLICATION DATA

Names: Kunnie, Julian E., 1958– author
Title: The Earth Mother and the assault of capitalism : living sustainably with all life / Julian E. Kunnie.
Description: Jefferson, North Carolina : McFarland & Company, Inc., Publishers, 2025. | Includes bibliographical references and index.
Identifiers: LCCN 2025019727 | ISBN 9781476683904 paperback ∞
 ISBN 9781476653846 ebook
Subjects: LCSH: Global environmental change—Economic aspects | Sustainable living—Economic aspects | Capitalism—Environmental aspects | Traditional ecological knowledge | BISAC: SCIENCE / Environmental Science (see also Chemistry / Environmental)
Classification: LCC GE149 .K86 2025
LC record available at https://lccn.loc.gov/2025019727

ISBN (print) 978-1-4766-8390-4
ISBN (ebook) 978-1-4766-5384-6

© 2025 Julian E. Kunnie. All rights reserved

No part of this book may be reproduced or transmitted in any form or by any means, electronic or mechanical, including photocopying or recording, or by any information storage and retrieval system, without permission in writing from the publisher.

Front cover image: photograph © Artem Oleshko/max dallocco/Shutterstock; layout based on a design by Veronica Rodriguez.

Printed in the United States of America

McFarland & Company, Inc., Publishers
Box 611, Jefferson, North Carolina 28640
www.mcfarlandpub.com

To our Ever-Living and Ever-Powerful Earth Mother and the Endless Spiritual Universe and all those Seeking Life in Wholeness, especially to the Women and the Children of all Creatures, particularly Indigenous people who Suffer Ethnocide, Genocide, Ecocide, and Violence, with over 16,000 in Gaza in 2023 alone, the Congo, Haiti, and elsewhere, and the Eternal Ancestors of the Indigenous Peoples of Turtle Island and Mother Africa

Table of Contents

Acknowledgments ix
Abbreviations xv
Preface 1

1. Indigenous Peoples' Creation Stories of the Earth and the Natural World 7
2. An Interactive History of Europe, Colonialism, Slavery, and the Emergence of Capitalism 26
3. Oil Production and Its Capitalist History: Ecocide, Earth Heating, and the Poisoning of Life 55
4. Chemical Geoengineering of the Weather: Corporate and Militarist Intervention and Effective Climate Collapse 97
5. Capitalism's Persistent Devaluing of Public Education and the Need for Restructuring and Reorienting Education 130

Epilogue: Whither Capitalism? The Earth Mother Prevails for Life 155
Chapter Notes 177
Bibliography 195
Index 223

Acknowledgments

This book is a text of constant giving thanks and signifies a thorough prayerful collective, organic, and solidarity effort, and thus extended thanks and gratitude needs to be accorded first and foremost to *Nahazdzáán, Pacha Mama, Umama Hlabathi, Diqui Muxin, Madzitateguru*, our Earth Mother and all of the Spirits of this sacred land, Turtle Island, *Abya Yala*, and of our *Mama Afrika*. As everyone knows, our Mother's support, protection, and approval are the basis of our existence because out of our Mother's Womb, we all have come to this present moment, still breathing invisible but indispensable Air, Spirit, *uMoya*, who permeates all life. We are so thankful that we are here, and regardless of past folly and selfish and contradictory deeds at whichever juncture, we can still pen these words even in the tongues of various relatives in this collective discursive and land-back struggle. *Ningadinwa Nangamso* and thank you so very much to the beautiful Earth Mother creatures who light up our daily journey each morning as we meditate and offer thanks in the desert, especially the hummingbirds and others who chirp and sing; lizards; ants; and tiny, barely visible bugs who remind us constantly that *Earth Mother is always alive and that we are all meant to live as One*.

Ahééhee' to our ever wise and gentle Teacher—the real *Professor*, Healer, Dancer, and World Champion Hoop Dance teacher—Hataali Jones Benally, and the Benally family, especially Berta and Jeneda, for your generosity, wisdom, and support through challenging times. To the broad Kunnie clan in so many different places, especially partner Kim for her editing and wonders Mandla and Sibu. To the Mathys clan, a special thank-you for support and persistence in this struggle. *Siyabonga kakhulu, Siyabamba ngazibini, Abahlali BaseMjondolo*, the *Landless Peoples Movement*, especially in Kwazulu, Azania, for your tireless defense and persistence of land repossession even in the face of deadly state violence against landless people, especially Sibu, Mazwi, and Thabo, who shared so much with us during our visit and during our presence at the state trial charging activists for restoring lands for the landless and homeless.

To Phoebe Farris, Powhatan-Pamunkey, long-time Indigenous art scholar, activist, and mentor, defending the rights of Indigenous people, *Baskonee,* for helping in so many ways to keep us on track so that we didn't steer off the path in a massive undertaking quite a few years until all was complete. To Blackfeet elder Betty Cooper, with whom we worked on Indigenous and Black familial support networks in Oakland, California, in the 1980s, and her daughter, Theda Wind Breast, *Nitsiniiyi'taki,* for sharing your ancestral teaching wisdom at the AIM west conference in San Francisco honoring the 50th anniversary of the Indigenous Occupation of Alcatraz in November 2019, some of which is shared here.

Acknowledgments

To Grandmother Isobel John, 94, from the Dene community in Tanacross, Alaska, who interviewed with us in 2016 during our historic 4,000-mile road trip from Arizona to Alaska via the Summer Canoe ceremony in Nisqually, Washington, and various Indigenous nation lands in Alberta and so-called British Columbia, including the Mikisew Cree First Nation reeling from the ongoing tar sands oil extraction at Fort McMurray, oil drilling at Athabasca, the Kwanlin Dun First Nation who hosted us in White Horse, and the Blackfeet First nation in Sturgeon Lake through which we passed, close to Montana, *Ahééhee' Sógá sénlá* to Brother George Donnessey and to Great-Grandmother Mida, renowned traditional Kaska elder, teacher, and healer, who passed to the Spirit World in December 2018 at 90, for your eagle feather gift of protection during our stop at Watson Lake in 2016. To Grandmother Margaret Nakak and grandson Kelvin, from the Bering Sea Yupiak village in Alaska, *Koyana*, for sharing your traditional wisdom and ancient food knowledge on video during the historic 25th anniversary of the *We the Peoples Before Conference* at the Kennedy Center, Washington D.C, in June and July 2022, where we did photography for *Cultural Survival*.

Heai Watapi to Senior Healer Dennis Khoibur, from West Papua, for your love, generosity, prayer mediums, and your work in training young people to carve in wood in honor of the Ancestors. And to Ezra, Bata, Socrates, Robert, Sister Finiranga, and all our relatives there … *Heai Watapi* … West Papua will be free from occupation!

To the AIM grandmothers Darlene, Debra, Sandy, and Beverly, at Wounded Knee, Pine Ridge Lakota Nation, for the beautiful sage bundles and the medicine pouch, and to Brother artist, Thurman Horse, and Raven Horse, a big *Wopila! Pilámaye! Pilámaye* to author, teacher, and counselor, Jessica Taken Alive-Rencountre, at Crazy Horse, and your family of dancers! Long Live the Struggle for Land Restoration in Lakota country in the spirit of Tantanka-Iyotanka (Sitting Bull) and Crazy Horse! *Ahééhee'* to Tony Gonzales, coordinator of AIM west, for reminding us about 2023 being the 100th anniversary of Gayogo̱hó:nǫ' (Cayuga) *chief* Deskaheh's visit to the League of Nations in Geneva, demanding full rights and independence for Indigenous peoples.

To Elder and Grandfather Matthew Bacoate, Jr., the historic "mayor" and mover of Asheville, North Carolina, thank you so much for hosting and educating us on local justice struggles 93 years in the making and still going strong in the summer of 2023. Thanks, too, to Ray Mapp, owner of *Purpose Publishing*, for the instructive Black Inventors poster, and to Mark Bolt at the Grind at the *Home of Black Wall Street AVL*. To Tata Fernando Santillanez, *muchas gracias,* for your life of simplicity in decades of consistent love of *Madre Tierra* and inspiring us to take care of her and all of the vegetation as we share the fruits of the Earth.

Ungadinwa Nangamso … Siyabamba Ngazibini, Umhlele Umfowethu Chief Editor Siyavuya Mzantsi, for all your support and consistently publishing our articles on historic events in the Azanian and Pan African struggle. *Siyabongile kakhulu, Udadewethu* Comrade Thandi Mkhize, Comrade Patrick's dream of a free Azania and *Mama Afrika* will be realized. To the Mathys and Kunnie clans, we are always grateful for your encouragement and support through difficult times. *Aluta Continua*! To sisters and brothers at Radio 786 in iKapa (Cape Town) for your interest in our work and consistent interviews on various issues and commemoration days in the Pan African liberation struggle, especially Ruqayyah, *shukran jazilan.* To *Mwalimu Dada Daktari* Alwiya Omar and the entire family, *Asante Sana*, for all your support and generosity over so many years.

Acknowledgments

To Sister Zari Sundiata and Brother James Stone, editors of the most relevant decolonization online site in the world, with particular focus on Pan Africanist liberation, *Conscientization 101, Àṣẹ, Odabo! Asante Sana, Kea Leboga,* for your conscientizing the world on neo-colonialism and eradicating false consciousness so our world can be holistic again. To Sister Kiilu Nyasha, *Àṣẹ,* for being the model tireless journalist for truth and epitome of courage and strength over decades amid challenges, on *Freedom is a Constant Struggle* in the Bay Area, and *Àṣẹ* to Sister Lesley Tiyesha Phillips, for your support and continued involvement in women's health struggles and the decolonization movement in Oakland going on for decades, too. To comrades in the *Black Alliance for Peace* who persist in demanding true peace and justice for Africans everywhere and for the world, *Asante sana, no compromise, no retreat!*

To Brother Samwel Naikada, protector of the Nyakweri Forest and the community at Transmara, *Asante Sana,* for all your consistent work to protect the sacred lands and the forest preserve for future generations in the wake of the devastating erosion and struggle from climate changing effects. To sisters and brothers at Africa University, the leading Pan African university in Mother Africa, who were so kind to me during our sabbatical in 2017, especially Victor, Munyaradzi, Sam, Alex, Trevor, Victor, Samson (who sadly passed away), Maganda, Tsitsiwana, and the Furusa family, especially Sister Zanele and all those in the community there who experienced the tragic passing of Vice Chancellor Munashe Furusa in 2021, *Maita Basa. Asante sana* to Abisalom Oluoch, Pan African activist and innovator in Kenya, *Melesi* to you, Daniel Bahati in the DRC, and *Muito Obrigado* to Belmiro Jotamo in Mozambique, working always to advance the decolonization struggle.

To our wonderful Indigenous resistors persevering in the global decolonization and self-determination movement and connecting struggles everywhere, the *Indigenous Peoples Movement for Self-Determination and Liberation*, based in the Philippines, Beverly Ingrid, Paul, Romeo, and to the *Cordillera Peoples' Alliance* in Baguio, *salamat,* to Indigenous land freedom fighters Windel Bolinget and Bestang Dededke, who have been subjected to deadly threats by the Philippine regime and its military presence there. To the heroic resistors from the Truku Indigenous nation against the Asian Cement Company in Hualien, Taiwan, Lowking Yirdaw and Miyai Yirdaw, thank you for staying the course and overcoming the conglomerate's destruction of the Taroko Gorge and Truku ancestral lands.

Mahalo to our Indigenous relatives resisting colonization of the Indigenous lands and people of Hawaii, especially Samson Kama, president of *Hawaiian Sovereign Mint* on Oahu, Elder Liko and Sister Laulani protecting sacred Mauna Kea against the colonialist-desecration telescope on the Big Island of Hawaii, and Leon Kaulahoa Siu, Minister of Foreign Affairs, *Le Aupuni O Hawai'i Ko Hawai'i Pae Aina* (*The Kingdom of the Hawaiian Islands*), whose work on coordinating the decolonization of Hawaii and many other places for decades continues, critical work as explained at our meeting at the UN Permanent Forum on Indigenous People (Issues) in New York City in April 2023.

Domo arigatou gozaimasu, Tomoko, Gil, and Jonathan, for hosting us at the Asian Rural Institute in Tochigi, Nasushiobara, Japan, in October 2017, and sacrificing family and well-being to stay on amid the Fukushima-TEPCO nuclear disaster in neighboring Fukushima in 2011 as we heard from people there. *Gamsahabnida* to Eunja Lee, long-time colleague and fellow activist-professor, for hosting us for lectures and discussions at Kwansei Gakuin University in Nishonimiya in 2017, and for continuing to advocate for the rights of all colonized and oppressed people, especially the Korean-descended people in Japan.

To the people of Haiti, *Bondi beni w (Creator bless you)*. We honor particularly members of Lavalas and the sixty liberation organizations persisting in the decolonization struggle since 1804, resisting the nefarious onslaught of the U.S. colonial war machine, and consistent rejection of military intervention by its surrogate neo-colonial states, Kenya, Jamaica, Barbuda and Antigua, and the Bahamas, *Haiti Mayibuye*, Haiti will return. *Mesi anpil*, to former Haitian president Jean-Bertrand Aristide and Mildred Aristide, for your ongoing inspiration, courage, and support of the University of the Dr. Aristide Foundation (UniFA), especially the medical and nursing schools that have trained with Cuban assistance thousands of graduates in medicine, public health, education, agriculture, dentistry, architecture, and physical therapy over the past two decades in Tabarre, Haiti, even in the wake of floods, earthquakes, state repression, and armed banditry from thuggish gangs.

To Indira Serrano, from Barranquilla, in Atlántico Department, Colombia, for your tireless defense of Black and other women of color, especially challenging racist images in media and film, as you demonstrated during your outstanding presentation at the *Association for the Study of Worldwide African Civilizations* (ASWAD) in Seville, Spain, in November 2017. To the Indigenous and African defenders against paramilitary state-protected groups in Choco, Cali, and Cacarica, Colombia, with whom we met during the *Communidades de Autodeterminación Encuentro de la Vida y Dignidad* in 2002, that has seen fruition with the election of radical peoples' rights champions, including vice president Francia Márquez from Cali, the first Black female, and president Gustavo Petro in June 2022, *muchas gracias, hermanas y hermanos*.

Bures and *Giitu* to Jan Erik and Benta Henriksen for your invitation to present our research talk at UIT Nordiges Arktiske Universitet in Alta, Norway, in the fall of 2017, and to colleagues at *Sámi allaskuvla*, Sámi University of Applied Sciences, in Kautokeino for hosting us that November. The result was an instructive collective text on decolonizing social work, *Recognition, Reconciliation and Restoration: Applying a Decolonized Understanding of Social Work and Healing Processes* by Orkana Akademisk in 2019, which led further to our participation in the 5th International Indigenous Voices in Social Work Conference in Hualien City, Taiwan, in August 2019, drawing Indigenous scholars globally.

Kuzuzanpola and *Kadrincchey* to our dear sisters and brothers in Bhutan during our historic visit to the "land of gross national happiness," where we learned so much from the silence within the Buddhist understanding of the pervasive sacredness of life, especially to informants, Karchung, Rinchen, Tashi, Dangyup, Chimba, and nuns and monks in Paro, Punakha, especially at the monasteries at Baylangdra and at Taktsan (Eagles' Nest), and Thimphu, including from colleagues at Royal Thimphu College. Truly a special land and stay.

S@nlAk'yaSOTa (Many thanks) to Sushpa, director of the Yuchie Project in Glenpool, Oklahoma, for hosting us in late July 2023, a remarkable ancestral language project where he and his daughter are now fluent in Yuchie and his grandchildren are learning their real language, along with parents and elders fluent in Yuchie teaching the language to young elementary school children. The beautiful 160 acres with bison, horses, and pigs are a reminder of the power of recovering community's ancestral language particularly in the critical Indigenous requirement of self-determination in land guardianship, language, and culture, a major decolonization subject in this book. Special thanks, too, for your presentation on reclamation of Indigenous languages in Russia at the April 2023 meeting of the UNPFII in New York, Sister Kseniia Bolshakova, making us realize how all Indigenous peoples understand one another through our inter-linguistic respect and caring for our Earth Mother.

We'd be remiss without expressing *Feichang ganxie* (thank you very much) to our dear colleagues in China, Jianping Yi, of the Chinese Academy of Social Sciences, who introduced us to the breadth and depth of Chinese academic work and society during stays in 2011, 2013, and 2014, and for introductions to various colleagues in social sciences, especially Zhongyang Lu from Shaanxi Normal University in Xi'an, where we were hosted and taught in 2017 and 2018, and his professor wife, Sasa, and growing daughter, Lingyi, and to cultural teachers in Xi'an, like Zhu Lee, Li An, Jing Jing, and Dean Chen Weiguo, and award-winning scholar Jiaoushou Liang, and Jiaoushou Juiping Wang, for our collaboration in academic research and publishing, including connecting to this book project. *Feichang ganxie* to Jiaoushou Yu Xuemin, Guozho, graduate teaching assistants Pengxi and Hui, and cultural teachers Zao Ma Ying and Ziyang from China University of Political Science and Technology, where we were invited to do a graduate class in the summer of 2019. And special thanks to Wang Hao, who was so moved by our graduate class that he facilitated a meeting with administrators and colleagues in Ecological Sustainability at Beijing Forestry University, a journey of constant love for protection of *Diqui Muxin* (Earth Mother).

To our colleagues at Vietnam National University for Social Sciences and Humanities in Ho Chi Minh City, professor and former president Pham Quang Minh, Professor Nguyen can Chinh, and students, for hosting us in the summers of 2017 and 2018, *Câm ơn*. To our friends in Ventiane, Laos, Pham and Alex, for arranging a wonderful visit in 2018, and Dr. Khounbally for meeting with me at the National University of Laos in beautiful historic Ventiane, *khob chai*. To Umakanta Meitei and Indigenous land defenders and dam resistors like Jiten Yumnam in Manipur, colonized northeast India, *Tha gut chari,* for your ongoing defense of Indigenous occupied lands and peoples, as we witnessed during our presentations at the University of Manipur in Imphal in 2012.

To our mentors of many decades—Percy Hintzen in Florida; Mark Juergensmeyer in California and Hawaii; "Tink" Tinker, Osage scholar in Colorado; and Chuck Tatum in Arizona—thank you for your consistent support and wisdom over the years.

To Sharon Streater, lead staff at Hillsborough Organization for Progress and Equality (HOPE) in Tampa, Florida, *muchas gracias* for your faith and perseverance in supporting the marginalized, including those from the United Farm Workers in California and the Farm Labor Organizing Committee from the 1980s, that has inspired many and will always serve as a reminder of the power of organizing among the least of creation.

Medase, Shukran jazilan to Akhi Adel Gamal and Yuxuf Abana, for your consistent support of our decolonization and liberation work for the past two and a half decades. *Merci beaucoup* especially to Sister Julia Wright, liberation activist daughter of iconic writer Richard Wright, whose defense of political resistors like Mumia Abu-Jamal and Leonard Peltier and other freedom fighters incarcerated in the U.S. prison industrial complex for decades persists, and whose poetry is transforming for all of us involved in liberation struggle. And to brother Yves Kounougous, traversing Africa and France, *merci beaucoup* for your consistent scholastic decolonization insights and concerns.

Special thanks and gratitude to our wonderful, prolific, and brilliant Brother Mumia Abu-Jamal, who was on death row (still uplifted from our historic visit along with Martha Conley in May 2009) and remains unjustly incarcerated on life without parole (LWP) in this system's dungeons and who frees us all from apathy and indifference to struggle and suffering. You will free us all, as will Leonard Peltier from AIM, finally free from life imprisonment for 47 years due to fabricated murder charges, *Pilámaye!* So, too, *shukran jazilan* to Brother Jahid Shabazz confined in Florence, Arizona, for so many years, yet still braving the storms of repression and resisting denial of fundamental life rights through writing.

Finally, to our technical collective co-producer team: Veronica Rodriguez, for all the years of supporting our video and book publications from various parts of the world; Casey Ontiveros, for your excellent graphic designing and image construction in this book and our 2015 text amid a very tight work schedule; Margaret Pearce, outstanding cartographer and geographer from the Potawatomi Indigenous nation, in generously offering your map constructions of Indigenous lands, especially "land-grab" universities documented in chapter 3 of this book; Cassidy Brandt, for your permissions to use your Big Ten College presidents' salaries graph; and, the mover behind it all, Layla Milholen, outstanding executive editor, operations at McFarland, who stuck with us throughout, thank you so very much for your patience and for enabling us to produce a successful book by one of the best publishers known, McFarland & Co., Inc., Publishers!

Abbreviations

AAWORD—Association of African Women for Research and Development
ADX—Adams Securities Exchange
AFRICOM—U.S. Africa Command
AIDS—Acquired Immunodeficiency Syndrome
AIO—Americans for Indian Opportunity
ALEC—American Legislative Exchange Council
APEC—Asia-Pacific Economic Cooperation
APOC—Anglo-Persian Oil Company
ARI—Ayn Rand Institute
BAP—Black Alliance for Peace
BJS—Bureau of Justice Statistics
BRICS—Brazil, Russia, India, and South Africa (also includes Egypt, Ethiopia, Indonesia, Iran, and the United Arab Emirates)
CARE—Citizens Against Ruining the Environment
CAVA—California Virtual Academies
CBD—Center for Biological Diversity
CCAP—Canadian Climate Change Adaptation Project
CCCNS—Canadian Climate Change Scenarios
CDC—Centers for Disease Control and Prevention
CFCs—Chlorofluorocarbons
CIA—Central Intelligence Agency
CNY—Chinese Yuan
CONAIE—Confederation of Indigenous Nationalities of Ecuador
COVAX—COVID-19 Vaccine Global Access
COVID-19—Corona virus disease–2019
CSO—Civil Service Organizations
DBCP—Dibromochloropropane
DNA—Deoxyribonucleic Acid
DOD—Department of Defense
DRC—Democratic Republic of the Congo
DTRA—Defense Threat Reduction Agency
ECCB—Eastern Caribbean Central Bank
ECLAC—United Nations Economic Commission for Latin America and the Caribbean
ECOWAS—Economic Community of West African States
EIA—U.S. Energy Information Administration
EMR—Electromagnetic Radiation

ENMOD—Convention on the Prohibition of Military or Any Other Hostile Use of Environmental Modification Techniques
EPA—Environmental Protection Agency
EPI—Economic Policy Institute
EPZs—Export Processing Zones
ERCOT—Electric Reliability Council of Texas
EU—European Union
FBI—Federal Bureau of Investigation
FDA—Food and Drug Administration
FDI—Foreign Direct Investment
FEMA—Federal Emergency Management Agency
FMLN—Farabundo Martí National Liberation Front
FSLN—The Sandinista National Liberation Front
GAVI—Global Alliance for Vaccines and Immunisation, Gates Foundation
GDP—Gross Domestic Product
GEO—Group on Earth Observations
GMO—Genetically Modified Organism
GWP—Global Warming Potential
HAARP—High-Frequency Active Auroral Research Program
HANDS—Humans Against Nuclear Dumping
HIPC—Heavily Indebted Poor Countries
HPV—Human Papillomavirus
ICA—International Coffee Agreement
ICAN—International Campaign to Abolish Nuclear Weapons
ICE—U.S. Immigration and Customs Enforcement
IDP—Internally Displaced Person
IEA—International Energy Agency
ILO—International Labour Organization, United Nations
IMF—International Monetary Fund
IPCC—Intergovernmental Panel on Climate Change
IPMSDL—International Indigenous Peoples Movement for Self-Determination and Liberation
IPR—International Property Right
ISCST—International Centre for Sustainable Tourism, APEC
IUCN—International Union for Conservation of Nature and Natural Resources
KTUU—TV station in Anchorage, AK
LGBT—Lesbian Gay Bisexual Transgender
LWOP—Life Without the Possibility of Parole
MAI—Multilateral Agreement on Investment
MOSOP—Movement for the Survival of the Ogoni People
mRNA—Messenger Ribonucleic Acid
NAFTA—North American Free Trade Agreement
NAGPRA—Native American Graves Protection and Repatriation Act
NASA—National Aeronautics and Space Administration
NATO—North Atlantic Treaty Organization
NEA—National Education Association
NOAA—National Oceanic and Atmospheric Administration

Abbreviations

NPR—National Public Radio
NSC—National Security Council
NYCLU—New York Civil Liberties Union
OACPS—Organisation of African, Caribbean, and Pacific States
OCC—Office of the Comptroller of the Currency
OCO—Overseas Contingency Operation
OECD—Organisation for Economic Co-operation and Development
OPEC—Organization of the Petroleum Exporting Countries
PLO—Palestine Liberation Organization
PRD—Party of the Democratic Revolution (Mexico)
PTSD—Post-Traumatic Stress Disorder
RT—*Russian Times* newspaper
SHU—Special Housing Unit
TBI—Traumatic Brain Injury
TEPCO—Tokyo Electric Power Company
TGNP—Tanzania Gender Networking Programme
TRIPS—Trade-Related Aspects of Intellectual Property Rights
UAE—United Arab Emirates
UNCTAD—United Nations Conference on Trade and Development
UNFCCC—UN Framework Convention on Climate Change
URNG—Guatemalan National Revolutionary Unity
USAID—United States Agency for International Development
USDA—United States Department of Agriculture
US DOE—U.S. Department of Education
WB—World Bank
WEF—World Economic Forum
WHO—World Health Organization
WIC—Women, Infants, Children
WIEGO—Women in Informal Employment: Globalizing and Organizing
WMO—World Meteorological Organization
WTO—World Trade Organization
WWF—World Wildlife Fund
WWNO—Public radio station in New Orleans, LA

Preface

Yágháhookáán, Nahasdzáán, the Earth, and the beginningless, endless Spiritual Universe, are beautiful and the epitome of both beauty and spirit power.[1] In the coyote metaphor used by Michael E. Marchand from the Confederation of Colville Indians in the northwestern part of Turtle Island (so named by the Indigenous Haudenosaunee peoples of the northeast to refer to North and Central America, which this book will use throughout), the coyote essential to the creation narratives is the trickster figure capable of both "good" and "bad" and neither, as we humans typically are.[2]

This book is about the beauty of the Mother of us all, Earth, and source of life, the "good." The beauty of the Earth is best expressed in music and silent observation, as *Wuauquikuna*, the Indigenous Inca-descended musical group, performs about Earth Mother.[3] Yet, on the flip side, we have the "bad," which is signified by the escalating wars and violence against our Earth Mother throughout the past millennium, most pronounced in the last five centuries. Reggae icon Bob Marley, known for his mind-and-body moving lyrics by billions on all continents, sang about "So much trouble in the world" in the 1979 album, *Survival*:

> So much trouble in the world....
> The way earthly thin's are goin'
> Anything can happen

This book is also about the trouble in the world and the most important source of life for everyone: *the Earth Mother*, and our most valuable of beings, our *Children*. Though we are all made of air, wind, fire, carbon, and soil from the Earth, never before in human history has the Earth been so taken for granted. Life comes from the Earth and life returns to the Earth for all beings. How do we restore our reverence for our Mother? What are the fundamental reasons for us veering radically off the path of reverence and respect for the Earth and rejecting who we really are like the rest of the natural creation, and challenging and vying against the One who brought us here? These are the core questions which warrant detailed responses through especially understanding history and particularly the central role that Indigenous people have always played, still play, and need to continue to play in assisting us to return to the path of wholeness with the Earth.

Right at the outset of this project, this author needs to acknowledge the serious limitation of embarking on a manuscript that covers major global issues and diversity of knowledge. Yet all elements in the text remain focused on the Earth and preservation of life in all of its fragility and vulnerability. This book is not intended to single out any particular group or people, but to examine and interrogate the history of human beings, particularly through the cultural and historical lenses of Indigenous peoples from ancient creation narratives into the present, and to understand how, in the words of Hataali Jones Benally, beloved 90+-year-old Diné/Navajo healer, teacher, and world champion hoop dancer in northern Arizona, "we all have the same roots." Though the book is written by one person, it represents the voices and struggles of many, particularly the Earth Mother and the most vulnerable creatures who have shared and given the most, their all, their lives, but who are consistently derided and crushed into the Earth. It highlights Indigenous ancestral knowledge as a medium of instruction for living in the manner that the Earth and the revered ancestors have always instructed. In this sense, the book intends to convey important facts and critical analysis, yet is very open to ongoing critique, dialogue, and multifaceted perspectives on protecting our Earth Mother and all repressed and obliterated life since so much of the dominant news that we receive is distorted and untruthful. In the vein of Indigenous cultures and philosophies, while ancestral knowledge is foundational in every respective context, no singular entity or tendency has the solution to the unending problems and crises in our world, particularly in a world that is heating dramatically, drying up and burning down, and being flooded out of existence in the 21st century. For example, the Colorado River now suffers its driest level in 1,250 years, losing 10 trillion gallons over 20 years, while Lake Powell and Glen Canyon Dam in northern Arizona are still 127 feet below capacity and original levels as the drought from 2000 still takes a serious toll on life in the region.

Ultimately, we all need each other in resolving the unfolding irreparable catastrophes that are fast depleting life as we have traditionally known it with little linear time in the timeless universe.

Lake Powell and Glen Canyon Dam, July 2023 (author photo).

World War III Has Already Begun

This war has been waging for some decades now. This war does not involve armaments and high-tech cyber warfare visible in so many parts of the world today but is the most devastating and annihilating. It is not over oil or "natural" gas. Neither is it over nuclear weapons and which regime will use them. Nor is it over ethnicity, religion, or culture as we are constantly ideologically bombarded daily by the huge monopoly news conglomerates. Or even the horrific Covid-19 pandemic that has claimed the lives of over five million the world over. WW III is over *Water*. *The Water of Life*. *Mni Wakan, Mni Wiconi* ("Water is Holy, Water is Life"), was the clarion cry of Indigenous Lakota people in the persistent struggle to protect and defend the *Water* at Standing Rock in the Dakotas, as thousands converged upon the area to protest the desecration of sacred lands by the Dakota Access Pipeline in 2016. There was a halting of the pipeline in 2021, yet it persisted in 2023 with Line 3 of the Enbridge Pipeline in violation of the sacred Mississippi River on Indigenous Cree, Ojibwe, and Anishinaabeg lands. Water is indeed sacred, a Divine Spirit, hence it is *Water* in this book, as Indigenous peoples understand the term.

The reason why history is important in a book on the Earth and the demise of capitalism is precisely because the Earth is a living species unto Herself, *Mother of all life*, and has been around since the beginning of time; capitalism, on the other hand, is a relatively late system in Earth time, evolving from the 17th century. It's important to note that capitalism went hand in hand with colonization of the Americas from the Columbian invasion in 1492, of Africa, Asia, and the Pacific subsequently. One cannot engage in any discussion of capitalism without seriously interrogating its precedents, colonialism and enslavement.[4] This disturbance of the peace of the world launched an

Drying Lake and Parched Earth (redcharlie-Unsplash).

unprecedented course of genocide, ethnocide, ecocide, and femicide, which have a bearing on the world of today, including climate change and impoverishment. These systems of oppression, violence, war, militarization, and dehumanization were and are rooted in *fear*—fear from the consequences of historical extermination, land theft and dispossession, exploitation of the peaceful and vulnerable women, children, and honest working classes, and fear of the ruling oppressors losing their ill-gotten dominant and privileged elite minority possessions and status in the world now fast eroding with increased resistance to injustice from the subjugated peoples. The persistent, growing organized protests and movements demanding justice, particularly by young people and students in this era, are often met with fierce police hostility and violence. The fear of the natural world and diversity and difference from centuries ago has unleashed legalized repression, banning, prohibitions, censorship, censure, electronic surveillance, and monitoring in so many quarters today, including in and by governments, corporations, and institutions, especially in the U.S. empire. Libraries, schools, colleges, and legislatures in the U.S. are restricting access to diverse forms of knowledge and perspectives on the world and continue to marginalize such content, dictating and curtailing fundamental human and constitutional rights. Such actions, as we will come to see in this book, have everything to do with the failure of these systems and those operating them in understanding our deep organic roots in our Mother, Earth, and realizing that ultimately we own nothing; everything we have, including ourselves, belongs to the Earth and the beginningless and endless Spiritless Universe, to whom we all inevitably return upon the ending of our physical existence. Nothing is permanent except the Universe, of which our Earth is the basis of life eternally. In the face of ongoing dispossession and colonization, Indigenous people (specifically in Turtle Island) repeat, "We are still here … as long as the Grass Grows and the Rivers Run." Indigenous people everywhere will always be here to protect our sacred ancestral lands from further destruction.

How and why did such cultural shifts occur from the feudal-colonial era? What are the questions that need to be pondered from such a discursive journey into history? How do we understand these questions with crystallized clarity, so we don't persist in repeating the catastrophic and lethal errors of history and recall well-grounded Indigenous wisdom and culture to collectively chart a sustainable path in the living present and the future so that future generations are preserved? How do we rededicate and reeducate ourselves to restore our broken relationship with the *Earth*, and her struggling children in the sea, on the Earth, in the Air, and everywhere? Most importantly, how do we protect our most vulnerable beings, the Little Ones, the Children of all life, so that they in turn can live in wholeness and joy in their present and future existence? These are part of the central themes of this book so that we can understand the gravity of these crises and how to resolve them. It is not knowledge about the world that we lack since we generally have access to information globally through variegated forms of swift communication across thousands of miles in minutes; what we often lack is the ability to understand both reality in depth and its interwoven and complex nature. Dismal levels of historical knowledge permeate our existence since our dependence on knowledge transmitted via the same electronic media communication forms has overwhelmed us to the point of making us shoddy in-depth readers. Such saturation of information with reduced discernment has denuded our innate strength, moral perseverance, and in many instances, consciences, in personally persisting and collectively reshaping the same reality, often unhealthily, in every way. Many formally educated people are not necessarily well informed and possessors of grounded and rounded knowledge to fundamentally challenge the dysfunctionality around them since they inadvertently have become part of such. It is precisely here that Indigenous peoples, ancestral knowledge, and cultures are extremely helpful in the journey toward self-realization and community empowerment.

The introductory chapter describes some of the instructive Indigenous creation narratives that define Indigenous cultures worldwide, demonstrating the basis of Indigenous life and culture. Indigenous histories and cultures have complex cosmogonic, cosmological, and epistemological stories of creation, some of which can be shared outside respective cultures, and some not. We thus make a limited and humble attempt in this chapter to portray some Indigenous creation stories and understandings of life's origins in one form or another, realizing that not all can be included within the scope of one chapter and realizing that original stories were not written down, were in Indigenous ancestral languages, and were transmitted through oral tradition. This chapter is devoted to understanding the way that various Indigenous peoples understand the Earth, the universe, and life in all of its diversity and complexity in balance. It will recount some Indigenous creation stories, those which can be shared with others, illuminate the fluidity of gender, and explain how these are connected to the sacredness of the Earth, and the endless Spiritual Universe and all life. Such stories can assist in explaining so many of the conflicts and divisions among us that have much to do with our failure to understand our deep common roots in the Earth. Indigenous peoples' understandings and histories of creation and evolution are axiological for addressing the urgent crises of life and existence, particularly the perilous path on which we humans find ourselves.

Chapter 2 delves into the history of Europe, especially the region's religious, ecological, political, economic, militaristic, and social evolution in the post-medieval era, the Renaissance and its challenges. It highlights Karl Marx's pointed and critical observations of the exploitation of workers, children, and women under early capitalism, and delineates the history of slavery and colonization that shaped the emergence of globalized industrialist capitalism in the world that is now fast eroding, especially as these matters unfolded in the western hemisphere with decimating and enslaving Indigenous peoples, Africans, and other peoples of color.

Chapter 3 discusses issues of oil's history, including whaling, oil production and consumption, and the inextricable interwovenness with capitalist extraction, violence, and war, the resulting ecocide, hydraulic fracturing for fracking oil and liquified natural gas (LNG), the Oil Pollution Act of 1990 and its flaws, and ecological and environmental decimation from repeated toxic spills from 1990 through 2020. It will highlight Indigenous cultural philosophies and contexts rejecting oil drilling and fracking.

Chapter 4 examines and discusses the widely discussed subject of Earth heating and perennial climate chaos and instability, even collapse, by illuminating the history of industrial chemical use, U.S. weather monitoring, chemical engineering of weather, and climate modification. It will highlight the collective impact of these facets on ongoing daily existence, unprecedented sea level rise, prolonged and entrenched drought, unheard-of extremes in heat and cold with constantly spiking temperatures high and low, unimaginable floods and consequential destruction, and devastation of plant life and soil fertility that has exacerbated impoverishment and "climate refugees" from the rapid disintegration of island countries around the world, especially in the Pacific.

The chapter discusses Indigenous cosmological perspectives on the climate change crises and how Indigenous knowledge can function as the basis for Earth-centered and Earth-friendly solutions to this escalating and unfolding life collapse, in concert with all environmentally and ecological defending allies.

Finally, chapter 5 discusses how we've all been conditioned by an educational ethos that is Earth-ignorant, anti-impoverished and anti-working class, patriarchal, elitist, money and profit-oriented, and capitalist engineered, shaped, and controlled, and has denied the diversity of knowledge systems and philosophies of Indigenous people, Black people, and other communities of color that can assist us in collectively addressing crises and perennial global and societal problems. It highlights the roles of billionaire and ruling class capital in defining what we call "education" that has had repercussions for every segment of our society, from the little ones to the elderly and struggling-to-retire folks, many now saddled with substantial debt from college loans. The distinctive element of this chapter is the manner that Indigenous cultural and cosmological perspectives view such developments and crises and how Indigenous knowledge can function as the basis for Earth-centered and Earth-friendly solutions in addressing even educational issues.

The concluding epilogue urges a reorientation and return to and restoration with the *Earth*, our eternal, indomitable Mother, and to wholeness and real living so that we can all live in mutuality and reciprocity, caring for and sharing with others, including the vulnerable creatures of life, because we all need each other in addressing the life questions facing us daily.

This is the essence of this book, dear reader, which we hope you will read with spiritually inspired patience and understanding.

1

Indigenous Peoples' Creation Stories of the Earth and the Natural World

"As you know," Jenny Leading Cloud said, "we Indians think of the earth and the whole universe as a never-ending circle, and in this circle humanity is just another animal.

*The buffalo and the coyote are our brothers; the birds, our cousins. Even the tiniest ant, even a louse, even the smallest flower you can find—they are our relatives. We end our prayer with the words **mitakuye oyasin**—'all my relations'—and that includes everything that grows, crawls, runs, creeps, hops, and flies on this continent. White people see humanity as nature's master and conqueror, but Indians who are close to nature, know better."*[1]

Introduction

The Earth is considered one of the eight planets within our solar system. There are interesting, powerful, and instructive accounts of the origins of the Earth in the scientific theories of how the Earth was formed, generally subscribing to the Big Bang theory, in which scientists like Tim Flannery and Stephen Hawking and other noted physicists describe the complexity of processes of evolution based on the views of theorists like Charles Darwin and Richard Dawkins.

Indigenous histories and cultures have complex cosmogonic, cosmological, and epistemological stories of creation, some of which can be shared outside respective cultures, and some not. We thus make a limited and humble attempt in this chapter to portray some Indigenous creation stories and understandings of life's origins in one form or another, realizing that not all can be included within the scope of one chapter, and realizing that original stories were not written down in non–Indigenous languages, and were transmitted through oral tradition.

Indigenous historical and cosmogonic accounts from every tradition recount the organic nature of life in all spheres, the sacredness of life, and the spiritual and physical interdependence of all life. They also reflect teachings and histories that have everything to do with the evolution of the world, the violence against the Earth and all life that has reached unprecedented destructive proportions, and the consistent economic, political, cultural, and social instability experienced particularly in the last five centuries. The ecological crisis, social conflicts and divisions, and exploitation we suffer daily have much of their roots in the repeated failure to understand our deep common roots in the Earth. Indigenous peoples' understandings and histories of creation and evolution are axiological for addressing the urgent crises of life and existence, particularly the perilous path on which we humans find ourselves. The land continues to suffer expansive degradation everywhere, drinking water is mostly contaminated and scarce in all regions of the Earth, floods are rampant and destructive, most of the trees and forests have been decimated, deserts are spreading their arid land base, and apocalyptic fires, accompanied by scorching heat, wind, and lightning, are normative even in the supposedly "economically secure" countries of the west.

This chapter is devoted to understanding the manner that various Indigenous peoples understand the Earth, the universe, and life in all of its diversity and complexity in balance. It will recount some Indigenous creation stories (those which can be shared with others); illuminate the fluidity of gender; and explain how these are connected to the sacredness of the Earth, the endless Spiritual Universe, and all life.

Some Indigenous Creation Stories That Can Be Shared with the World

Many Indigenous peoples do not share the intricate stories about creation because such sacred knowledge is protected and off-limits to those from outside such communities. For Indigenous peoples the world over, life is organic, integrated, and interdependent. As Diné (Navajo) cosmologies describe it, the Earth, the Sun, the Moon, and the Universe (including nine major creation deities) are all involved in the creation of life, of which humans are a fragmentary part, albeit significant.

Since original creation stories were generally unpublished as part of the necessary preservation of Indigenous sacred knowledge and were transmitted transgenerationally via elders' memory and oral tradition practices, there are various accounts over time. The sacred roles of Sky Woman and the turtle (female) are central in all the major northeast Turtle Island Indigenous nations' creation stories, including the Oneida, Haudenosaunee, and Mohawk accounts. These accounts explain the basis of what many of our Indigenous people here call Turtle Island. Varied accounts exist with nuanced differences over time and source. One such Oneida account describes how a long time ago above in the sky, a beautiful world teemed with plants, game, fish, fowl, diverse vegetables, creatures, and human-like beings who were very happy and had neither disease nor terminal existence, neither storms nor tempests. Passions of jealousy, anger, hatred, and hostility were unknown. Due to a couple's separation, a pregnant woman fell from the sky into the dark world below. As she fell, some accounts describe her pulling strawberries and tobacco plants with her; others state that she fell with very useful elements like corn, deer meat, and other food preparations. A muskrat from below noticed the falling Sky Woman, and other creatures of the water identified the falling one as being human. They then conferred and referred the matter to muskrat, who in turn opted to entrust the developing situation to Turtle (female). Turtle then proposed that all of the creatures work collectively to dive deeper into the waters below and bring mud back up so that the Sky Woman's fall would be cushioned. They brought up the soil to Sky Woman and laid it on her back, so that she fell comfortably on the island. Thus, Turtle Island was formed. A Haudenosaunee account describes Sky Woman giving birth to twins, one of

1. Indigenous Peoples' Creation Stories 9

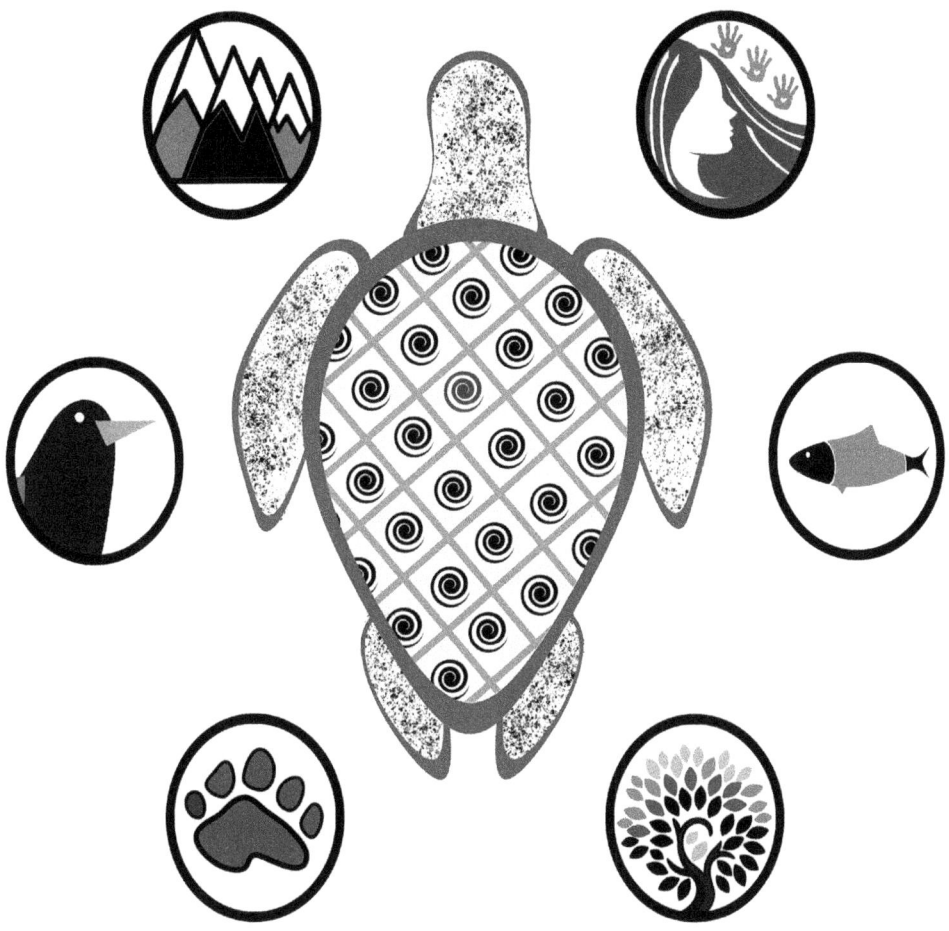

Turtle Creation Story (Veronica Rodriguez).

whom followed the "good way" and the other the "evil way." Thus it was that good and evil became present in the world and continue to be with us today.[2]

For the Hopi Indians of Arizona, Turtle Island's (North America's) origins are in the Sun divinity, *Tawa,* and *Huzruiwuhti,* Spider Woman Earth Goddess. The ancient Egyptians also viewed the Sun and the astral beings as deeply involved in the creation, with the sun god, *Ra,* a form of *Atum,* who created himself out of *Nun* by will and mouthing his name. Humans were created out of the eye of *Ra,* the eye of wholeness, when the eye separated from *Ra* and was unable to return. Thereupon, *Shu* (the offspring of *Atum,* who signified dry air) and *Tefnut* (offspring of *Atum,* who represented moist air) went to look for the lost eye, but the eye refused to return and instead struggled. During the course of the struggle, the eye shed tears that subsequently led to the birth of humans.[3]

In the creation story of the Indigenous Tohono O'odham nation, whose ancestral lands are now occupied by the state of Arizona, the two Creator Beings are *Tcuwut Makai* (Earth Maker) and *I'toi* (Elder Brother), whose sacred home was in the

Babaquivari mountains, south of *Schuk Toak* (meaning Black Mountain), *Ts-iuk-shan*, anglicized as Tucson.[4] Among the Tohono O'odham people (meaning the "desert people who have emerged from the Earth"), who are related to the Akimel and Salt River O'odham people of Arizona, there are diverse creation stories, all of which are sacred and to be narrated at certain times of the year and seasons. These stories cannot be narrated by outsiders or published. The Tohono O'odham live mostly in southern Arizona and Sonora, northern Mexico. The mountain located in the west side of Tucson has been desecrated by a huge "A" sign made out of red, blue, and white stones, which demonstrates a legacy of colonial disregard, violation, and refusal to recognize and respect sacred places of Indigenous people here and globally.

Very importantly, for the O'odham people, like with all Indigenous peoples, the design of all created beings was and is circular, as with the Sun, Moon, Stars, Earth, and other celestial beings. The great flood on the Pinacate millions of years ago (when the volcanoes were very active) shaped the geology of southern Arizona; during this time, the Elder Brother, *I'itoi,* created the O'odham people from the clay of the Earth and lived in a cave in *Waw Kiwulik* (rock in the middle) known as Baboquivari. Creator created the prickly pear to sustain the people and life in the hot desert. This area, with the highest mountain in southern Arizona, over 7,000 feet high, is the most sacred point of the region, and a place where Indigenous peoples pray and make offerings of feathers, plants, and ceremonial mediums to the Creator. Tragically, *Waw Kiwulik* has been dispossessed from the traditional Tohono O'odham ancestral lands as the result of the U.S. colonial–Mexican War and the "Gadsden purchase" in 1854, where the U.S. government paid Mexico, another colonially orchestrated nation, $10 for the transfer of 29,670 square miles of Arizona and New Mexico, total anathema to Indigenous nations there. Indigenous lands can't be bought and sold in Indigenous histories and cultures. The land is administered by the U.S. Bureau of Land Management, a desecration reflected in many Indigenous sacred places around Turtle Island, including *Chi Chil Bildagoteel* (Oak Flat), sacred ancestral lands of the Chi Endé (Chiricahua Apache) nation outside the San Carlos community and near Globe, southern Arizona, which is now environmentally and culturally threatened by Resolution Copper, a subsidiary of mining conglomerates Rio Tinto Zinc and BHP Billiton.

The Maidu Indians of California describe stories of *Earth Starter* coming down a rope of feathers and lowering into a raft where *Turtle* and *Pehipe*, two divinities, were, in a realm where it was all dark and only water. *Earth Starter* asked for some dry land when he landed in the raft, and *Turtle* asked *Earth Starter* to tie a rope around his leg so he could dive underwater and fetch dry land, which he did. However, he was gone for six years. Later, when *Turtle* returned, he was covered with slime and had a tiny portion of dry land under his nails. With the little land scratched from *Turtle's* nails, *Earth Starter* made a small pebble of earth and left it in a corner of the raft. Each time he would come to look at it, it grew in size, until on the fourth look, the pebble became the world. Then there was land, and the raft ran ashore on the land. *Turtle* was glad that there was land but complained that there was no light, to which *Earth Starter* said that he would call his sister, the *Sun*. Thus, the *Sun* appeared, and *Earth Starter*, *Turtle*, and *Pehipe* were all happy to see the *Sun*, but later *Pehipe* began crying when the *Sun* went down. *Turtle*, unsatisfied even then, asked *Earth Starter* whether that was all that was going to be made. *Earth Starter* called the stars, and they shone and made a huge oak tree, under which they sat for two days. While they all beheld the beauty of the newly created beings, *Coyote* appeared with his dog, *Rattlesnake*, but they were not permitted to enter *Earth Starter's* house. Later, *Earth Starter* took mud and called the birds of the air, the trees, and the first deer, followed by other animals. *Turtle* wasn't happy again and asked that *Earth Starter* make some other being; *Earth Starter* said that he would make people and proceeded to make a man and woman from the red earth combined with water, except for their hands. He laid on his back and sweated all night. Early in the morning, the woman tickled him on his side, but he sat still and never laughed. When *Coyote* saw the woman and man without hands, he asked *Earth Starter* to make their hands like his paws. But *Earth Starter* declined and said that he would make the hands of woman and man like his hands so that they could climb trees if chased by bears. Thus, the first woman was called *Morning Star Woman*, and the first man was *Ku'ksu*.[5]

For the Yoruba creation account in West Africa, there were four major deities involved: *Olodumare*, the executive Creator; *Ifà*, the Creator of wisdom; *Obatala*, the Creator of form; and *Eshu*, the Creator of morality, who ensures humans behave in acceptable ways that would otherwise evoke serious consequences.[6] In fact, in the *Ifà* literary corpus, the authoritative religious scriptures of the Yoruba tradition, there are 600 +1 divinities, reflected as such:

> *Irinwó o mólé ojùkòtún,*
> *Igba mole ojùkòsì òwúrò.*
>
> *Four hundred primordial supernatural powers of the right,*
> *Two hundred primordial supernatural powers of the left.*

In the cosmogonic and cosmological universe of the Yoruba tradition, the universe is divided into the right hand and the left hand, with the 400 right-handed benevolent *Orisà* (divinities), who sometimes inflict punishment on evildoers, and the 200 left-handed malevolent *Ajogun* (anti-gods-warriors-warmongers), who cause violence to the *Orisà* and to humans. The *Ajogun* are constituted by *Ikú* (death), *Àrùn* (disease), *Òfò* (loss), *Ègbà* (paralysis), *Òràn* (big trouble), *Èpè* (curse), *Èwòn* (imprisonment), and *Èṣe* (affliction).[7] Interestingly, there are two powerful divinities with "supernatural" abilities who are in-between the two sides: *Àjé* (incorrectly translated as "witches"), and *Èsù* (a universal "guardian" of morality), who maintains neutrality and is a mediator between the forces of good and evil. Thus, *Èsù* is considered an *Orisà*, and is part of the right-handed side of the universe. Interestingly, too, humans are part of this right-hand side, and though not divine, wield the potential to become a divinity. In this sense, the Yoruba pantheon is composed of 400 +1 divinities, with the +1 viewed as a flexible deity where humans who have lived spiritual lives are included in the pantheon. The complexity of the Yoruba cosmology is found in most other Indigenous traditions of the world, underscoring openness, unfolding, dynamic, and constantly renewed beliefs, thought systems, and deities, all of which cater to the needs of the constantly changing environment and ethos within which the people live. Anthropology generally considers such systems fixed in time and space and thus has often misunderstood Indigenous religious and cultural complexity and overlooked Indigenous cultural fluidity and openness.

Among the ancient Dogon people of Mali (northwest Africa), there's an interesting creation account of Amma, the Creator, who deemed originally that all life should be characterized by complementarity and mutuality, as with female and male. Each being that Amma created had two spirits—twins, female and male—at birth. This principle underlined all phenomena in the universe. Unfortunately, at the time of creation, one of these twins, a male spirit, Ogo, was impatient and did not wait for his full gestation period to materialize; instead, prematurely, he emerged and opted to compete with Amma, the Creator. He believed that he could create a better world than the one Amma had created. With his incomplete placenta, he proceeded to create the Earth. Soon he realized his folly and inadequacy and went to Amma to acquire his female spirit twin, but it was too late since Amma had imparted his female Creator spirit away. Thus, Ogo was destined to be forever deficient and incomplete, and the Earth he had created would also be incomplete. Beings from the Earth would thus suffer the consequences of this original Earth creation defilement and always experience impurity. Ogo's incestuous act of procreation with his own placenta, which represented his mother, lives on in the struggle of life to realize wholeness and complementariness.[8]

The Indigenous people of the Marshall Islands in the eastern region of Micronesia have creation stories that describe a time when there was just the ocean, until one day, *Lowa,* the heavenly God, came down and, through a humming sound, created the Marshall Islands. After Lowa returned to heaven, four men came to the island of Ailing-lap-lap, and these four people went in four cardinal directions, west, east, south, and north. The person from the east brought light (Sun); the one from the west brought Life, since all things were made through him; the person from the south brought Wind; and the final person from the north brought Death. After each of these co-creators found their directions, Lowa sent a fifth person, who then arranged the Marshall Islands in a basket and placed them in the respective order found in the ocean today. Thus, Sun, Life, Wind, and Death are the divinities determining all life on Earth, and the Indigenous people there perform worship rituals always in reverence of these divinities.[9] It is indeed catastrophic that the U.S. has never respected the sovereign inalienable rights of the Indigenous Marshallese people to their lands and cultures, but instead have occupied these islands, committing ethnocide by secretly using them for nuclear weapons testing, missile weapons delivery, and the "Star Wars" program, detonating 66 thermonuclear and atomic bombs from 1946 to 1958 that caused most of the residents of these islands to suffer from death and radiation poisoning.[10] Many were thus forced to leave their traditional homeland.

In the Indigenous Blackfeet cosmogony of Turtle Island, Creator Sun is the Source of all life and has always existed. Living alone in a spiritual aura for timeless eons, Creator Sun felt lonely and one day gathered space dust and formed this into clay, with future creation in mind. Creator Sun molded the clay into a mud ball and blew into it, and it was filled with air and remained suspended in air: Earth. When Creator Sun became tired of playing around with the mud ball in the air, Creator Sun gathered the dust again and spat on it, resulting in a long, slim form: a snake. The Creator Sun allowed the snake to multiply, and soon they overgrew their numbers on the Earth. They covered all of the Earth, and being so numerous, grew arrogant to the point that they ignored Creator Sun's commands to listen and follow Creator Sun's ways. This stubbornness compelled Creator Sun to sadly resolve to destroy the rebellious creatures, by making the mud ball, Earth, boil out from the bottom. The snakes were unable to escape and were thus consumed. However, miraculously, a small female snake escaped into a narrow crevice where the boiling liquid was strangely absent. When Creator Sun noticed the small snake, Creator Sun decreed that the small snake would multiply, but was unconcerned about the snakes becoming too powerful since other natural causes would limit the snake population. Later, feeling for something soft on the mud ball, Creator

Sun invented the soft green grass, but even after this creation, Creator Sun felt unsatisfied and searched for a being that would be similar to the Sun. Creator Sun spat on the mud ball and created a figure that resembled the Sun in many ways. This was the Moon and Creator Sun's mate. Later, the Moon would give birth to seven sons, the Big Dipper stars. Things were going smoothly until the Moon became enamored with a snake being who lived in a den in a brush where the Moon woman had regularly gone looking for food. The snake being could change form to look like a being of Creator Sun and was very deceptive, and gradually drew the Moon woman into an intimate relationship, of which Creator Sun was fully aware. Instead of destroying the Moon woman, Creator Sun decided to flee to another place with the seven sons, Big Dipper. Moon woman, possessing virtually similar powers of Creator Sun who bestowed these on her at creation time, attempted to destroy Creator Sun and the Big Dipper seven sons as they fled, but just as Moon woman was catching up with Rawman, one of the sons, Creator Sun threw a hatchet that severed the leg of the Moon woman. Moon woman was stopped, and Creator Sun still gently attempted to talk with her, but she remained feisty and challenged Creator Sun. It was then that Creator Sun, whose powers are infinite, decreed that Moon woman would see only half the time, and during the other half when Creator Sun and the Big Dipper sons rested, she would not be able to see at all. Thus, we have separation of night and day, and the story develops into various elements of creation, with Earth becoming central as the wife of Creator Sun, and human beings created by Creator Sun blowing into the nostrils of the mud form made by the Sun. The snakes that emerged multiplied and continued to remain rebellious, forcing Creator Sun to cause it to rain heavily for many, many days, resulting in water forming all over the Earth and making it muddy. The multitudinous snakes and other reptiles—even those with legs that grew to a giant size—were all washed away in the great floods that covered the Earth, with those remaining in smaller form moving to higher ground. The Earth is thus undulating and uneven throughout as the consequence of these floods.[11]

In the Huichol/Wixárika creation story from the Sierra Madre Mountains of central northwest Mexico and parts of the southwest in Turtle Island, so-called animals and human beings could move back and forth in the New World. The Goddess of Growth and Germination, *Takutsi Nakawe*, who came through the tunnel of light from the sky world, and *Kauyamari*, the Deer Spirit who left green circular disks on the ground, *Hiykuri*, peyote, used for visions, together with *Tatewari*, the Grandfather Fire, are all the creation deities. *Takutsi Nakawe* admonished the young, lonely boy of the mountains, *Watakame*, to build a boat so that he and his dog, *Tusi*, would live safe from the floods. Reluctant at first, he eventually did exactly as *Takutsi Nakawe* instructed and stayed with his dog, *Tusi*, and took many seeds with him. After the flood, Goddess *Takutsi Nakawe* continued creating the animals and plants. Later, because of jealousy from the other beings, *Watakame* decided to stay by himself in the mountains that he knew so well. He found a place with good water. He then found an ideal spot for planting his corn seeds and began clearing the brush with his sharp machete. Not focusing, he swung his machete and the blade hit his leg, so he covered the graze with his bandana. In anger, he swung the machete at a rock, and it became blunt. Tired and frustrated, he returned home and fell asleep. Returning the next day to resume clearing the undergrowth, he was shocked to see that his previous day's work was not evident because the growth appeared like nothing had been cut. He realized that something was amiss and couldn't understand who caused such a thing. Apologizing to his machete, he sharpened it and continued cutting the undergrowth. However, the following day, he returned only to see that all of the growth remained, and the same continued each day, even after *Watakame* had cut the vegetation. Finally, exhausted, he decided to stay at the place to see who was undoing his clearing work. He lit a fire, fueling it with the cuttings from the undergrowth. In the flame of the fire, he saw many nature beings and flowers of the mountains. He was sure that the fire would show him who was responsible for restoring the undergrowth and reversing his hard work of cutting. He was happy that he had a good fire. He followed Goddess *Takutsi Nakawe's* instructions about planting seeds, and these grew into corn, squash, and beans, and he was taught how to care for these food crops of the Earth. So, life grew in different ways and forms. The instructiveness of the Huichol/Wixárika creation story is that we need to respect and revere all life and take care of the plants, animals, and even the tools that we use that enable us to grow food and live so that they, in turn, can take care of us. It is a profoundly holistic, integrated, organic, mutualistic and interdependent creation story, scientific in its core as Tewa Nation scholar Greg Cajete notes. He explains that creation stories of all Indigenous people are relational and inextricably intertwined with all beings of life in the ever vast natural world, with whom we humans are instructed to harmoniously live and share life.[12] We were created to be *in and one with* "nature," in close interrelationship, as opposed to living in isolation from the complexity and diversity of the rest of the natural world.

In many ways, the preceding cosmogonic accounts recall some elements of the Genesis story of the flood after creation in the ancient Eastern and north African world, reminding us that while there are significant differences in creation accounts by peoples around the world, there are also close connections in describing powerful spirit forces—Sun, Moon, Earth, Sea, and other beings who participate in the creation and ongoing creating of all life, with humans, while important, being neither the center nor the pinnacle of the mosaic of life.

In the Indigenous Okanagan people's account of creation from northwest of Turtle Island, the Earth was once a human being made by the Creator, Old One, who declared that Earth would be the mother of all people. The soil of Earth is her flesh, the rocks her bones, the wind her breath, and the trees and grass her hair. When she moves, earthquakes result. Originally, Old One took the woman and took some of her flesh, the soil, to make the ancient beings. The ancient beings were two-legged and four-legged, and some could fly as birds, while others could swim as fish and sea creatures. All these beings had the gift of speech and communication. Only the deer remained as an "animal." The Indians were made from the last of the mud balls by Old One, who made them and all creatures female and male, and thereupon people and "animals" grew numerous and spread.[13] Sadly, many of the ancient ones became selfish and grew into monsters, and not knowing the difference between deer and humans, they often ate humans in error. Old One was very disturbed because the monsters were wiping out the humans, and thus sent Coyote to destroy the monsters among the ancient beings, and Coyote then taught the Indians how to live.[14]

The Okanagan creation story clearly demonstrates the centrality of *Earth as Mother,* and all life on Earth as being derived from her. Further, it indicates the interwovenness of humans and the rest of the creatures and life on Earth and the sea, with each having its own respective manner of speech. No life is considered inanimate, but rather all are alive with spirit, including trees, rocks, wind, mountains, rivers, lakes, and air. This is the principal reason that all Indigenous people in the world, particularly in the Americas, revere the Earth as Mother of life and have consistently resisted capitalist industrialism, which has commodified all life on Earth and in the Sea.

To move to another part of the Earth, the ancient Indigenous Jawoyn people of colonized Australia describe creation as *Buwurr* (Dreaming or Dreamtime). Jawoyn Elders Lily Bennett, Margaret Katherine, Sybil Ranch, and Liz Thompson explain:

> One of these Dreaming figures was known as *Nagorrgo*, a tall spirit who came from saltwater country in northern Arnhem Land. *Nagorrgo* taught of many things and he gave the **law** about *mowurrwurr* or clan groups, as well as showing what foods (plant and animal) clans could and could not eat....[15]
>
> There are strong **laws** relating to food and people's relationship to it. People are connected to plants and animals through particular ones' ("totemic") relationships and have responsibilities to look after those plants and animals. Different communities have different laws for foods and how they should be collected and prepared. These laws were given by the **Creation Ancestors** of the **Dreamtime**.[16]

Among the Jawoyn, like all Indigenous peoples, food is spiritual and a gift of the Earth and Creation, and harvested seasonally with particular ceremonies conducted to show gratitude to Creation for rain that enables life to grow. Women and men have complementary yet different responsibilities, with the former gathering seeds and berries and other food, and the latter hunting and fishing.

Creation for the Jawoyn people is a very sacred event that constitutes the basis for the living of all timeless generations, and 60,000 years ago as the time for such creation is "simply a number," as we learned from Elder Willie Gordon in August 2014.[17] Elder Gordon is keeper of the sacred serpent caves and Nugal clan member of the extensive Guugu Yimithirr people in the vast area between Annan River and Princess Charlotte Bay in the northern region of Queensland, Australia.

Indigenous people there and everywhere *do not subscribe to a linear concept of time* (unlike Western epistemologies, which use zeros meaninglessly in an attempt to age the stars, suns, and celestial and terrestrial beings), since it's hardly concretely possible to imagine the substantive difference between a hundred million and a billion and two billion years ago. When Indigenous people narrate creation stories beginning with "a long time ago," it implies a timeless universe, which, in fact, the universe scientifically is. Anthropological, archaeological, and paleontological methodologies employed with carbon dating are but one parochial system to understand historical happenings and dates. While these may be relevant for recent historical construction (with some caveats as Vine Deloria cautions and critiques), they really are extremely limited in understanding ancient histories that span millions and even billions of years. Willie Gordon notes:

> Aboriginal people believed that there was a Spirit Giver for all living things. And when that life finished, they questioned what happened to that spirit. Where did it go? Did it go back to the Creator? This is why they protected the bones of people in the *dubai* (cave adjacent to the deceased's birthplace, to show the Great Creator where he came from, so that when the decision was made to complete the Circle, his spirit would remain in the place he desired), preparing them for the completion of their life's circle.
>
> Because life comes to us all from the spirit-giver, Aboriginal people believe that we are created equal. The differences between people only occur with religion, culture and environment; but spiritually we are all the same.[18]

It certainly was no coincidence but rather spiritual convergence to see a large poster on the way up to the 8th floor of a small hotel in Sydney, Australia, in 2013, that read:

"We are here, to learn, to love, and grow, and then we return home" (Aboriginal Proverb). Indigenous ancestral wisdom crystallized.

In the Maori creation story, *Papatuanuku,* primeval Earth Mother, and *Rangi-nui-etuiho-nei,* primeval celestial Sky Father, become united into one deity. They enjoyed the union and gave birth to many children, and some of these children embodied traits of their mother and others of their father. *Papatuanuku and Rangi-nui-etuiho-nei* enjoyed having their children very close to them, but some of them became resentful at being so close and desired moving away to discover what life was like outside of their mother's and father's close embrace. The *ikai-whakangirai* (kindlers of fires or initiating ones) then conveyed their discontent with being confined and informed their shocked parents of the intention to leave. Some of the other children slipped out of the saliva of *Papatuanuku*. Once these children were outside the purview of their heavenly parents, they communicated with the rest of the children still with their mother and father, and this revelation caused further disruption in the family.

Papatuanuku and *Rangi-nui-etuiho-nei*'s most powerful son was *Tane,* who possessed divine powers and was known best for his transmission of the seed of life from *Io-matua-kore,* the Supreme Spiritual Being, creating birds and trees, but also the one who tossed his father, *Rangi-nui-etuiho-nei,* upward with his feet and split the union of his mother and father, who had always been one. He then asked the gods about *ite uhai,* "the female element within a woman," and after much persuasion, two of these spirit beings showed him the first woman known by many names, including *Hine*. After seeing her, he was moved by his physical instincts, even though he had spiritual power to meet his spiritual needs. When *Tane* attempted to engage in his first male attempt with *Hine*, he hurt her eyes and tears formed. He then interfered with her ears and a discharge resulted. His attempts at entering *Hine's* nostrils produced mucus, and he was unsuccessful entering *Hine's* mouth since saliva came out. He then moved downwards, and as he moved around the armpits, sweat formed. Finally, he found the pubic area, and when he connected with *Hine* here, he felt a great force from within her. He was finally fulfilled, and from this union, humanity emerged. This union of woman and man persists today and is the basis of new life. *Iwahinei* is the Māori word for woman, and *Taneî* is the word for man. Had *Tane* not interfered with *Hine*, human beings would not have experienced the end of physical life.[19] This account is part of a more detailed and complex Māori creation story.

The Indigenous Shona people of Zimbabwe, like other Indigenous peoples and nations, have many cosmogonic stories to explain creation. One such story is known as "*Mwedzi* and his Two Wives," described as follows:

> Mwari (God) made man whom he called Mwedzi (Moon). He then created a woman whom he called Hweva (Morning Star) and gave her to Mwedzi as wife. Mwedzi was going to live with her for only two years after which Hweva would return to heaven. At night in Mwedzi's hut the two slept together and the following morning Hweva's belly was hugely swollen. She gave birth to vegetation of all kinds. When the two years were up Mwari called Hweva back to heaven. He then sent another woman, Vhenekeratsvimborume (Evening Star) to be Mwedzi's wife with whom he was again going to live for two years. After sleeping with Mwedzi, Evening Star's belly became swollen and she gave birth to cattle, goats and sheep. The following morning she gave birth to boys and girls. Mwari came and said the two years were finished and Evening Star had to return to heaven, but Mwedzi slept with her and she gave birth to lions, leopards, snakes, scorpions and other dangerous creatures. Mwedzi became king of a large realm.[20]

Another creation story from the Shona tradition notes how the *Madzetateguru,* ancestral spirits, and the natural landscape and ancestors evolved from the land, how the sacred Nyangani Mountain in the east, the highest in Zimbabwe, is home of the ancestral spirits that protect the people of the land. Any desecration and disrespect when visiting the mountain results in death and mysterious disappearance.

In Africa, where over 1,435 billion people lived in 2023, various creation stories emerged, representing a plethora of ideas about the universe:

> As they went through life, African peoples observed the world around them and reflected upon it. They looked at the sky above with its stars, moon, sun and meteorites; with its clouds, rain, rainbows and the movement of the winds. Below they saw the earth with its myriad life-forms, animals, insects, and plants, and its rivers and lakes, rocks and mountains. They saw the limits of human's powers and knowledge, and the shortness of human life. They experienced and witnessed the processes of birth, growth, procreation and death; they felt the agonies of the body and mind, hunger and thirst, the emotions of joy, fear, and love.[21]

It was the wonder of the natural world and the complex and dynamic life-giving processes of life that compelled African reflection on the source, and thus the ancient Africans conceptualized a Creator of the Universe figure, known as God in the Anglicized western tradition. Though there are thousands of creation stories among the range of diverse peoples in Africa, there is unequivocal unanimity in affirming the eternal and all-powerful Creator being the cause of everything, often working with other deities as the Yoruba account described. What's evident from this summary on Indigenous African views of creation is that African people, like all other Indigenous people, understand the limitations of human beings, particularly regarding capacities and knowledge, underscoring the role of higher spiritual powers on Earth and in the universe that ultimately determine destiny and the manner in which the world functions. Hence, various life sequence ceremonies from birth to passing away are normative for Africans and all Indigenous peoples.

Lest many consider Indigenous creation stories centered around male deity figures and imaginary constructions, Indigenous Turtle Island scholar Paula Gunn Allen explains creation gender inclusiveness in her classic, *Grandmothers of the Light: A Medicine Woman's Handbook*, noting the words from John Gunn's book on the Laguna and Acoma,

> Their theory [concerning Thinking Woman] is that reason (personified) is the supreme power, a master mind that has always existed, which they call Stich-tche-na-ko. This is the feminine thought for thought or reason. She had one sister, Shro-tu-na-ko, memory or instinct [nako means woman]. Their belief is that Stich-tche-na-ko is the creator of all, and to her they offer their most devout prayers. [Gunn, 1971][22]

Allen explains further,

> Spider Grandmother, the major deity of the Keres, is weaver and thinker: she thinks, therefore we are. Though she is "supreme"—the thought sounds wrong put in those terms and read from a Western perspective where "supreme" means king or pope or dictator—she is not alone. There isn't an "only," just as there isn't a beginning as such. Surely, the Western mind inquires, something comes before her, something made her. Surely the universe has a beginning, and an end. But like these stories, which go on and on, Indians seem to believe that life itself does not have endings. And if that is so, then what use is there of beginnings?[23]

Allen portrays Spider Grandmother as a Great Goddess with infinite medicinal powers akin to the power of the wind, since she is constantly in motion and is the basis of all movement that impels life. Such is similar to the ancient Mayan Goddess creator spirit whose aura spans the land, mountains, clouds, corn, and deer, all sacred for Mayan people. Allen cites these examples as the foundation of the sisterhood of life that encompasses the universe, and as the Laguna Indians practice, always humble. Creation is multifaceted and based on multiple deities, reflected in Keres cosmogony:

> Thinking Woman (*S'tsi'naku, Tse-che-nako, Sussistinako, Stich-tche-nako, Tsi'its'nako*) and her cohorts *Nau'ts'ity* (*Nautset, Naotsete, Nautsete*) and *Ic'sts'ity* (*Ic'ts'ity, Icsity, Utset, Iyatiku*) create and recreate the cosmos giving shape, form, and meaning to all that is. As she thinks, so we are. Spider lives everywhere and presides in *Shipap*, the underground source of life of the people, where she sits on *Iyatiku's* shoulder and advises her.[24]

Similarly, *Agawelu* (Old Woman or Crone) or *Selu* (Corn) is the well-known female deity of the Cherokee, Allen notes, with corn and gynocracy going hand in hand in ancient Indigenous cosmogonies that may have begun to erode with the decline of the classic Mayan and Pueblan civilizations in the 12th century.[25] The roles of female deities like *Asdzaa Nádleehé* (Changing Woman) and *Yoolgai asdzáá* (White Shell Woman) in the Diné creation narrative, with every ceremonial song honoring *Asdzaa Nádleehé*, echo similar concepts of female deities in creation. Such are akin to many other Indigenous traditions connecting the southwestern region of Turtle Island like the Zuni, Hopi, and Keres and other Indigenous cultures around the Earth. While there are striking similarities, there are also inevitably marked differences—interrelated, yet equally distinctive.

Though there are instructive descriptions of the role of female deities in many Indigenous creation stories, it is important to note that Indigenous languages at their root did not and do not have specific gender pronouns. For example, in most Indigenous Turtle Island, African, and Asian languages, there are no specific words for "she" and "he" predicated on biological attributes as we know them from the English language. Gender inclusiveness is normative in Indigenous languages, and the dualistic binaries that we see extant in European languages, for instance, do not exist in the former. This is a critical note because it serves as a reminder that the current and recent historical context in the world was not and is *not* the norm for most cultures and histories of the world. There may have been shifts toward "avuncularchy," "an uncle-centered system" following the decline of the classic Mayan and Pueblan civilizations, Allen suggests.[26]

Yet it is equally instructive to realize how gender hierarchy based on the supremacy of males has become the norm of most societies rooted in feudal, colonial, or other patriarchal-based societies, especially capitalist, even influencing Indigenous societies. Indigenous creation stories generally refer to creation figures and deities by individual names rather than ascribing gender. Diné Hataali Jones Benally from the Diné tradition often refers to such figures in the plural and not singular, recalling the nine divinities who were all collectively the Creation figures of the beginningless and endless creation, timeless, and genderless. Ifi Amadiume notes issues of gender complementarity in the foundational illumination of the ancient Indigenous Umueri clan and the Igala people of the broader Igbo nation in Nnobi, Nigeria, which she studied and where she lived. She avers:

> The flexibility of Igbo gender construction meant that gender was separate from biological sex. Daughters could become sons and consequently male. Daughters and women in general could be husbands to wives and consequently males in relation to their wives.[27]

In the discourse on women and gender, Amadiume insightfully emphasizes the point made in Maurice Godelier's 1981 book on the massive discrepancy between the vast number of societies in the world and the relatively low numbers of anthropological studies on relations between women and men. She explains that anthropologies have engaged in shoddy research on the subject of gender and sex among Indigenous societies, citing Godelier's observation that anthropologists cumulatively may have studied at the most 800–900 societies in the world, yet there are as many as 10,000 societies globally. She reiterates that the proposal of African and other women of color collecting information themselves for the construction of gender narratives at the *Association of African Women Research and Development* (AAWORD) workshop in Dakar, Senegal in December 1983 is critical in the decolonization of gender scholarship.[28] Similarly, Oyeronke Oyewumi's book, *The Invention of Women: Making an African Sense of Western Gender Discourses,* is relevant because she challenges the western biological determinism of gender and stresses that social organization in Igbo society was based on relative age and accumulation of knowledge over extended time and other cultural factors.[29]

Though most non–Indigenous people consider many Indigenous creation stories "folk myths," these stories are viewed as factual in all Indigenous societies and serve as the basis of existence. They are verifiable—for example, the recalling of floods in many Indigenous evolving creation narratives, like the Diné who understand creation from the time when the Grand Canyon in Arizona was replete with water billions of years ago. Indigenous creation stories are important instructive teaching mediums (particularly for children) that provide the basis for understanding roots in the Earth, ancestral lands, and the sacredness of life in the universe. Indigenous peoples' understanding of creation, often particular to the specific culture or clan, is all encompassing of two-legged and four-legged creatures from the sea, and various natural landscape formations that are intrinsic and interwoven with each other. Humans are not exclusively accorded the pinnacle in a hierarchy over other forms of life as in the Hebrew scriptures' Genesis account in the Abrahamic religions. While there are various levels of ability, diversity of numbers and presence, and multiple functions and relevance, all life forms are very much alive as in stones and mountains, hills, rivers, deserts, savannahs, seas, etc. convergent in that each form possesses the creative and creation spirit and is essential to the sacred thread that holds all life together. These beings do not always necessarily function in accordance with human intellectual impulses and calculations. Indigenous ancestral and contemporary knowledge of the universe includes such beings in the unfolding of life, and the relationships with these beings provide fundamental direction, purpose, and meaning in and of life.[30] One of the most classic illustrations of this fact is the manner in which ants, as beings and cultures, live their lives in total constructive harmony with each other in pluralistic environments all over our Earth Mother, relentlessly taking care of each other and never in conflict with each other, operating independently of our expectations. This seemingly simple but generally overlooked observation was noted and accentuated by renowned ecological biologist E.O. Wilson, who commented that after writing 27 books and retiring from his academic vocation, studying the ants and "saving the natural world now ... was more important than winning any intellectual fights."[31]

Indigenous Creation, Creating, and Cyclical Views as Opposed to Linear Views of Reality

The Earth never makes progress, is seasonal, and simply rotates on its axis every twenty-four hours, completing the revolution around the Sun every 365 days. The ideology of "progress" as an "exceptionalist" tendency in the train of human history has, in fact, caused much suffering and ecocide. It is one of the reasons that the Indigenous peoples insist on the protection of sacred sites and places in all parts of the world to this day. It is particularly at this juncture that Indigenous creation stories and accounts are telling: the sacredness of all life and the Earth and the rest of the universe and all females as life-givers are indivisible. Human life is but one tiny, albeit significant, element in the diverse mosaic of creation.

Indigenous creation and spiritual knowledge are sacred and, when violated, have lethal consequences. Siy'aka, the Hunkpapa spiritual leader who revealed his dream to the anthropologist Francis Dunsmore, is an example. The violation of silence on sacred Indigenous knowledge as required by elders in that situation resulted in Siy'aka's passing away. Certain sacred knowledge is necessarily off-limits to outsiders to the community and essential in certain cases to protect the integrity and power of such knowledge. Vine Deloria correctly noted that "many times, the thoughtless scholar … [was] the cause of great tragedy among Indians."[32]

Everything is organic and interwoven, like the elements in the ecology and the environment found within us: Earth, wind, air, and fire. Nature is part of us, and we are part of Her inextricably. There is no objectification or separation. Nature is all encompassing. Thus, imposing extra-natural solutions in every situation can be problematic. For instance, in 2021, as the world faced the lethal pandemic, Covid-19, the founder of the mRNA gene therapy as an antidote to the Covid-19 virus, Robert Malone (who was later erased as the pioneer of the gene modification therapy), warned the U.S. government that there were significant health risks due to the unpredictability of inserting an artificial mRNA spike protein into the cytoplasm. He found a close interrelationship between natural, unmodified RNA with the DNA of individuals, and that such unnatural processes involving artificial GMO proteins could have unknown, long-term health consequences, particularly increasing risks of blood clots among the young and elderly, myocarditis (heart inflammation), and miscarriages among pregnant women.[33] Mae-Wan Ho, director of the Institute of Science in Society, a leading scientist, prolific author, and winner of the 2014 Prigogine Medal, explained the complexity of RNA and DNA interactions in her early caveat about the dangers of genetic modification and engineering that was based on very inaccurate research about the human genome, a fact scientifically confirmed today:

> The original rationale and impetus for genetic modification was the "central dogma" of molecular biology that assumed DNA (deoxyribonucleic acid) carries all the instructions for making an organism, which are transmitted via RNA (ribonucleic acid) to protein to biological function in linear causal chains. This is contrary to the reality of the "fluid genome" that has emerged since the mid–1970s. In order to survive, the organism needs to engage in natural genetic modification in real time, an exquisitely precise molecular dance of life with RNA and DNA responding to and participating fully in "downstream" biological functions.[34]

High technology, while useful in the short term, comes generally at the cost of the lives of other related forms of life including four-leggeds, birds, insects, plants, trees, and all beings, and the source of life, Water. It was and is continually anchored in a distorted ideological assumption of radical dichotomies between "living" and "inanimate," one that considers other forms of life expendable, reflected, for example, in caging, experimenting with, injecting, and torturing other creatures in the supposed pursuit of medical cures for humans. This profound scientific disregard is also a factor in the current ecological, environmental, climate, water, land, and food crises permeating all sectors of life on Earth as noted in the preface of this book, with extinction of many forms of life, including us, as probabilities in the wake of Earth heating and ecocide. This peculiar human hubris and apotheosis has been lethal. Once sustainable environments have turned into zones of death and suffering has been caused by protracted drought, lack of consistent rainfall, mostly contaminated water for drinking, agricultural biodiversity extinction of 75 percent, species extinction of between 150–300 daily, and land degradation of 75 percent and growing due to chemical use and industrial pollution since the industrial "revolution."[35] For Indigenous peoples and Indigenous science, plants and other forms of life are closely interwoven with us and are sacred because they are like us, created with divine purpose and never expendable for gluttonous human wants, desires, profit, and materialistic accumulation. The timeless instructions and wisdom of Indigenous peoples echo, *We are the plants, we are the "animals," we are the Water, we are the Land and the Earth, we are Nature*—eternal scientific truths captured in the diverse Indigenous creation accounts described earlier in this chapter. The unequivocal truth is that any destruction and annihilation of our relatives will ineluctably lead to human life being obliterated, too, since we can't survive as a separate species.

Indigenous oral traditions "represented not simply information on ancient events but precise knowledge of birds, animals, plants, geologic features, and religious experiences" of particular peoples, all violently disrupted by colonization, echoing Black Elk and all Indigenous medicine people who have held from time immemorial that oral tradition accounts are authentic.[36] A cardinal point that Deloria highlights that many scholars elide is the distinctive nature of Indigenous languages that differ markedly and radically from European language systems in foundation, orientation, structure, meaning, cultural nuances, and complexity:

> Storytelling was a precise art because of the nature of Indian languages. Some tribal languages had as many as twenty words to describe rain, snow, wind and other natural elements; languages had precise words to describe the various states of human emotion, the intensity of human physical efforts, and the serenity of the land itself. If the stories began "Once upon a time…." they quickly gave the listener a complete accurate rendering of a specific experience which Western languages could not possibly duplicate…. In some of the larger Indian nations elders functioned pretty much as scientists do today.[37]

Taiaiake Alfred also stresses the complex and distinctive differences between Indigenous and non–Indigenous languages, or what he refers to as Onkwehonwe and European languages, citing his teacher, Leroy Little Bear, who explained that Indigenous languages are *verb-based* and "communicate through descriptions of movement and activity," underscoring dynamism, fluidity, and resilience.[38]

Deloria points out that apprehending and acquiring knowledge for Indigenous people is a lifetime process, thus certain persons were provided with specific mediums to gain access to specialized knowledge through vision quests, fasts, and other ceremonies, which was then selectively shared with communities as determined by spiritual teachers and knowledge guardians. "Once upon a time" is an expression used in many children's story books, usually couching the story in mythic and fictional terms, unlike in Indigenous oral transmission stories. Further, while the west uses the human intellect to "discover" origins and explanations, Indigenous peoples depend on the rest of the natural world and powerful nature forces to understand the Earth, life, and the Universe:

> The difference between non–Western and Western knowledge is that the knowledge is personal for non–Western peoples and impersonal for the Western scientist. Americans believe that anyone can use knowledge; for American Indians, only those people given the knowledge by other entities can use it properly.[39]

Deloria elaborates:

> The major difference between American Indian views of the physical world and Western science lies in the premise accepted by Indians and rejected by scientists: the world in which we live is alive. American Indians look at events to determine the spiritual activity supporting or undergirding them. Science insists, albeit at a great price in understanding, that the observer be as detached as possible from the events he or she is observing. Indians thus obtain information from birds, animals, rivers, and mountains which is inaccessible to modern science. Indians also know that human beings must participate in events, not isolate themselves from occurrences in the physical world.[40]

The principles of interwovenness of the creation of humans, with the rest of the trees, plants, mountains, rivers, four-leggeds, birds, insects, stones, and all the life on Mother Earth, lie at the foundation of Indigenous histories, heritage, and cultures that is the basis of preservation of Creation from the original times.

> The Indians developed as comprehensive analysis of the world as has Western Science, but the goals have been very different. Western science is based on Roger Bacon's command to pry nature's secrets from her, by torture, if nothing else. The medicine men (and women-ours) sought alliances with other entities and, according to some tribes accumulated both spiritual power and cumulative knowledge over several lifetimes. The Muskogees thought much knowledge had been given at the beginning of the world, and we should examine their beliefs in this respect.[41]

Deloria explains that the Muskogee believed that the world was originally *mind*, from which matter emerged, and that all life, including plants, four-leggeds, and other creatures, connect all forms of life with this preexisting nature. The Omaha Nation, too, held that an Invisible Energy generated life in visible and physical form and that all life was thus connected to the Invisible Energy, so that there were no radical distinctions between that which can be seen and that which cannot, and between "the dead" and the living. All emanated from *Wakonda*, the Invisible Power and Energy.[42]

This is the core teaching of Indigenous creation, histories, and cultures that western societies have refused to concede until the present: that we are all sacredly endowed with the spirit of the Creator, the Spirit-Giver, as Willie Gordon, from Gurrbi in Australia, asserts. The ancient Indigenous creation narratives and philosophies function as a constant reminder that the unjust, militarized, exploitative, polarized, and institutionally conflicted world is certainly *not* what the world has always been and not what it was meant to be.

Trying to imitate natural ways or improving upon the ways of the Earth unnaturally has been destructive, if one thinks, for example, about Water today needing chlorination for drinking purposes. Hataali Jones Benally expressed his revulsion at all these illusory attempts at "change" because he explains that the more humans insist on "changing" things, the more things "become messed up." He avers that the Sun, Moon, Earth, Stars, and all the Spirit Powers of the Endless Universe at work on the Earth are the ones who really possess power and that everything has been set accordingly, requiring understanding and respect of these natural, powerful spirit ways. He has urged human beings to back off from their egotism and delusional belief that they have power to "change the Earth." Similarly, well-known religion scholar Charles Long intimated some years ago at an American Academy of Religion meeting that it was absurd that people thought that they can "save the Earth." It is only the Mother Earth that can save anybody because all life is borne of Her/Them and returns to Her/Them as articulated earlier in this chapter. Very importantly, for all Indigenous cultures, the design of all created beings was and is circular, as with the Sun, Moon, Stars, Earth, and other celestial beings.

The drying up of rivers around the world—for example, the Colorado River, which sustains over 40 million people in the western part of Turtle Island and is now reduced to a trickle—combined with unprecedented heat and over 100 fires in the summer of 2021 and 400 fires in the northern part of the continent in 2023 are just some signs of the looming catastrophes facing the nation and the world. Capitalist industrialist technology and monetary investment by the trillions are powerless to halt the drying up and depletion of the very limited fresh water accessible on Mother Earth in many regions of the world, illustrated in the tar sands oil extraction industry in the boreal forests in Alberta, where freshwater is being depleted for such extraction.

There is great and pervasive Spiritual Cleansing occurring all over the Earth. Air conditioners and "green energies" like electric, lithium battery charged automobiles and even solar and wind energies only exacerbate the problem, according to Hataali Jones Benally. He states that such desperate measures are rooted in a philosophy and culture that refuses to understand that we are specks of dust that have no moral right to tap into the spiritual powers of the Sun and Wind while focusing essentially on meeting our own, often industrial, human material wants. Instead, Indigenous medicine persons performing Earth and spirit ancestral ceremonies and dances and songs are the key to understanding how we all need to work, live, and heal with what's unfolding daily, including addressing issues of drought and lack of rainfall. Rain dances, for example, are still practiced in many Indigenous communities so that the dryness of the Earth is fulfilled by falling rain, as among the Ohlone, Lakota, Cherokee, Diné, and Mayan peoples. The essential inclusion of other beings and forms of life is required for understanding the regeneration of life on the Earth, principally mediated through Indigenous healers, teachers, and practitioners from around the world who bring the deepest and broadest knowledge and experience to bear on such regeneration.

Summary and Conclusion

Creation stories of Indigenous peoples all over the world are united in their understanding of the sacredness of all life and the foundational spiritual bond that is rooted in the overarching and underlying spiritual and sacred nature of life. These wisdom-instructive Indigenous creation stories underscore how Indigenous people are rooted in Creation/Creating cosmologies, the foundation of harmonious living on the Earth, and practice ceremonies in honor of such Creation/Creating, since without these events, we would neither be here on our Mother Earth, nor understand our purpose in life: living sustainably and in reciprocity with all our relatives of every shape and form in the world. The closing Lakota prayer for each ceremony and prayer is *Mitakuye Oyasin*, "All our Relations." Capitalism has catastrophically never grasped the power of this spiritual anchor of all life, hence the perennial and perpetual conflicts, divisions, and violence from its systemic function and operation. Rooted in the Earth implies a belonging to the Earth so that we realize we really own nothing; instead, we are "owned" by our mother, Earth.

Renowned Lakota healer Lame Deer crystallizes and summarizes this timeless Indigenous understanding, way of life, purpose of life, and complexity of life when he says:

> The Great Spirit is one, yet he is many. He is part of the sun and the sun is part of him. He can be in a thunderbird or animal or plant. A human being, too, is many things. Whatever makes up the air, the earth, the herbs, the stones is part of our body.... The Great Spirit ... only sketches the path of life roughly for all the creatures on earth, shows them where to go, where to arrive at, but leaves them to find their own way to get there. He wants them to act independently according to their nature, to the urges in each of them.
>
> All creatures exist for a purpose. Even an ant knows what that purpose is—not with its brain, but somehow it knows. Only human beings have come to a point where they no longer know why they exist.[43]

Lame Deer echoes what Malidoma Somé from the Dagara tradition of West Africa says about the western capitalist way of life in contradistinction to Indigenous knowledge and ways of being and living:

> Indigenous people find their rhythms in nature. Westerners, on the other hand, seem to seek meaning in the realm of the machine, where one finds neither peace nor wholeness, but ceaseless movement. In the West, people are always frenetically rushing somewhere in the countless lanes of the multiple highways of progress.[44]

Though these two medicine persons and traditional healers are from two different continents living decades apart, their common essence of understanding life and the interwovenness of nature of which humans are an indispensable part is unequivocal.

Contradictions abound in capitalist-driven societies. While progressive climate change activists and fossil fuel opponents like Green Peace and Extinction Busters in the west demand the immediate cessation of fossil fuel production and consumption, the U.S. federal budget earmarked $550 billion dollars in 2021 for the rebuilding of infrastructure that includes hundreds of billions for the construction of new highways and roads that presuppose the use of automobiles and other motorized means of transportation heavily dependent on fossil fuels or combustible electric battery charged vehicles. Factually, both do irreparable harm to the Earth, the ecology, and the environment, including us.

The confusion about identity and purpose in western cultures does not exist among traditional Indigenous peoples and cultures because Indigenous cultures are rooted in ancestral creation histories and accounts, many of which are recounted in this chapter that anchor our being and living in this world. As we embark on the next chapter on the history and evolution of capitalism, Lame Deer again has didactic lessons for all, albeit his words were recorded almost fifty years ago, narrated in his typical humorous tone:

> I made up a new proverb. "Indians chase the vision; white men chase the dollar." We are lousy raw materials from which to form a capitalist. We could do it easily, but then we would stop

being Indians ... deep down within us lingers a feeling that land, water, air, the earth and what lies beneath its surface cannot be owned as someone's private property. That belongs to everybody, and if man (humanity—this author) wants to survive, he had better come around to this Indian point of view, the sooner the better, because there isn't much left to think it over.[45]

There certainly isn't much time to ponder and reflect on *whether* we should all change course and return to Indigenous ancestral ways of living, but *how*; each waiting day pushes us confused humans further to the brink of imminent destruction and extinction because of our persistent and defiant egotistical and destructive violence against our only universal mother, Earth.

The next chapter provides a detailed illumination of the root cause of the crises we are living today, generated by capitalist industrialism, built upon the genocide of Indigenous and African people in particular, the enslavement of people of color, and all workers in general.

2

An Interactive History of Europe, Colonialism, Slavery, and the Emergence of Capitalism

> The medieval aristocracy was primarily a military aristocracy and the men who participated in its high medieval diaspora were trained fighters. They possessed a particular set of arms and equipment and had been brought up to fight in a particular way. As a consequence the spread of the Frankish aristocracy entailed diffusion of a military technology-armaments, fortifications and methods of waging war—from its place of origin.[1]
> —Robert Bartlett, *The Making of Europe: Conquest, Colonization, and Cultural Change, 950–1350*

> Weapons they have none, nor are acquainted with them, for I showed them swords which they grasped by the blades, and cut themselves through ignorance.... It appears to me, that the people are ingenious, and would be good servants and I am of opinion that they would very readily become Christians, as they appear to have no religion. They very quickly learn such words as are spoken to them. If it please our Lord, I intend at my return to carry home six of them to your Highnesses, that they may learn our language.[2]
> —Christopher Columbus, "Columbus' Journal Entries from August to November 1492"

> They forced their way into native settlements, slaughtering everyone they found there, including small children, old men, pregnant women who had just given birth ... the native population of the two islands was certainly over six hundred thousand (and I personally reckon it more than a million) ... fewer than two hundred survive on each of the two islands.[3]
> —Bartolomé de las Casas, *A Short Account of the Destruction of the Indies*

Introduction

This chapter will interrogate the historical, religious, ecological, political, economic, militaristic, and social evolution during the post-medieval period in Europe to understand the emergence of incipient capitalism—the system that birthed industrial capitalism following the processes of slavery and colonialism. Understanding history in its complexity and dynamism is essential for addressing the plethora of crises facing us in the world as outlined in the preface of the book.

Highlights of Post-Medieval Europe

The medieval period in Europe was characterized by polarization and domination by the feudal ruling classes and the Papacy, the latter signifying the unquestionable authority of the Christian church. The constant war between the French and the English, for example, from 1338 to 1453, also had a bearing on the evolution of these two nations. Much of the impetus for the expansion of learning, education, and social uplift, however, came from the Moorish occupation of Spain. Communities then were mostly rural small villages where the peasants and serfs worked as vassals for landlords who were connected to monarchical rulers in various parts of Europe. They enjoyed no freedom of movement while wars were being waged by the monarchs. In France, for instance, from 1471 to 1483, Louis XI consolidated his rule over the princes and dukes and expanded the military, reinforcing the total dominance of the central monarchy. This was followed by Charles VIII (1483–1498), Louis XII (1498–1515), Francis I (1515–1547), and Henry II (1547–1559). In England, where Henry VII ruled from 1485 to 1509, there was a strategic end to the widespread slaughter of nobles fighting for monarchical recognition of their nobility under Henry VI, drawing in the middle classes and lower gentry, even while the merchants still made money from their commercial trade and house manufacturing practices. But these were still small-scale practices in commerce.[4]

Spain subsequently saw major revolutionary changes. The accompanying Muslim educational, economic, and socio-cultural milieu transformed Spain into the major industrial, scientific, learning, technological development, and arts and cultural center of Europe. Prior to the Moorish invasion and occupation of Spain, the region was ruled by the Visigoths, a Germanic tribe that was radically disconnected from the general population. The Visigoths managed ongoing conflicts between nobles and the ecclesial establishment, both vying for political and economic power, compelling many to migrate from urban centers to the rural areas that were typically feudally structured. Catholicism was still the religious law of the day. With the Moorish invasion came Islam, and with it all the benefits of high Islamic culture that in a real sense could be considered the pre-enlightenment era of Spain and Europe. Arts, science, literature, medicine, mathematics, astronomy, botany, history, philosophy, and jurisprudence flowed and flourished. Cities in the fertile valleys of Guadalquivir and Guadiana grew, and city baths, street lighting, ship building, artisan guilds, and commercial transactions expanded to heights never experienced in the country or in any part of Europe.[5] John Jackson explains at length about the Moorish intervention and occupation of Spain:

> They, for instance, introduced the silk industry into Spain. Ibn-al-Awam and Abu Zacaria wrote learned works on agriculture and husbandry. A translation of a treatise on agriculture by Ibn-al-Awam (twelfth century) was published in Spain as late as 1802 for the instruction of Spanish farmers. Ibn Khaldun, another Moorish expert on agriculture, wrote a valuable treatise on farming and worked out a theory of prices and the nature of capital.[6]

Subsequently, mining of precious minerals and metals expanded—gold, silver, copper, quicksilver, tin, lead, iron, and alum. Maritime commerce boomed with the construction of hundreds of ships that provided a boon to manufacturing and continental and global trade. So, too, women surgeons and doctors from Andalusia were prominent, and female doctors in places like Cordoba were no exception.[7]

Yet, this era of flourishing prosperity in Spain and the rest of Europe was rudely disrupted with the onset of a pervasive cultural despair stemming from the Biblical scriptures' reference to the great consummation and the end of the ages. This overall pessimistic era was due to the endemic corruption and brutal suppression of dissent through the Inquisition by Christendom led by the Catholic church and accompanying social crime. The intense irruption of disease, magnified by the black plague that hit several times from the 14th through the 17th centuries, decimated hundreds of thousands. Famine caused by repeated crop failures, particularly grain, took the lives of just as many.[8] Kirkpatrick Sale notes this important historical shift in Europe that changed the ecological, socio-economic, and cultural landscape of the continent permanently, the legacy of which lingers in the evolution of Europe's "rational" and "scientific" culture, where it was seriously believed then that the end of the world would occur "7000 years after creation," a belief cherished by none other than Christóbal Colón (Columbus), as part of the fulfillment of Biblical prophecy.[9]

It was in this conflicted world that the "European Renaissance" was born, with its striking distinction of the humanism tradition and elevation of "scientific thought" and rationale to absolute proportions. The "Renaissance" humanist tradition was closely associated with Erasmus, the Dutch Catholic theologian who emphasized enhancement of the self and education as the key to such improvement. Petrarch, another devout Catholic scholar from Italy who excelled in translating ancient Greek scripts like Cicero's and in poetic and lyric composition, shared Erasmus' humanist beliefs. Petrarch held firmly that the human potential of creativity and excellence was part of divine design, though women were excluded. In the late 15th century, Italian humanist philosopher Marsilio Ficino, in his formative work, *Theologica Platonica*, described this ideology as striving "to be as God everywhere."[10] In his analytical work on the Italian "Renaissance," Lauro Martines explains that humanism was a very aristocratic movement and all of the major figures of this philosophy and theology that emerged in Europe in the 15th century and those following prided connections to the ruling aristocratic classes of ancient Greece and the Roman empire. Cicero was held up as a hero to Petrarch the humanist, Aristotle was eulogized for being tutor to Alexander the Great, Plato was elevated as teacher to Sicilian kings, and Caesar was commended for writing his "Commentaries while on campaign."[11] Martines rebukes the humanism of the "Renaissance" as a tool of the ruling classes and the extension of empire:

> All humanists, whatever their stripe, made a candid alliance with power. They plumped for the ruling classes, empires, and luminaries of past *civil* times.[12]

Martines cites the cases of two renowned Venetian humanists of the 15th century, Bernardo Giustiniani (1408–1489), who observed that all great cultures followed an empire, and Lorenzo Valla, author of six books of his *Elegantiarum*, echoed such.[13] Humanism as conceived by European humanists was in fact conceived as legitimation of the rule by privileged elites destined to preside over the unfortunate working classes, so that iconic humanists "spoke for and to the dominant social groups."[14]

As the perverted insistence of humanism advocated "a rebirth of the human in the image of God" and simultaneously and contradictorily "promoted rather a new secular pragmatism" that was unabashedly materialistic at its core, it also focused on preserving the unassailed dominance of the male over the female and the general exclusion of women, reflected in clerical patriarchy and the dictatorial father at the head of family and principality.[15]

Along with humanism, as Kirkpatrick Sale elaborates, followed the ascendancy of the faculty of human reason and the development of rationalism that became embedded in every sphere of life in science, aesthetics, culture, religion, commerce, and economics that sowed the seeds of the ideology of atheism, again particular to normative European cultural evolution of the 15th century and following.

> The task of rationalism, through science, was to show—no better, to *prove*—that there was no sanctity about these aspects of nature [trees, streams and rivers, forests, rocks, etc., mentioned earlier by Sale], that they were not animate or purposeful or sensate, but rather nothing more than measurable combinations of chemical and mechanical properties, subject to scientific analysis, prediction, and manipulation.[16]

Sale adds that with nature being "degodded," rationalism deified humans so that they were the "masters and possessors of nature," in classical Cartesian terms. The merger of science and technology on the one hand and humanism and rationalism on the other culminated with materialism and the emergence of early commercial capitalism, followed by colonization.[17] Mechanization, standardization, routinization, and de-individuation were and are core elements of this ideology.

Sale notes that the distinguishing features of the early technological development of Europe were the printing press and the gun, the latter that transmogrified into "'armaments' ... 'corned' gunpowder from the 1420s and more sophisticated gun bores and firing mechanisms from the 1460s ... followed by the arquebus ... allowing individual soldiers to have powder-fired weapons, and then with the perfection of mobile and large-bore cannon in the 1480s (proven by the French in the field in 1494)."[18] The large-bore cannon invented by post-medieval Europe was the single most lethal and devastating weapon used against the Indigenous peoples of the world. The inventors of gunpowder, Chinese monks, who had discovered the technology in the 9th century CE, originally pursued "a life-extending elixir" for healing purposes, combining saltpeter (potassium nitrate), sulfur, and charcoal, which would, in admixture, explode.[19] Though the Mongol empire used gunpowder to expand its geographical conquest across west Asia in the 13th century, the use of the cannon by colonialism was unprecedented. Commercial trading companies like the English East India Trading Company, the Dutch East India Company, the Dutch West India Company, the French "Compagnie des Indes Orientales," the Royal African Company (England), the Company of Royal Adventurers, the Company of Merchants, the Muscovy Company (Russia), and the Turkish Company (which later became the Levant Company) depended on such overwhelming firepower in their conquest of the world and the chattel slavery enterprise.[20] African scholar Chinweizu highlighted how this military industrial development had disturbed "the peace of the world" and has been responsible for occupying Indigenous lands globally.[21]

Ecocide was part of the early elements of this post-medieval culture, with the "Greek empires ... [that] destroyed the once wooded hills of the Mediterranean through deliberate fires and urban encroachment, careless herding and overgrazing, ignorant planting and relentless cultivation," and the Christian ideology of the subjugation of the rest of the natural world that resulted in irreparable loss of the ecology and vegetation. "England, for example, was significantly deforested as early as the eleventh century, with probably no more than 20 percent of it still wooded (and not even 2 percent virgin) by the time of the Domesday Book in 1086," Sale asserts.[22] Historian Keith Thomas notes that the ecological preservation and environmental consciousness movement in the west is a relatively recent phenomenon, with "the profound shift in sensibilities … which occurred in England between the sixteenth and late eighteenth centuries"[23] moving away from the injunction to "replenish the earth and subdue it," and shifting to the injunction "to level the woods, till the soil, drive off the predators, kill the vermin."[24] The subjugation ideology insisted on land cultivation as part of the ingredients for "civilized" men. It became the basis for the violent occupation of British dispossession of Indigenous lands in Massachusetts, where colonizers argued that Indians had failed to cultivate the land and had thereby surrendered it, akin to the British colonial occupation of Ireland.

Today, the United Kingdom (as it's now known) has less than 50 percent of its biodiversity remaining, far less than the 75 percent of most of the world, the most intact being Indigenous-occupied lands and territories in Asia, Africa, the Americas, the Caribbean, and the Pacific. The need for utilitarian wood elements in everyday existence resulted in many of Europe's forests being erased as early as the 16th century in Spain.

Animals became associated with being "derived from beasts" in the 16th century and were thus viewed as distinct and radically inferior. In this regard, animals served as entertainment through cruel sports: bull fighting in Spain; bear baiting in northern Europe, where the bear was chained and forced to defend itself against predatory dogs; cockfights that required that wings be clipped and feet fitted with razor-sharp spurs to cause optimal injury to opponents; and hunting that was oriented toward senseless killing, with food a secondary concern. The elimination of over 1,000 deer and 1,000 wild boars in a single hunting expedition was not uncommon for feudal rulers.[25] Overfishing of herring in the Baltic Mediterranean from the 11th century intensified, so that by the end of the 15th century, these had disappeared. Other fish, too, like barbell, bream, dace, and flounder were radically reduced in number, and the right whale was eliminated from its plenitude in the eastern Atlantic from the 16th century.[26] The common sturgeon has become extinct in the Lower Rhône River Basin due to nineteenth-century shipping navigation, pollution from industrialism, and urban expansion and commerce.[27] Wolves, which are connected to the forest spirits of the world, had disappeared in England by the 15th century and in France by the early 16th century, and bears had become extinct in England by the 13th century.

It's unsurprising that sports teams in Turtle Island have been named after or associated with "animals" or mascots mimicking Indigenous people and Indigenous historical figures, rejected wholly by the latter: the Chicago Bears, Cleveland Indians, Florida Seminoles, Atlanta Braves, Washington Redskins, etc. Some 2,129 sports teams' names include references to Indigenous people, such as the Braves, Chiefs, Indians, Orangemen, Raiders, Redmen, Reds, Redskins, Savages, Squaws, Tribe, and Warriors, as well as tribe names such as Apaches, Arapahoe, Aztecs, Cherokees, Chickasaws, Chinooks, Chippewas, Choctaws, Comanches, Eskimos, Mohawks, Mohicans, Seminoles, Sioux, and Utes.[28]

The carnivorous diet was fully developed in England in contrast with other cultures, especially in the East, which was predominantly vegetarian with occasional meat.

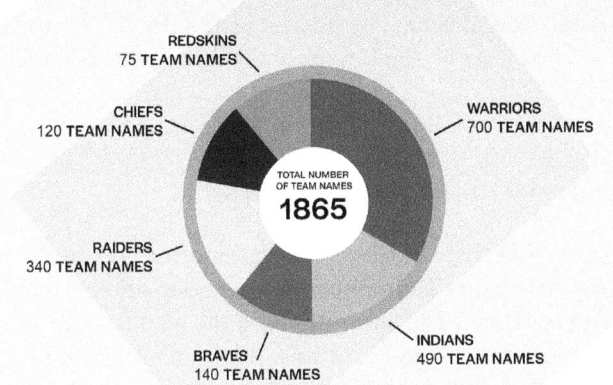

Sports Teams at College, High School, and Professional Level Using Native American Names (2013) (Veronica Rodriguez, data from Hayley Munguia, "The 2,128 Native American Mascots People Aren't Talking About," *FiveThirtyEight*, September 5, 2014).

The Amazon rainforest has now been transmogrified from a carbon sink to a net emitter of carbon dioxide, equivalent to that emitted by the U.S. each year, due to deforestation and expanded beef production, causing even more emissions of carbon dioxide, nitrous oxide, and methane that may contribute up to 14 percent of world emissions according to the UN Food and Agriculture Organization (FAO).[29] Fifty million buffalo, sacred to the Lakota Nation of the Midwest, were killed by colonial forces and beef was imposed as a staple of the Lakotas' diet.[30] Keith Thomas notes that flesh eating was believed to strengthen male virility.[31]

"Europe's technophilia, its unchecked affection for the machine, also distinguished it among world cultures," Sale explains.[32] Lewis Mumford, the philosopher who observed that it was in Europe that all of life's processes and functions revolved around the machine, noted in his book, *Technics and Civilization*, that the Cretans, the Egyptians, and the Romans were formidable engineers and had machines, but it was Europe that developed "the machine" so that life became structured around the capacities and pace of the machine.[33] He notes that the development of the mechanical clock by a monk named Gerbert (who later became Pope Sylvester II) for the purpose of meticulously keeping disciplined time as part of the rigidity that Benedictine and Catholic monastic routine demanded marked the prelude to the emergence of capitalism. Forty thousand monasteries "helped to give human enterprise the regular collective beat and rhythm of the machine," with the clock "synchronizing the actions of men."[34] This mechanical idolatry from the 17th century was castigated by Karl Marx in his critique of the routinization of capitalism.[35]

The concept of "time" that was hitherto open and undefined—and among Indigenous cultures, cyclical—saw linear concepts conducive to profit maximization in the industrial revolution.[36] The apotheosis of the machine from the late medieval period signified an ominous harbinger of the deification of high technology in the late 20th and early 21st centuries. Electronic mediums like the computer and the cellphone now have assumed dominance in the industrial, economic, financial, medical, health, social, educational, and even religious spheres. The internet, high speed computers, wireless technologies, drone warfare, and, of late, "artificial intelligence" or "AI" are part of this trend. "AI" is now being diffused into the civilian sector with mail delivery and weather observation, for instance, and is heavily present especially in U.S. universities and colleges. It is also used for testing by the U.S. military through its AFRICOM military arm in Africa and Syria.[37]

A History of Slavery and Colonization

The conquest of the Caribbean and South America was initiated by the Italian enslaver and pirate Christopher Columbus, serving none other than the Spanish monarchs Ferdinand and Isabella I. The latter imposed the bloody Inquisition on the people of Spain, causing massive suffering of Muslims, Jews, and other non–Catholics who had enjoyed such religious and cultural liberties under Moorish rule in the country. There were burnings at the stake of perceived recalcitrants in England, France, and Germany.[38] What distinguished the Inquisition in Spain was the established monarchical tribunal and trials of those considered heretics that were accountable to the Spanish monarchy and not the Catholic Papacy. Noted Latin American writer Eduardo Galeano explained that the militaristic nature of the Inquisition and the unholy marriage of monarch and papacy, which caused 150,000 Jews to be expelled, had Queen Isabella serving as the "patroness of the Holy Inquisition."[39] He stressed that the "discovery" of the western hemisphere was intrinsic to the "Christian" crusade against Islam, and he adds that Spanish Pope Alexander VI "ordained" Isabella as "proprietor and master of the New World [sic]." Isabella in the papal view was the instrument of the "extended God's reign over the earth."[40]

Pope Alexander VI, a head of the Catholic church during the 15th century, pursued personal wealth and provided the papal and religious authority for Columbus' conquest of the western hemisphere, sanctifying exclusive rights to the Spanish monarchy and colonizers. He issued Papal Bull "Inter Caetera" on May 3, 1493, which declared that under divine Catholic authority traced to Peter and Jesus Christ, all "discovered" and "undiscovered" lands in the western hemisphere and beyond were to become the possession of the Spanish invaders.[41] This Papal Bull became the ideological basis for the widespread "Doctrine of Discovery" that has yet to be denounced and formally nullified by the current papacy, notwithstanding repeated efforts by various representatives of Indigenous nations in the western hemisphere at the UN Permanent Forum on Indigenous Issues (Peoples) in New York City in 2013. Indigenous legal scholar Robert Miller notes that colonial England added another fabricated principle in line with its history of the expropriation of Indigenous lands:

> This was the principle called *terra nullius* (a land or earth that is null or void), or *vacuum domicilium* (an empty, vacant or unoccupied house or domicile). This element stated that lands that were not occupied by any person or which were actually occupied but were not being used in a manner that European legal systems approved, were considered to be available for Discovery claims.[42]

Luis Rivera notes that "discovery" and "expropriation" became "concurrent acts," and expropriation a "formal and juridical act."[43]

Galeano notes that three years after Columbus launched military campaigns against the Indigenous peoples of Haiti (known as Hispaniola then), about 200 Spanish

Columbus' Initial Landing and Colonial Encounter with the Indigenous Taíno of the Caribbean on Hispaniola (Wikimedia Commons).

foot soldiers employed dogs to annihilate Indians and then shipped 500 Indians to Spain as enslaved captives. Spanish conquistadors read a long papal *Requerimiento* that urged the Indians to accept the Catholic faith and Spanish sovereign as divine rulers, just before they were executed, burned, hanged, or thrown to the dogs as food, and demanded allegiance of the survivors under threat of death if disobeyed.

The estimates of the Indigenous nations who were annihilated by the brutality of the Iberian invasion and enslavement of the Indigenous people run into millions, around eight million, a real genocide and ethnocide, reducing Indigenous Arawaks, Taínos, Caribs, and other Indigenous peoples to hundreds. Ward Churchill asserts (citing the work of Troy Floyd and Stuart B. Schwartz) that the Taíno nation was reduced to a couple of hundred by 1542, and that the entire Caribbean Basin Indians were virtually exterminated from numbering over 15 million prior to the Columbian invasion.[44]

James Loewen notes that Columbus sent the first 5,000 enslaved Indians across the Atlantic. Portugal followed suit and launched ethnic cleansing of Labrador, forcing Beothuk Indians to become slaves in Europe and Cape Verde. The Pilgrims and Puritans made profits from selling the survivors of the Pequot War into slavery in Bermuda in 1637. The Natchez Nation was kidnapped and shipped in shackles to the West Indies in 1731.[45] Cynthia Chambers, an educator from the northern part of Turtle Island, describes the need for *Treaty Education* as part of all core curricula in this land, similar to science and math.[46]

The rape of very young Indigenous women by Columbus and his crew was foundational in the slavery regime, precursors to the child sex trafficking rampant in the world today. Like subsequent British colonial opium trafficking protected by the British Royal Navy in the 19th century after it was confiscated from Bengal in India during the British occupation there, sex trafficking has deep colonial and enslavement roots. The confiscated opium product was then militarily imposed on China, making it the most prized commodity of the 19th century, along with chattel slavery, which financed the construction of railroads and Ivy League colleges and universities in North America.[47] Colonial enslavement was the province of pirates on the open seas, as Gerald Horne writes, like "Sir Henry … and Charles-François d'Angennes, Marquiz de Mainenon, by 1680 the largest sugar magnate in Martinique—and a notorious freebooter," with 2,000 corsairs (French pirates acting on behalf of the monarchy and who were thus spared the death penalty for trafficking enslaved Indians and Africans). Horne stresses that capitalism at its core was enslavement based, white supremacy grounded, and gangster executed.[48]

Peter Martyr d'Anghiera, a 15th-century historian who served the Spanish monarchy and was chaplain to Ferdinand and Isabella, described the high mortality rates of enslaved Indians forced to constantly work with little periods of rest and women inducing abortions to avoid children becoming enslaved.[49] He concluded that the original number of 12 million Indigenous people in the Caribbean was radically reduced, but had no idea by how many. Las Casas documented similar information.[50] Eric Williams, drawing upon Las Casas' *Historias de las Indias* (1559), highlighted how Indians, young and old, female and male, were confined by the hundreds below deck in overcrowded slave ships, causing suffocation and death, akin to that of enslaved Africans in transatlantic slave ships.[51] Bodies were dumped into the sea and served as map pointers for ships. Andres Reséndez notes that when Columbus invaded the Caribbean ten years earlier, he had a model of slavery from São Jorge de Mina on the coast of Guinea, a Portuguese-built fort to ensure ownership of enslaved Africans in contemporary coastal Ghana. The site subsequently became notorious with the largest port transporting enslaved Africans, Elmina.[52]

The western hemisphere, America, however, was not named after Columbus; instead, it was named for Amerigo Vespucci, another Italian colonial who insisted in his letter to Lorenzo di Piero de' Medici that he had "discovered" "the New World," a continent that was "more densely peopled and abounding in animals than our Europe or Asia or Africa, and in addition a climate milder and more delightful than in any other region known to us."[53] For Vespucci, it was only "capital" that mattered as he wrote. Over 100 million Indians were exterminated as a result of the Columbian and post-Columbian invasion, as a result of slavery, violence, and unknown diseases.

There was a continuity of slavery from the Caribbean through South America. Peru, Bolivia, Venezuela (in particular), and Mexico were areas rich in silver, tin, and gold, and thus slavery grew early in the 16th century. The Iberian conquest unleashed by Columbus had huge ripple effects through the likes of Cortez and Pizarro. Eduardo Galeano writes that in Bolivia, today considered the most impoverished nation in the Americas after Haiti, eight million Indians were killed or died from exhaustion after working in the colonial mineral mines, where "one of the diamonds encrusted in a rich caballero's shield [a sword-wielding Spanish cavalier] was worth more than what an Indian could earn in his whole life under the *mitayo* [a forced tribute system]."[54]

Citing Celso Furtado's work on this subject, Galeano underscores the fact of how external Spanish colonial trade was limited and did not allow for any serious accumulation of wealth; it was private Spanish traders (pirates) in the Americas who provided the bulk of early capitalist accumulation in the "home country, while undermining the accumulation of wealth and viable economic development in the colonies." He observes that though there were feudal dimensions to Latin America's early colonial economy, "it functioned at the service of capitalism developing elsewhere."[55] Furtado's research revealed the gross imbalance of trade between Spanish imports and exports, providing the basis for capital accumulation through export of vital metals like gold, tin, and silver indicated earlier:

> The external trade of the Spanish colonies was subject to strict control by the metropolitan authorities.... The averages over the long periods indicate that the value of the precious metals shipped by the private sector was about four times that of the total imports ... the foremost objective of the work carried out in the Americas was to create a flow of resources for accumulation in Spain.[56]

Spain pillaged South America and the Caribbean of its vast wealth. The French colonization of Haiti, Martinique, and Guadeloupe had similar objectives, as did the British in their early colonization of islands in the Caribbean like St. Kitts in 1624, Barbados, Montserrat, and Antigua in 1627, Nevis in 1628, and Jamaica in 1655 after expelling the Spanish colonizers. Sugar became the basis for the spawning of France's and Britain's capitalist industrial economy. Sydney W. Mintz observes in *Sweetness of Sugar*:

> though sugar cane was flanked by other harvests—coffee, cacao (chocolate), indigo, tobacco, and so on—it surpassed them all in importance and outlasted them.[57]

Haiti had become the leading sugar producer in the Americas, if not the world, with enslaved African forced labor. Trevor Burnard and John Garrigus emphasize how the sugar-based plantation economies of the Caribbean, specifically Saint-Domingue (Haiti and the Dominican Republic) and Jamaica in the 18th century were central to the emergence of Atlantic capitalism that was at the foundation of Europe's unsurpassed ill-accumulated wealth. Sugar production became the basis of what we know as "whiteness" and white supremacy, with Jamaica and Saint-Domingue (Haiti) being key in profit maximization and at the core of capitalist accumulation. Wealth thus became equated with whiteness and "supplemented class standing as a primary basis of social standing core ideologies of capitalism then and now."[58]

The enslavement of Africans, especially in the French and British colonies of Saint Domingo and Jamaica from the 17th century respectively, and the enslaving of Indigenous Indians in the Caribbean and colonization of peoples in Africa, Asia, and the western hemisphere were the preconditions for the expansion of capitalist accumulation.[59] Abuse and violation of enslaved Black bodies in the Caribbean during chattel slavery was unprecedented, and that legacy lingers in the unresolved and persistent contemporary police brutality against and repression of Black people in the U.S., Brazil, Haiti, Colombia, Western Europe, and other parts of the Black diaspora.[60] The African-Caribbean slave regime based on the sugar industry in the 18th century was pivotal in connecting Africa and Europe in the latter's formative capitalist development.[61] The hyper greed of European merchants and pirates intensified, escalating the trade in human flesh.[62] Sugar thus became a "de facto currency" and "sugar was not just used to sweeten tea and coffee, but was for years seen as a marker of sophistication and refinement, ideal for a rising capitalism that exploited such a trend with merciless dedication."[63]

Though France today possesses no formidable major gold mine, it possessed 2.436 tons of gold reserves worth $203.65 billion euros in December 2024, the fourth largest in the world. France looted sugar in Haiti through enslavement in San Domingo then and Haiti now and made billions from the 18th century through 1791 when the African Haitians overthrew the yoke of French slavery led by Toussaint L'Ouverture (1743–1803) and his successor, Dessalines, sparking the first successful Black revolution that uprooted slavery in 1804. When Jean-Bertrand Aristide, a popular leader of the Lavalas Movement there in the late 1980s, was elected for the second time in 2001, he was overthrown in a U.S.-orchestrated, French- and Canadian-supported military coup, similar to his first presidential coup in 1991.[64] In 2004, during the bicentennial of the historic Haitian revolution overthrowing slavery, he called for reparations of $21 billion by France ($30 billion in 2021 value) for the 90 million in gold francs Haiti was forced to pay France for its "confiscated property," enslaved Africans, from 1825 to 1947. He was subsequently overthrown.[65] Haitian workers still toil for slave wages for multibillion dollar transnational corporations like Walmart, Gildan, Hanes, Palm Apparel, Levis, Gap, Nike, Fruit of the Loom, Old Navy, H&M, JCPenney, Zara, and others. On February 9, 2022, thousands of textile workers (80 percent of the formal labor force) went on strike, demanding livable wages and an increase from 500 Gourdes ($3.66) to 1,500 Gourdes ($10.98) per day (still below the Solidarity Report's recommendation then of 1,750 Gourdes daily [$12.81]), resulting in state repression.[66]

France terrorized and enslaved Africans in Haiti in the 18th century for massive profits and wealth accumulation, and it continues the neo-colonial rape of Africa today.[67] France exploits 860 gold mines in Mali and extracts uranium for nuclear power from Niger that supports 70 percent of French power needs, hence the Indigenous hostility to France in Francophone West Africa. Currently, military coups in Niger, Mali, Burkina Faso, and Guinea, former French colonies, have demanded the departure of French troops and an end to mineral extraction. While three of every four light bulbs in France are powered by Nigerien uranium, 10–20 percent of urban Nigeriens enjoy electricity access and only 2–3 percent in rural villages do, while over 60 percent of the people are forced to eke out an existence on $1 per day.[68]

Eric Williams notes how the emergence of the European Industrial Revolution and the triangular character of the trade in enslavement of Africans were axiological for the British capitalism that spawned in the 17th century. British ports and cities benefited momentously from slavery. Products manufactured by British industries whose production depended on enslaved African labor in the Caribbean were shipped for sale to Britain's West African colonies, generating gigantic profits that enabled the purchase of more enslaved Africans, an endless triangle of torturous human trade.[69] Historian Philip Foner notes that such trade was the ground of the English cotton industry with "the manufacture of cotton goods for the purchase of slaves" providing "the initial stimulus for the emergence of Manchester as the great cotton-manufacturing center of the world."[70] Joseph Inikori's groundbreaking work on slavery in Africa explains that it was the same process by which British palm oil imports from Africa expanded and fueled its machines, leading to further imports of rubber, cocoa, tin, coal, timber, cotton, and ground nuts, and subsequently gold from the Gold Coast (Ghana). The ripple effect was an unprecedented economic boom for Britain from which it heavily benefited, with the correlative colonial extraction of raw materials and the undermining of indigenous West African economies.[71] Inikori adds that slavery sparked the least expensive cotton production in the Americas performed by an enslaved African population and facilitated British manufacturer expansion, while arresting the independent evolution of African economies globally in both the 19th and 20th centuries.[72]

In the 1950s in British-occupied Ghana, the latter was humiliated by being coerced into borrowing its own money, 200 million pounds expropriated by Britain, with interest.[73] African nations have been forced through neo-colonial dependency for their livelihood on "foreign aid" as the supposed path to economic independence and stability. The net consequence is African nations become mired in mammoth unpayable debts of hundreds of billions of dollars—$800 billion today—forcing them to be forever beholden to hegemonic financial behemoths like the World Bank (WB), the International Monetary Fund (IMF), and the World Trade Organization (WTO). Debt accumulation is core to capitalism.[74]

Chattel enslavement of Africans, along with the Columbian invasion and European colonial settling of the Americas, was the catalyst for European commercial and industrial capitalism, with 12.5 million Africans kidnapped from African homelands from 1450 to 1850, leaving the African continent depopulated and destabilized ever since, with the permanent arrest of Africa's autonomous and Indigenous socio-economic development. Though there were forms of coerced indentured labor and forced servanthood in many parts of the world, the 15th through 18th centuries marked the first time that chattel slavery of Africans and Indigenous Indians in the Americas was deployed as the principal form of economic exchange, production, and intercontinental finance. The very intentional separation of kinship groups among enslaved Africans, depriving people who spoke the same language and derived from the same culture and religion from being together, and the practice of splitting up African families—children from parents, wives and husbands, and siblings—as well as denying the right of Indigenous religious practices and mother tongues were acts of ethnocide. The violent prohibition of the African drum and use of Indigenous languages constituted hallmarks of perpetual cultural fragmentation and disintegration that persist in troubled and ossified familial and cultural life in Black communities in the African diaspora, especially in the U.S. and in places like Brazil today. Outrageously high and disproportionate rates of Black incarceration perpetuate such inhumanity. The forced imposition of European Christianity during chattel slavery robbed Africans of vital ancestral and linguistic connections, resulting in the abandonment of ancestral traditions foundational to every culture. On the other hand, diverse European groupings during the slavery era became unified under the rubric of "whiteness."[75] Yet enslaved Africans and Indians persistently rebelled and resisted their enslaved condition, marked by the regular and hundreds of revolts through the 17th and 19th centuries, never surrendering their inalienable right to be human and free.[76]

Two million of the 12.5 million Africans shipped in 34,934 documented transatlantic journeys never made it to the western hemisphere because they died from the inhumane conditions, an unremovable bloody blemish in history. This practice was indispensable to the development of industrial capitalism.[77] Capitalism was borne in the crucible of brutality and violence against people of color.[78] Gerald Horne asserts,

> From the sixteenth century through the nineteenth centuries nearly 13 million Africans were brutally snatched from their homelands, enslaved and forced to toil for the greater good of European and Euro-American powers, London not least. Roughly two to four million Native Americans also were enslaved and traded by European settlers in the Americas, English and Scots not least.
>
> From the advent of Columbus to the end of the 19th century it is possible that five million Indigenous Americans were enslaved. This form of slavery coexisted roughly with enslavement of Africans leading to a catastrophic decline in the population of indigenes.[79]

Horne explains that a genocidal colonialist onslaught against Indigenous Indians in the Caribbean Basin, Gulf Coast, northern Mexico, and the U.S. southwest reduced the numbers of these nations and communities by about 90 percent.[80] He echoes the basis of sex trafficking highlighted earlier:

> The majority of the enslaved were women and children, an obvious precursor, and trailblazer, for the sex trafficking of today. But for the massive revolt of the indigenous in 1680 in what is now New Mexico, the toll might have been much worse.[81]

Following the incipient capitalist cash-generating crop, sugar, tobacco, indigo, and rice became the major staples that caused the enslavement of Black labor in the U.S. northeast and the Carolinas, where Indigenous African methods of cultivating multifarious varieties of rice were exploited to ensure maximum profits from crop diversity.[82] Then it became cotton from 1790 through the pre–Civil War, with New England being the major cash slave cow, while Indians and Africans were subjected to the worst forms of coercion and labor regime brutality. Eli Whitney's invention of the cotton gin was attributable to an enslaved African known only as Sam (enslaved Africans were never accorded credit for any invention or innovation). The machine revolutionized cotton manufacturing and production in the 1790s and signified a radical upswing in expanded cotton production from 300,000 bales in 1820 to 700,000 bales in 1830 and 4.5 million bales in 1860.[83] Cotton production didn't just provide plantation owners with wealth in the southern region of the U.S.; northern white industrialists and commercial capitalists, too, were the largest beneficiaries of southern slavery, sparking food production in the Ohio Valley locality. One million Africans were forced to migrate from the upper to the lower South as a result.[84]

The enslavement of Indigenous Indians—some three million in the middle of the 16th century in Mexico, Central America, and Venezuela, with around 200,000 in Mexico by 1555—aggravated by the obsessive pursuit of gold from the 1520s, sparked a boon in wealth accumulation for colonizers.[85] A small number of Africans in Mexico were also enslaved alongside the Indians. While Hernando Cortés is portrayed normatively as a pioneering Spanish "discoverer" and "innovator" on behalf of Spain and credited with being key for the conquest of the Aztec empire (as Pizarro was in the destruction of the Incan empire), few are taught that Cortés was the catalyst for the institutionalization of enslavement of Indigenous Indians through the *encomienda* system. He was a leading slavery beneficiary, just as U.S. founders Thomas Jefferson and George Washington were. Cortés acquired gold mines as well as enslaved Indians with his ill-accumulated wealth, paying 6,230 pesos on November 20, 1536, for 70 Indians, and eventually claiming 120 Indians as his enslaved "property" with 30,000 pesos on one day. He became the wealthiest owner of enslaved Indians, and the Spanish colonizers hurriedly followed in his footsteps.[86] From the 16th and 17th through the 18th centuries, the terror of mining predation and slavery persisted with the looting of silver from Spanish colonial mines in Mexico, so that for most of the 19th century, Mexico was the most important silver producer on Earth, producing half of global fine silver. This "silver revolution" was made possible through the blood, sweat, tears, and destruction of Indigenous Indian bodies as enslaved chattel, resulting in phenomenal quadrupling of silver production from 1,403 tons in 1701–1710 to 5,378 tons in 1801–1810.[87]

Closer to home, well into the latter part of the 19th century, during the "gold rush" in California, Indians were used for all kinds of labor on farms, constructing railroads and waterways, clearing land for white colonial invaders and gold rush occupiers, while being progressively eroded, exterminated, and destroyed from the relentlessness of racial oppression. Indian children were kidnapped at very young ages and bought and sold as enslaved labor property of white farmers and settlers. California, which was born in the crucible of genocide and slavery, boasts the fifth largest economy in the world, the home of giant high-tech corporations in Silicon Valley, and the center of the film and modeling industry, Hollywood. M. Kat Anderson asserts that following land dispossession by colonists, Indigenous people in California were forced to work for the miner-settlers, and those who were not recipients of low wages were forced into "actual slavery."[88] So too, "California Indians sheared sheep, grew crops, threshed domesticated grains, rowed boats, dug irrigation ditches, cut timber, maintained railroad tracks, washed clothes, rounded up cattle, built structures, ferried travelers across rivers, panned for gold and served meals for white settlers," making them the cornerstone of the economy in the 19th century.[89] Many wealthy white families like the Murphy brothers (Murphys), Andrew Kelsey (after whom Kelseyville is named), John Bidwell from Chico Creek, Peter Lassen from Deer Creek, Pierson Reading from Cottonwood Creek, and James Savage participated in the forced dispossession of Indians from traditional ancestral lands. Horace Bell, a ranger, wrote in his memoirs that the "cultivators of vineyards commenced paying their Indian peons with a*guacaliente*, a veritable fire-water and no mistake," and like New Orleans, Los Angeles "had its slave mart" where the enslaved Indigenous person was sold each week for one to three dollars, and then paid with the same fire-water that destroyed the Indians fully.[90]

By August 1848, Indians in California made up more than half of the four thousand miners in the gold fields.[91] They were routinely and systematically cheated by white miners, which was the case of all Indigenous-European colonial invader encounters, as recounted by J.H. Carson, a miner settler from Virginia:

> In the early days of gold digging, these Indians looked on in wonder at the exertions of the white men to procure from the rivers and the gulches things not to be eaten. Indians were at work for miners and others, receiving in payment for their week's work an old shirt or handkerchief.[92]

History repeats itself, so we are constantly told. In Tulare County today, home of the Tule River Indian Nation that consists of the Mono, Tūbatulabal, and the Yokut (the latter of which had sixty different groups prior to European invasion), over 30,000 Mexican farm workers living in the same county still pick fruit and vegetables for the country's food needs, earning exploitative wages and subjected to pesticides and herbicides with few farmworker rights and little fair wage protection.

Yosemite National Park's formation as the first national "nature" park entailed forcing thousands of California Miwok Indians out by white settlers and "environmentalists" who were determined to enjoy the untrammeled "appreciation of nature" through the establishment of the National Park Service. Yosemite sparked national parks and nature reserves in Indigenous lands around the world that made (and continue to make) lucrative profits for tour companies from "ecological tourists." There are tens of thousands of sacred sites and places in these dispossessed Indigenous ancestral lands in Turtle Island and globally. Some of those experiencing desecration are noted here:

- Yellowstone National Park in Montana and Wyoming, which was built upon the extermination of the Nez Perce Indians;
- Glacier National Park in Montana that dispossessed the Blackfeet Nation;
- the Grand Canyon National Park, which confined the Havasupai in northern Arizona;
- the Everglades National Park, which dispossessed the Seminole Nation in Florida;
- *Paha Sapa* (the Black Hills) in South Dakota, which is still being defended by the Lakota today against mining desecration;
- *PeeHee Mu'huh* (Thacker Pass), the site of the massacre of over 30 Shoshone people, is now facing lithium mining desecration by General Motors and Lithium Americas Corp.;
- *Dook'oos'liid* (the San Francisco Peaks) in Flagstaff, which is occupied by Arizona Snowbowl making snow from sewage water for recreational skiing;
- Chitwan National Park in Nepal, which dispossessed the Sherpa people;
- the Ukuhlamba-Drakensberg Park in Kwazulu, Azania (South Africa), which dispossessed the ancient Indigenous San people;
- Nairobi National Park, which, along with Amboseli, Maasai Mara, Serengeti, Ngongoro, and Lake Manyara parks, dispossessed the Maasai people in east Africa;
- Machu Picchu, Peru, where the Quechua-speaking Indians were dispossessed;
- the Acre region of the Nukini Indians in the Brazilian Amazon, who still face mining and industrial invasion threats to their lands;
- the Tacana Indians of Bolivia, who still face threats from mining invaders on their sacred lands; and
- Uluru National Park outside Alice Springs, Australia, which dispossessed the Indigenous Anangu people and was only returned in 1985 as *Uluru-Kata Tjuta* traditional land.[93]

Ojibwe writer David Treuer, in reflecting on his extensive visits and stays in Indigenous lands dispossessed by the U.S. park system all over the western part of Turtle Island in 2021, captures this ethnic cleansing in modern parlance when he writes, "From the perspective of history, Yellowstone is a crime scene."[94] He recalls the history of Chief Joseph and the Nez Perce band seeking refuge there after a 1,500-mile journey from their traditional ancestral Walowa Valley homeland in 1877. But refuge was not realized because the U.S. Army, headed by John Gibbons, stormed the area and killed women, men, and children while they slept. Today, Indians live in the U.S. on almost 90 million acres of lands, just five million more than all of the U.S. national parks' land area of 85 million acres.

Indigenous historian Roxanne Dunbar-Ortiz has delineated the far-reaching history of Indigenous people of the western hemisphere, describing how Indigenous peoples from the Caribbean through Meso-America revolutionized the world with the *maize* (corn) plant, sacred for all Indians, and for 70 other globally known foods.[95] The same Indians who built towering mounds of Georgia still evident in Turtle Island today were agricultural specialists in the north and southeast from millennia ago, as are the ancient fisherfolk of the west, and irrigation canal builders like the Anasazi from over a millennium ago, descendants of the O'odham people of the southwestern desert. The canals' preservation of water and food silos still benefit U.S. residents in the west today.

This historical recounting of slavery and extermination echoes Marx's assertion on the emergence of industrial capitalism:

> The discovery of gold and silver in America, the extirpation, enslavement, and entombment in mines of the aboriginal population, the beginning of the conquest and looting of the East Indies, the turning of Africa into a warren for the commercial hunting of black-skins, signalized the rose dawn of the era of capitalist production.[96]

The European Colonization of Africa and Asia: Critical Highlights

Just as slaver Columbus is still portrayed by millions of Europeans and Euro-Americans as a "discoverer" of "America" in 1492, "Prince Henry the navigator," a Portuguese colonial invader and plunderer, is also portrayed in a positive light. Henry invaded and occupied Ceuta, a port in the Moroccan region of north Africa in 1415, that has since been occupied illegally by colonial Spain into the present. The Portuguese made excursions further south and landed without permission on the coast of west Africa, where they sought gold, ivory from the tusks of dead elephants (setting off the ivory trade that persists in the world today!), spices, and African people for the enslavement enterprise. African leaders in that region were foolishly enticed and bribed through petty trade for selfish gain, collaborating with Portuguese colonial slavers, so typical of colonial collaborators from the colonized nations the world over. Portugal swiftly and strategically colonized Brazil in the southern part of the western hemisphere, to which Portuguese ships carried their precious "cargo," Africans, to launch the chattel slavery trade. Similarly, the Portuguese invaders set sights on Angola, where they forcibly occupied Indigenous lands, and then on the maritime trade of the Indian Ocean zone along with the Swahili east African coast and further east toward Malaya, Indonesia, and the Philippines in the pursuit of the lucrative spice trade. They pressed on toward south Asia and then Africa, and back to Brazil. In India, they established trading ports for spices in places like Goa and Cochin; Jaffna in then Ceylon; Macau off the coast of China; Elmina on the Cape Coast of Ghana; Luanda in Angola; São Tomé, an island off the central-west African coast; Mombasa in east Africa; Sofala in Mozambique; and then Bahia, Rio de Janeiro, and Pernambuco in northeast Brazil. Brazil has the largest number of African people in a single country after Nigeria, with over 100 million people.

Not to be outdone by their Iberian neighbors, the Spanish launched their Columbian invasion of the Caribbean in 1492, but upon initially realizing that these were not the spice islands originally conceived, set their sights on east Asia. They penetrated the long transpacific route by journeying across the Atlantic desperately coveting the Portuguese accumulation of wealth from enslaved Africans and the spices trade of the east Indies. Ferdinand Magellan launched a "pioneering expedition" to the East Indies, circumnavigating the Atlantic and the Pacific oceans, and was hailed as a "discoverer" engaged in "valuable trade." From the vantage point of the Indigenous peoples of the Philippines, for instance, Magellan was no discoverer, but an invading plunderer whose colonization on behalf of the Spanish monarch Charles I has since destroyed the Philippines, which was occupied by Spain from 1565 through 1898. Almost half a million lives were lost over those three centuries of brutal occupation. This carnage was followed by the invasion and occupation by the U.S. that then cultivated elitist-imposed and established client regimes so that it could establish some of the largest foreign military bases on Earth, at Subic Bay and Clark Naval Base, where 200,000 Filipino women were forced into working as prostitutes to "serve" the sexual needs of U.S. military servicemen.[97] Though the Philippines, a nation of over 100 million today, was formally granted "independence" from the U.S. empire in 1946, the large U.S. twin military bases were not closed until 1992. The regime of Rodrigo Duterte, preceding that of the current Bongbong Marcos, son of the tyrannical rulers Ferdinand and Imelda Marcos, gained notoriety for killing 25,000 people, including many preemptively, as his strategy of ostensibly erasing crime and thuggery. The majority of the people remain marginalized and impoverished, with Indigenous lands being dispossessed and militarized, especially in the southern region of Mindanao and the outlying islands of Basilan and Jolo, all the while enriching the large mining corporations and the fringe wealthy elite.[98]

The Spanish empire, thanks to Spain's military and naval dominance from the 15th century, "augmented in 1580 with that of Portugal through a merger of the Portuguese and Spanish crowns, stretched from Macau to China to Potosi in Peru, girdling the globe."[99] Yet this hegemonic role was soon challenged by the emergence of Holland and then Britain, vying for the looted wealth of the lands and nations of color.

Yet tiny Holland found its niche. Gerald Horne explains in detail that many of the Pilgrims entering the Americas in the early 17th century spoke Dutch because they were migrants in Holland from England. Owing to Holland opening itself to Jews fleeing from the Spanish inquisition, they were the real colonialists of Turtle Island since they formatively developed "overarching racial identities in order to facilitate colonialism, a process that took the name of 'whiteness' on the west bank of the Atlantic."[100] Horne notes further that the foundational role played by the Dutch in launching mercantile capitalism and expanding the arms industry as a result of the Dutch war with Spain from 1569 to 1648 paved the way for what could be described as the early "military-industrial complex."[101]

Dutch colonists fought for possession of Portuguese and Spanish colonial "possessions" of lands in the western hemisphere, southern Africa, and the East Indies. Walter Rodney, the noted Guyanese scholar assassinated in 1980 by members of the Peoples National Party led by the U.S. and British-supported Forbes Burnham from 1964, documents in his instructive book *How Europe Underdeveloped Africa*, "In 1634, the Dutch captured the vital Brazilian port of Penambuco and 'the director of the Dutch West Indian Company'—paved the way for further enslavement for what came to be notoriously known as the 'Slave Coast.'"[102]

By the 1660s, the Dutch, through forced wars of attrition against the Spanish and Portuguese, came to possess in entirety the spice trade in the east and the West African trade in enslaved Africans, ivory, and gold. The Dutch also invaded Indonesia, and after a series of defeats in Bali in 1846 and 1848, finally subdued the Balinese in 1849, and occupied Indonesia in 1900, after nationalizing the Dutch East India Company that had earlier invaded and occupied the Cape Bay in Azania/South Africa in 1652. The

Dutch East India Company exploited the spice trade with the east and greedily sought to monopolize the trade spoils for the Dutch economy, particularly to finance its colonial military machine in the Caribbean and the U.S. east coast. The Dutch were some of the earliest colonial invaders and enslavers of Indigenous peoples in the world, enslaving Indonesians in the 17th century and transporting them to the Cape in Africa. They also occupied part of the West Indies: Tobago in 1628, Tortuga and St. Martin in 1631, and St. Croix in 1634.

The British subsequently extirpated the French in the Caribbean and Turtle Island by the 18th and 19th centuries, asserted colonial control of lands in India, and forced the opium trade on China. The British conveniently suffer amnesia today when it comes to acknowledging their historical national hostility toward China and other peoples of color that forced Chinese people to become drug addicts, anathema in Chinese culture:

> The Viceroy of Hupeh and Hunan estimated in 1838 that there must be over four million opium addicts.... Another Chinese scholar reckoned that the 40,000 chests due to enter China that same year would supply eight and a half million smokers. Toogood Downing, a well-informed English physician with experience in Canton had calculated two years before that the quantity of opium imported in 1836 would make when prepared for smoking about 32,200,000 taels of mixture enough for the needs of 12,500,000 smokers.[103]

In 1830, the opium trade was controlled through personal royal countenance of Queen Victoria, to whom the Chinese government through Viceroy Lin Tse-hsu appealed in 1836 to end the decimating opium trade for the mutual benefit of both China and Britain. The Viceroy was unaware that opium was legal in Britain then.[104]

Under British and French colonialist occupation of Indigenous lands in Turtle Island from the 17th century, Dutch settlers occupied the Hudson Valley, the French took over St. Lawrence and the lower Mississippi basin, and the British occupied the eastern Atlantic coast of what came to be known as colonial North America. As the Spanish empire drew to a close at the end of the 17th century, the beginning of the 18th marked the spawning of British imperialist dominance. The French colonizers were forced out of their strongholds in the St. Lawrence and Mississippi regions, including Louisiana, and from the West Indies that included the islands of Martinique, San Domingo (Haiti), Guadeloupe, Dominica, Guiana, St. Lucia, Grenada, and Tobago, and retreated to India where they colonized portions of the Coromandel Coast at Pondicherry, Karikal, Yanoan, Malabar, and Chandernagor in Bengal. However, their colonial occupation was soon displaced by the British who occupied India and once again, in 1760, the French were compelled to flee the land and return to home base. The insatiable British colonial desire for more land to exploit and more people to enslave for profits then colonized Australia and Aotearoa (New Zealand). In 1769, British Captain James Cook and his crew of prison convicts invaded *Aotearoa* of the Māori Nation and the British expanded their colonial invasion of the land from 1841. Subsequently, Cook and company invaded Australia and occupied Kamay Botany Bay on the east coast in April 1770, sparking a genocidal colonial occupation from 1778 that enslaved Indigenous people and resulted in their near extermination. In the island of Tasmania, European invaders formally proposed "extermination" to address what was called the "Aboriginal problem." At a meeting held in Hobart, Tasmania in 1830, colonial solicitor general Alfred Stevens, who subsequently became a chief justice in New South Wales, Australia, demanded that the colony protect its stolen farming areas (previously Indigenous hunting grounds) against Indigenous attacks. If the colony could not do it without extermination of the Indigenous people, he retorted, *"then I say boldly and broadly exterminate!"*[105] (italics ours).

From 1760–1858, Britain occupied India, forcing the English language on its inhabitants and raping the cotton production of the land to finance its imperialist colonialist expansion, reaping ill-gotten wealth of $34 trillion in today's value.

Subsequently, the U.S. set its sights on Japan in the 1850s and demanded access to sales of U.S. industrial products. In 1885, the General Act of Berlin was signed, granting illegitimate *carte blanche* rights to European colonists to split the momentous African continent into partitions divided among the invaders, with France occupying major areas of west Africa; Britain occupying Ghana, Nigeria, The Gambia, Sierra Leone, Egypt, Kenya, and Azania/South Africa; the Portuguese colonizing Angola, Mozambique, Sao Tomé and Principé; the Spanish occupying and expropriating the wealth of the Canary Islands; and the Germans occupying Tanzania. The Italians illegally held on to the port of Assab in Ethiopia in 1869, but though they attempted to fully occupy Ethiopia, they were thoroughly defeated by heroic Ethiopian resistors at the Battle of Adwa in 1896—Europe's first major defeat in Africa—only to retake portions of the country subsequently. In 1898, though the Sudanese Madhists fought valiantly and defeated the Anglo-Egyptian invaders in 1883 and 1885, they were finally forced to abdicate at the Battle of Omdurman in 1898. Italy also reoccupied Libya, dropping the first aerial bombs ever used in November 1911 in the war between the Ottoman Empire and Italy. The Belgians occupied the Congo from the late 1800s into the early 1900s and were responsible for the deaths of 10 million Bakongo people in the latter, cutting off peoples' hands and ears for inadequately extracting rubber.[106]

In 1898, Filipino revolutionaries expelled Spanish colonizers of the islands from 1511, only to be replaced by the U.S. empire that illegally occupied the country in 1898. In the same year, the combination of U.S. and Cuban forces on the other side of the Pacific defeated the Spanish there, with African Cuban Antonio Maceo leading a successful guerrilla struggle in eastern Santiago de Cuba against the Spanish in 1895. The U.S. then occupied Cuba from 1899 to 1902, passing the Senate Platt Amendment in 1901 that authorized military intervention in Cuba "to protect US interests" represented by the thousands of U.S. businesspeople and personnel in various capacities living in Cuba. This path paved the way for the installation of the U.S. client state regime of Batista in 1940, despised by the Cubans for his repression. Batista was finally successfully overthrown in the Cuban socialist revolution of 1959, through a fierce guerrilla war led by Fidel Castro and Che Guevara, even after an initial unsuccessful attempt by Castro in 1953. The shocking result was U.S. imperialism's complete economic blockade of the island nation and the colonization of Guantanamo Bay with a permanent military base there until the present, in total violation of international law. This international economic embargo of the worst kind has caused untold death, suffering, and economic and social hardship for the Cuban people, with the loss of hundreds of billions of dollars in today's value due to the prohibition of free maritime trade and commerce and cheaply priced food, medicine, technology, agricultural products, etc.

Yet Cuba has struggled defiantly amid the U.S. blockade and still established free education, free health care, free housing, and subsidized food programs for all of society, the best in the western hemisphere and among the top in the world, including during the Covid-19 pandemic, in stark contrast to the U.S. Cuba has sacrificed much to assist in the decolonization of Southern Africa in the struggle to liberate millions from Portuguese colonialism in Angola and Guinea-Bissau, and apartheid in Namibia/South Africa, and in health care programs on the continent and in Iran, Venezuela, Haiti, Grenada, and other parts of the world. Though Cuba has made phenomenal achievements in numerous areas, the nation still struggles with the vestiges of racism, slavery, and genocide.[107]

On the other side of the world, Russia as a large, feudal, Euro-Asiatic state became prominent in the early 18th century and industrialized quickly, making western European colonial regimes uneasy. Russia broke the chains of feudalism with its watershed Bolshevik revolution of 1917 that shook the corridors of western imperialism. Changes toward a mixed market economy began in 1989 with the collapse of the Soviet Union, yet the state command of the economy is still prominent. During the Russian defense against Nazi Germany's military invasion to expand Nazism, the Russians fought and resisted to the point of losing over 20 million people. The U.S. and the European Union have conveniently forgotten this horrific history, during which Russia sided with the European Allies and the U.S. against Nazi Germany. Russia was extremely apprehensive about the violence suffered by Russian speakers in neighboring Ukraine at the hands of the Nazi sympathizers in the areas of Donetsk and Donbass on its borders in the 2000s. Russia struggled repeatedly to convey its security concerns to the EU and the U.S., to no avail, calling for adhering to the Minsk I and II Accords from 2014 to 2015, which were designed to establish peace in the Ukraine-Donbass region.[108] The failure of such adherence by Ukraine resulted in Russia's incursion into the Donbass region and the bloody war today involving tens of thousands on both sides. Diplomacy between the U.S., NATO, EU, and Russia—a course the U.S. rejected in March 2022—could have prevented this bloodshed.

The major events of the 20th century were, of course, the first and second major wars in Europe in the second and fourth decades respectively. Britain and France sided with Russia against the alliance of Germany, Turkey, and Austria, marking the first major conflict. In the late 1930s, Germany, Italy, and Japan were the leading imperialist powers, along with the U.S. earlier on from the 1820s with the Monroe Doctrine and military interventions in Latin America and the Caribbean. Germany sought to expand its territory across eastern and western Europe, triggering the second major intra–European conflict. It was also the period of Japanese incursions into China and Korea, occupying parts of these countries in the 1940s, and marking Japan's colonialist designs—the only country in Asia to have them.

The collapse of the Ottoman Empire was finalized in 1922, after it had sided with Germany and was defeated. Subsequently, it was "carved up" into Turkey, Italian Libya, and the British and French Protectorates of Palestine, Transjordan, Iraq, Kuwait, Lebanon, and Syria. All these states continued to suffer territorial assault by western militarism following oil discoveries in Saudi Arabia, Iraq, the United Arab Emirates, Syria, and Iran from the early 1900s. Tragically, this region has been under siege with continued military occupation through U.S. military bases like in Bahrain and Saudi Arabia and constant destabilization and military intervention by forces in places like Iraq, Syria, and Iran. The British Balfour Declaration of 1917 gave Britain colonial occupation rights over Palestine, in which it expressed its intention to create a Jewish state on Palestinian land, with no consultation with Indigenous people living there for millennia. This colonial charter led eventually to the horrific dispossession and *Nakba* of

Indigenous Palestinians and their occupation by Israel from 1948 through today under the auspices of British and U.S. colonialism, an occupation marked by daily bloodshed of Palestinian people resisting occupation of their homeland. Ongoing Israeli military incursions into the occupied territories of Gaza and the West Bank and arrests and killing of children, youth, and women in particular have persisted for the past 50 years of a 75-year-long colonial occupation. The 2000s (particularly 2006, 2007, 2012, 2014, 2021, and 2023) saw intensified Israeli military bombing and house demolitions of Palestinian homes and lands. The occupation culminated in the genocidal annihilation of over 15,000 people, including over 6,000 children and 4,000 women, from Israeli bombing in October-November 2023 on the pretext of fighting the Hamas movement. Repeated UN resolutions calling for Palestinian national independence and Israeli withdrawal from territories occupied by the regime since 1967 have been consistently vetoed by the U.S. at the UN Security Council. This is an unfolding deadly tragedy to both Palestine and Israel and needs an immediate humanistic and just solution so these peoples can live peacefully together, but the U.S.-funded Israeli occupation violence must end as a precondition.[109]

The deaths of millions in the Americas, the Pacific, Asia, and Africa in resisting colonization can't be understated since effects linger in impoverishment and political instability today. Some of these are:

- the promulgation of the Monroe Doctrine of 1823 that declared that the U.S. considered the lands of Latin America and the Caribbean as its "backyard" and illegally set precedents for the overthrow of governments that were considered oppositional to the interests of U.S. imperialism;
- Indigenous Tasmanian and Australian annihilation by British colonizers deliberately dispossessing Indigenous lands;
- the overthrow of Queen Liliuokalani, sovereign leader of Hawaii, in flagrant violation of international law in 1898, followed by the brutal occupation and systematic dispossession of Indigenous Hawaiians by the Dole company and the U.S. military through the present;
- the genocidal killing of the Bakongo in the Congo perpetrated under the Belgian colonialists in the early 20th century, during which rubber and other minerals were forcibly extracted, persisting in the violence from coltan mining for electronics today, giving Congo possession of 80 percent of the global supply;
- the genocide of 80 percent of the Herero and Nama in Namibia from 1904 to 1910, who were forced to leave ancestral homelands with water sources poisoned in the desert by German colonists;
- the lethal exposure of the Shoshone people of Nevada and Utah to 900 aboveground and underground nuclear tests on ancestral lands from the 1940s, the most bombed area on Earth, and resulting in tens of millions of people being afflicted with cancer through today[110];
- water and land poisoning of the Chi Endé (Chiricahua Apache) lands of Arizona by Agent Orange testing in the 1950s prior to it being used in Vietnam;
- the spraying of 20 million gallons of Agent Orange on food producing areas, waterways, forests, and people in Vietnam by the U.S. military during the late 1960s and early 1970s that caused four million deaths and maimed tens of thousands;

- the incineration of parts of the Marshall Islands by atomic testing while people were still living there from 1946 through 1958;
- the extermination of five million people in Korea in the 1940s by U.S. military bombing;
- the killing of millions in Laos and Cambodia as the result of the violence of the U.S. military onslaught from 1954 through 1974, including carpet bombing;
- the occupation of Indigenous lands by France (Guiana), Britain (Caymans, Virgin Islands in the Caribbean), and the U.S. in the Caribbean (Puerto Rico, Virgin Islands) and the Pacific (Tahiti, Bikini, Samoa, Mariana Islands, Guam) in the 1900s;
- the killing of almost two million and forced dispossession of hundreds of thousands in Iraq from the initial U.S. bombing in 1990–1991 and full-blown invasion and overthrow of the Iraqi government in 2003, including almost one and a half million children, from sanctions and the use of 300 tons of depleted uranium, a banned munition[111];
- the killing and permanent maiming of 200,000 to 300,000 people in Afghanistan from NATO and U.S. military occupation from 2001 through 2021;
- the deaths of 5,000 to 10,000 and thousands more wounded and hundreds of thousands displaced by the NATO bombing of Kosovo, Yugoslavia, and Serbia in 1999; and
- the relentless and illegal bombing and destruction to the ground of Libya in 2011 by NATO, the U.S., and France that sought to destroy the nation's attempt to introduce the gold standard and forge a unified Africa.

The aforementioned is certainly not the complete litany of such war-orchestrated suffering and death.

Three centuries of sanguinary brutal invasion, occupation, and enslavement of Indigenous peoples by the various western colonial countries and the accompanying looting of the gold and silver bullion and precious metals extraction provided the foundation for primary capital accumulation for the industrial revolution. The European bourgeoisie were thus able to emerge dominant in their respective feudal societies, and the industrialist and nationalistic revolutions (like the French Revolution of 1789) shifted their focus from mineral extraction to securing materials from the colonized for Europe's booming factories. Subsequently, manufactured goods from these raw materials were in turn sold to the residents of the colonies and former colonies to reap mammoth profits for the corporate industrial business owners in Europe. The methods of colonization were radically changed, from outright colonialism to "neo-colonialism," as coined by the first president of independent Ghana, Kwame Nkrumah.[112]

Africa, with 1.374 billion people today, numerous like China and India, is firmly in the hands of her former European colonizers and serves as the surrogate for western capitalism and militarism. Yet, while the west confiscated the wealth from African nations and those of the underdeveloped world in Latin America, Asia, and the Pacific, the underdeveloped countries are still forced to pay illegitimate debts with exorbitant interest for "loans" from the IMF and World Bank. The rejection and ongoing expulsion of the colonial French military presence by regimes in Burkina Faso, Mali, and Niger in west Africa from 2021 to 2023 signifies a radical repudiation of such entrenched extractive neo-colonialism. These are very optimistic signs of the unfolding decolonization movements in Africa, languishing for decades under French neo-colonialist tyranny. The west has punctured and destabilized every major revolutionary movement and assassinated every nationalist independence leader who resisted imperialist oppression. The assassinations of Patrice Lumumba in the Congo in the late 1950s, Steve Biko in Azania (South Africa) in 1977, Thomas Sankara in Burkina Faso in the early 1980s, and Muammar Gaddafi in 2011 all triggered widescale carnage and displacement of millions. These movements in west Africa today serve as inspiration to the rest of Africa and all colonized and oppressed people yearning for liberation from long-standing neo-colonialist and capitalist subjugation.

Karl Marx's Analysis of Capitalism's Formation in Industrial Europe

In his extensive exposition and analysis of the formation of early manufacturing and agricultural capitalism in Europe in three volumes, *On Capital*, Karl Marx explains that the self-sustaining farmer in 14th-century England was forced to become dependent on the landlord as proprietor who provided the ingredients for agriculture like seed, cattle, and other tools. The police of the English state were responsible for ensuring that farmers, serfs, and peasants were controlled and coerced into supplying the illegitimate landed proprietor who expropriated land for the accumulation of profit. Common held peasant land was privatized for the emergence of the individual entrepreneurial farmer, marking the incipient stages of a nascent limited European capitalism. The early capitalist farmer was enriched through acquisition of more cattle stock, richer supplies of manure to enhance production, and with the corresponding rise in prices of materials like wool, corn, and meat, earned significant wealth while he paid progressively low rent to the state over a 99-year lease in the 16th century. Essentially, as with capitalist conglomerate CEOs and corporate bankers today, the capitalist farmer made lucrative profits lazily through state-protected economic and social benefits and privileges "without any action on his part."[113]

Marx importantly provided extensive detail on the difference between use-value and exchange value of commodities and how money metamorphosed into commodities at the foundation of incipient capitalism's spawning. He noted, "The first chief function of money is to supply commodities with the expression of their values, represent their values as magnitudes of the same denomination qualitatively equal, and quantitatively comparable," serving as a *"universal measure of value."* He asserted, "Money as a measure of value, is a phenomenal form that must of necessity be assumed by the measure of value which is immanent in commodities, labour-time."[114] He captured the essence of the post-medieval European *sacralization of money* and its global expression of wealth, while ignoring the fact that various Indigenous nations on Earth have never followed this trajectory of life value:

> Money of the world serves as the universal medium of payment, as the universal means of purchasing, and as the universally recognized embodiment of all wealth. Its function as a means of payment in the settling of international balances is its chief one. Hence the watchword of the mercantilists, balance of trade. Gold and silver serve as international means of purchasing chiefly and necessarily in those periods when the customary equilibrium in the interchange of products between different nations is suddenly disturbed.[115]

Yet, Indigenous people used various forms of exchange and arbiters of value from time immemorial. For example, cowrie shells were used in Africa, and *wampum* beads in the northeast of Turtle Island (derived from the Algonquian language) functioned as exchange mediums, the Narragansett Indians being the first makers. The latter shared in trade and binding agreements with other Indians from the region so that early European invading colonists accepted the Indigenous value system of *wampum*.[116] *Wampum* as part of mutual exchange among the Indians was radically displaced by objective monetary exchange. Other forms of Indigenous commercial exchange included corn, tobacco, furs, and wheat. For Indigenous peoples globally and other cultures of people of color, *land* was paramount and always ancestrally rooted, communally based, and accessible to all. Colonialism and capitalism disrupted this collective basis of land tenure and replaced it with the privatization of land as individual property. The fringe elites in capitalist countries followed suit for self-seeking class benefit and profit accumulation.

In essence, the capitalist system at its initial formation, Marx averred, and continuing into the present has imbued commodities, derived from monetary exchange through adding of labor value, with a pseudo-mysterious character "abounding in metaphysical subtleties and theological niceties," while in actuality, they originate out of the "functions of the human organisms."[117] Yet a commodity's "mysticism does not originate in its use value," and neither does it derive from the "nature of the determining factors of value."[118] Marx amusingly provided the example of wood that changes form when it's fashioned into a common table, yet its triteness changed the moment it became a commodity, something "transcendent." He retorts that the table "not only stands on its feet on the ground, but in relation to all other commodities, it stands on its head, and evolves out of its wooden brain grotesque ideas far more wonderful than 'turning table' ever was."[119]

It is this "fetishism of commodities" which Marx describes that dovetails into his critique of religion as illusory and a method of obfuscation, particularly Christendom. He caustically identifies the religious sanction of capitalist commodification by Christendom as correlational to the latter's ideology:

> The religious world is but the reflex of the real world. And for a society based on commodities in which the producers in general enter into social relations with one another by treating their products as commodities and values, whereby they reduce their individual private labour to the standard homogeneous human labor—for such a society, Christianity with its *cultus* of abstract man [humanity], Deism, etc., is the most fitting form of religion.[120]

Of course, Marx based his critique of "religion" and Christianity in particular on his careful observance of the workings of European Christianity and the latter's endorsement of the exploitation of human labor for early capital formation.

Marx's passionate concern was for the laborer whom he witnessed being dehumanized in the mines and factories of Europe, a consistent trajectory from the mid–14th century to the end of the 17th century. The average person had been reduced to the essential entrails of being a lifelong laborer, not for herself or himself, even in her or his free time; instead, the laborer was required to be "devoted to the self-expansion of capital" for the landowner and industrialist.[121] Marx lamented the fact that the human need for intellectual growth and physical rejuvenation—even "fresh air and sunlight"—became a wish for the laborer since she or he was constantly confined to either working underground in the mines in places like England and Wales or the factory in Britain and other parts of Europe. These exploitative practices were compounded during the era of Indigenous and African chattel enslavement in the Americas when maximum extraction of raw materials for intensified profit-making from industrialism became the core of European labor regimes. Marx contended:

> The capitalistic mode of production (essentially the production of surplus value, the absorption of surplus labor), produces thus, with the extension of the working-day, not only the deterioration of human-labour by robbing it of its normal, moral, and physical conditions of development and function. It produces also the premature exhaustion and dearth of this labour power itself. It extends the labourer's time of production during a given period by *shortening his actual life-time* [italics added].¹²²

Today, capitalist exploitation is classically illustrated in the repressive call centers globally in various regions that serve western and other capitalist transnational corporations. The leading call center country is the U.S., with over 2.4 million workers, followed by the Philippines, with 800,000 employed at call centers, of the 1.3 million employees in the outsourcing sector, the country's major employer. India has 1.1 million outsourcing workers. Expanded call center operations are growing in Latin America, Africa, and eastern Europe.¹²³

Marx observed firsthand, too, the manner that young children—girls and boys as young as 9 or 10—were exploited as coal miners in England. He was thoroughly disgusted with such child slavery, venting repeatedly his denunciation of capitalism and the commodification of labor especially exploiting the young. He noted that following the manufacturing era, "the division of labor" was based particularly on "the employment of women, of children of all ages, and of unskilled laborers, in one word, on cheap labor as it is characteristically called in England and in the capitalist world since the 19th century."¹²⁴ In 1861, Marx documented that at least 73,545 of the 246,613 people

Labor in 19th-Century England (Wikimedia Commons).

working in coal mines and 30,810 of 125,771 working in iron foundries respectively were under 20 years of age.[125]

Such horrific exploitation recalls the deaths of Indigenous children by Catholic authorities at residential and mission schools in northern Turtle Island (Canada) unearthed in 2021, as well as in Australia from decades ago, though not directly tied to child labor. Nevertheless, it signifies shameful exploitation involving dehumanization, brutality, and coercion of children. The brutal anti-child practices that were pervasive in such institutions could be viewed as a correlate of the British child labor regime from the 19th century.[126] With the introduction of machinery technology to "revolutionize" European industrial production, the exploitation of such cheap labor using women and children was magnified monumentally. Marx observed:

> The exploitation of cheap and immature labour-power is carried out in a more shameless manner in modern Manufacture than in the factory proper ... because the technical foundation of the factory system, namely the substitution of machines for muscular power, and the light character of the labor, is almost entirely absent in Manufacture, and at the same time women and over-young children are subjected, in a most unconscionable way, to the influence of poisonous or injurious substances.[127]

Marx noted that this labor exploitation was more shameless in domestic labor than in manufacturing because there's no organized labor resistance in the former. The domestic worker was always pressured from protesting because she or he needed to compete for employment with the factory worker in the manufacturing sector.

Additionally, the invention of the factory machine, Marx averred, while viewed by the capitalist as a revolutionary leap for industrial production, had devastating effects on the human laborer. The latter became dependent on engaging in such work as a means of making a living. The machine, a tool in the hands of the human being, distorted the concept and division of labor, forcing the laborer to sell her or his labor value as a commodity. Her or his intrinsic capacity for labor became degraded into the ability to master a particular tool belonging to the machine.[128] Women and children, even as young as ages four and six, were heavily employed in varied British industries like print binding, coal mining, tile and brick production, salt mines, candle manufacturing, looms for silk weaving, chemical production, and the strange industry of "rags-sorting." Marx made a critical point about the inhumane and essentially gendered female-exploitative nature of the development of the European industrial machine that deployed even more women and "over-young" children "unconscionably" exposed to toxic chemicals and poisons since mechanized inventions no longer depended on physical muscular ability associated with male labor.[129] He argued that "the capitalist buys with the same capital a greater mass of labour-power, as he progressively replaces skilled labourers by less skilled, mature labour by immature, male by female, that of adults by that of young persons or children."[130] He castigated the subjugation of the woman as wife in bourgeois European society, too:

> The bourgeois sees in his wife a mere instrument of production. He hears that the instruments of production are to be exploited in common, and naturally, can come to no other conclusion than that the lot of being common to all will likewise fall to the woman.
> ...it is self-evident, that the abolition of the present system of production must spring with it the abolition of the community of women springing from that system, i.e., of prostitution both public and private.[131]

Europe's rigidly patriarchal society exploited all women in all spheres, whether it was through capitalist production as heretofore explained or through marriage that was viewed as legalized institutionalized enslavement of women.

This misogyny persists in the contemporary capitalist system, albeit in various institutionalized and corporate forms. Women and children in the U.S. and other capitalist centers are the most afflicted by impoverishment, lack of health care, adequate compensation for comparative male labor, sexual violence, and social exploitation by the wealthy. Half of those who are within the ranks of deep poverty in the U.S. are children, and globally, they are the most victimized by war, forced difficult physical and manual labor, and environmental devastation and climate change. Women's labor as sweatshop workers, secretaries, receptionists, administrative assistants, nurses, health care workers, schoolteachers, street vendors in most parts of the underdeveloped world, in the imposed and forced sex trade, and as household laborers and cooks persists as women continue to be exploited and paid less than men in equivalent roles.

This selling of the forced cheapness and availability of unskilled labor is very much in use in the global capitalist world today, with over three billion workers worldwide who are forced to either sell their labor at the cheapest level or eke out a survival existence in what scholars call the "informal sector." Day laborers in many parts of the world are the most exploited and most dependent on this twenty-first-century system of modern slave labor and are forcibly beholden to the extractive capitalist. Sweat shops run by trans-national corporations, especially in Asia, Africa, and Latin America and the Caribbean, are pervasive in this regard, particularly in textile factories, precious metal mines, the electronics industry, and various sectors of the capitalist economies, with young women as the primary victims in many instances.

Farm workers in California, mostly Mexican/Latina/o, have been the most vulnerable as workers, possessing little rights of labor negotiation. They suffer inadequate health and social security benefits and are generally deprived of decent and adequate incomes to maintain families, even while being subjected to a host of pesticides and herbicides under the grower regime. Farm workers, agricultural workers, construction workers, roofers, landscapers, and other outdoor workers, most of whom are Mexican or Latina/o in the U.S., who toil in the 100+ degree Fahrenheit (37.8° Celsius) summer heat from sunup to sundown to ensure that we eat daily and have houses in which to live, even experience death in the unmitigated heat. They are now forced to labor to support their poverty-line families for 21 days of deadly heat levels each year, a number steadily growing. The U.S. Bureau of Labor Statistics (BLS) reported that 40 people working in outdoor jobs like agriculture, construction, and delivery die annually from heat exposure. The Occupational Safety and Health Administration (OSHA) responded that this was a severe undercount, because thousands of unreported cases and those dying in hospitals from "cardiac arrest and respiratory failure" are part of the 600 to 2,000 deaths and 170,000 heat exposure deaths each year.[132] Agricultural workers are 35 times more likely to die of heat stroke and exhaustion and other heat-related causes than other outdoor workers. The insouciance of capital was classically illustrated in the three-week-long, unprecedented heat conditions in the U.S. in July 2023 when temperatures remained consistently in the 100+ Fahrenheit (110–120 degrees) range in much of the country, especially in the southwest, yet Texas ruled that construction worker water breaks were prohibited in Austin and Dallas, and Nevada failed to pass a state law mandating employers to "require water, rest and shade … once temperatures exceed 95 degrees."[133]

Even raising the temperature threshold to 105 degrees Fahrenheit failed in that state. While numerous states have penalties and fines for leaving a dog outside in heat with no shade or water, the same level of protection and penalties only apply in California, Oregon, Washington, and Colorado. Dog and pet rights are rightly important, for the protection of any life, yet the burning question remains: Why not accord such rights to the people who feed us and often labor to construct our houses and gardens? Capitalism assumes that people of color, like those from Mexico and Latin America, are immune to the heat, as are Euro-Americans, and therefore can tolerate it. This inhumanity recalls colonial slavery where Indigenous Indians and Africans were subjected to the lash, the sword, and inhumane manual and physical labor conditions in the Americas discussed earlier. People of color, especially women workers and laborers, have always been and continue to be the primary victims of hypercapitalist greed, exploitation, and subjugation, a legacy of the slavery-colonial regime. All laborers and workers of the world, especially women, are similarly subjected to such economic predation.

The fasting practices of historic United Farm Workers (UFW) leader and organizer Cesar Chavez in the early 1980s as a protest against inhumane farm worker conditions (that eventually weakened him to the point of passing away) serve as a traumatic reminder of the gravity of the inhumanity of the capitalist enslavement system, entrenched but daily weakening in the world, particularly as led by the EU and the U.S. empire.[134]

Similarly, in the delivery sector, consumers enjoying the luxury of having purchased goods delivered to them by UPS, FedEx, and other carriers intensified during the 2020–2021 Covid-19 pandemic. Consequently, 340,000 UPS workers went on strike for raised salaries, humane working conditions like air-conditioned trucks and vans, and a reduction of the slave-regime delivery quota of 150–300 packages a day even in extreme heat conditions. UPS promised to raise beginning workers' wages to $21 per hour above any other already committed wage adjustments and install air-conditioners in all vehicles by 2024. Capitalism's profit maximization by any means necessary is the cardinal principle, all the while sacrificing the lives of the most vulnerable of workers, even in deadly climate conditions.

Retail conglomerates like Amazon and Apple and its leading electronics supply company Foxconn continue to exploit workers. Amazon owns 400 private label brands and refuses to accept worker unions, denouncing most Black and Latina/o workers voting for a union in Alabama in 2021, with hundreds of thousands of workers globally forced to work long hours with unreasonable packaging and delivery quotas. Worker strikes in Asia and the U.S. have grown resultantly. Apple, the largest conglomerate on Earth, with $3 trillion in capitalization worth, still runs repressive, cheap labor manufacturing plants in places like Zhengzhou, Hubei, where half of all iPhones are made. At the Apple plant, 200,000 workers are forcibly locked down, a huge crisis particularly at the height of the Covid-19 pandemic. Worker protests and suicides a decade earlier did not deter the conglomerate from its inhuman labor practices.[135] In the case of Apple in China, even after it insisted that it would investigate its subcontractor, FoxConn, for repressive worker conditions for those making iPhones, iPads, iMacs, and other Apple devices in 2011 after the inhumanity of labor conditions at its Chinese plants was exposed, it refused to change course. Such is the nature of capitalism: talking sweetly and promising the world economic security on the one hand, while extracting incessantly and entrenching economic insecurity and protracted impoverishment. The impoverished have suffered unconscionably. Abolishing capitalism in all of its globalized forms is a *sine qua non* for the world to breathe.

Summary and Conclusion

The colonialist construction of the world as we are forced to live through it today has a very short and checkered history, hence the preceding illumination of post-medieval Europe's history. Europe's inability to understand its own Indigenous roots in the Earth as a result of imperialist, patriarchal, and ideologically dogmatic Euro-Christianity interwoven within this region's monarchical and feudal aristocracy has produced perpetual violence. This chapter illustrates that far from European history being the "progressive" history of the expansion of "enlightenment" and "discovery," it has been one of great suffering for the peasantry and workers in rural and urban settings, particularly women and children as delineated by Karl Marx.

The deification of monetary profit, materialistic accumulation, and apotheosis of the industrial oil machine has ruined much of life on the Earth and destroyed most of the fragile ecological links between humans and other forms of life. Yet, the accelerating ecological, environmental, and chaotic climate change crises show no signs of abating given the obduracy of capitalist industrialism, aggravated by the obsession with oil production and its intrinsic partner, militarism, the subject of our next chapter.

3

Oil Production and Its Capitalist History

Ecocide, Earth Heating, and the Poisoning of Life

> I know that in the past there was a lot of focus on the polar bears. In my attempt to get attention on our own situation I draw a comparison that what happens to the polar bears will also be happening to us in our part of the world....
>
> I saw for myself with my own eyes the huge sheets of ice from the glaciers, and imagined that if that was to melt then obviously it's going to mean a lot more trouble for us and a lot of trouble for other people.... Because for the islanders, the debate is not about higher taxes, political ideology, or any of the other elements of the debate on climate change.... It is about culture, identity and life itself....
>
> I don't know how old I'll grow to be, but I can guarantee you, I will live and die here, and if I need to build my own island I will try to do that.
> —Anote Tong, president of Kiribati, 2014[1]

Introduction

No other issue is as urgent and pressing today as the current crisis of ecological and insect collapse and associated unprecedented Earth heating and climate chaos and instability. Yet many climate change activists with good intention omit reference to the root causes: capitalism and its twin, militarism.

This chapter will thus delve into issues of oil's history, including whaling, and the inextricable interwovenness of oil production and consumption with capitalist extraction, violence, and war, resulting ecocide, hydraulic fracturing for fracking oil and "natural" gas (a misnomer because chemicals undermine the "natural" core), the Oil Pollution Act of 1990 and its flaws, and ecological and environmental decimation from repeated toxic spills from 1990 through 2020. It will highlight Indigenous cultural philosophies rejecting oil drilling and fracking.

History of Oil Production Fueling Western Industrial Capitalist Production

Indigenous peoples who have roots in the ancient lands and cultures of the Earth, of whom we are all descendants, differ from westernized, globalized, and standardized peoples and cultures. The distinguishing demarcation of Indigenous peoples is the insistence on retaining traditional ancestral languages, cultures, religions, and ways of living on the Earth and in harmony and reciprocity with all forms of life, visible and invisible, in the Air, Water, and on the Earth. Indigenous peoples used oil incidentally millennia ago and perhaps even longer. The ancient Indians of the western hemisphere, ancient Sumerians (southern Iraq today) and Mesopotamians (Iraq, Kuwait, Syria, and Turkey today) used small portions of petroleum for medicinal purposes, for soaking arrows during wars like the one between Persia and Greece in the 5th century BCE, and other minor utilitarian purposes, like sealing certain work implements, walls, and floors. The ancient Chinese and other peoples used oil for lighting and heating. Yet oil extraction, production, and core utility were never the foundation of these cultures until industrialist capitalism's birth in the 19th century.[2]

In August 1859, Edward L. Drake, who previously worked as a conductor on the New York/New Haven Railroad, launched the first drilling operation from oil seeping into a field in Titusville, Pennsylvania, that then moved the production of coal for energy to oil refineries. Subsequently, German engineer Nikolaus Otto developed the first oil-run engine in 1861, and soon oil drilling and production became the singular most important event in the development of western industrialization and urbanization at the turn of the 19th century, paving the way for the spawning of oil wells, towers, railroads, refineries, and pipelines by 1869 so that the northeast would become the energy hub of the U.S. economy. The oil industry has grown steadily since with over 50,000 oil fields discovered globally.[3] In 1908, British geologist George Bernard Reynolds, through continuous geological digging, found oil in the southwestern province of Khuzestan of Persia (Iran today). In 1909, Burman Oil, a British company, opened a subsidiary, the Anglo-Persian Oil Company (APOC), which began substantial oil production by 1913. The British colonial navy became the principal client of APOC.

Whale oil originally provided lighting for whale oil lamps and was used for transmission and brake fluid, plant food, pesticides, and other industrial purposes, leading to the decimation and rapid depletion of whales (now under the threat of extinction due to warming oceans, the acidification of seas, and the lethal loss of aquatic and marine environments and ecologies through the ecocidal whaling industry led by North America, northern Europe, and Japan). Hunting operations carelessly ignore that whales are the largest absorbers and sequesters of carbon dioxide—some 33 tons per whale—and their radical depletion and near extinction in the ocean exacerbates heating of our Earth Mother.[4] The International Whaling Commission banned commercial whale killing in an advisory capacity in 1986 and lifted elements of the ban in 2010 when it was reported that the whale population was recovering.[5] The U.S. supported the ban, but was already a leading global commercial whale killer, especially from 1812 through 1859, when oil was first drilled in Pennsylvania, having initiated the bloody commercial whale hunting enterprise in Southampton, Long Island, in 1644.[6]

Though the actual numbers of whales killed over the nineteenth century is not fully clear, the U.S. killed hundreds of thousands from the early 1700s. It possessed 640 whaling ships for its whaling expeditions, three times the rest of the world combined (similar to it being the leading arms seller with 40 percent of all global sales and surpassing that of other arms sellers like Russia, France, China, and Germany, with India, Italy, and Israel being smaller global arms sellers).[7] Estimates of U.S. whaling deaths annually may have been around 4,000–6,000 or less; however, no accurate annual records were kept of such carnage, and whaling records mostly focused on ships, voyages, workers, profits, and products from the whaling industry.[8] The fact of whales numbering in the hundreds today provides evidence of such ecocide. Indigenous nations in Massachusetts where whaling developed following early colonization, like the Pequot, Pawtucket, Massachusetts, Pokanoket (Wampanoag), Narragansett, western and eastern Niantic, Quirpi, Nipmuck, Pocomtuc, Tunxi, Podunk, and others, were ecological defenders and unsupportive of such extermination for commercial purposes, although some Indigenous people may have been compelled to join to earn a living given colonial land and economic dispossession.

Whaling was a lucrative commercial enterprise and the fifth largest sector of the U.S. economy in the 19th century prior to oil's production beginning in 1859. It made New Bedford, Massachusetts, the wealthiest economy per capita in the U.S., "if not the world according to a 1854 newspaper," which stimulated the subsequent "railroad, oil, and steel" industries.[9] Cape Cod and Nantucket were other coastal areas where such whaling occurred. San Francisco, too, was a place of commercial whaling after the "boom" in New Bedford. In 1744, with depleted whales on the east coast, whalers began to move into deeper waters, and soon hundreds of ships were built for this growing whale hunting/killing oil economy. The result of this obsessive commercial enterprise is that whales began to decline in the North Atlantic so that just 350 right whales and only 100 West Pacific grey whales remain today.[10] Blue whales, the largest creatures on Earth, which can measure up to 100 feet and weigh up to 200 tons (over 300,000 pounds), have been decimated to less than 1 percent of their original numbers in the hundreds of thousands, substantiating the ecocidal nature of commercial whaling from the 17th century. Some scholars deny the gravity of the ecocide of whale deaths by arguing that there were still plenty of whales after the period of the initial decline in whaling in the east in the 1850s.[11] At least 114 whales have been killed since 2017 in documented cases; innumerable others underscore that this number is a gross undercount. Most whale species are in radical decline and face extinction in many quarters. The washing up of 55 whale bodies on the Scottish coast and constant washing up of whales on the west and east coasts of Turtle Island are indicators that the ocean has been contaminated and violated by commercial shipping and naval vessels that poison the ocean and strangle the diverse, complex, and beautiful mosaic of aquatic life in the seas and oceans.[12]

When the U.S. invaded and occupied Iraq in 2003 to control and expropriate the land's vast oil reserves, a culmination of 12 years of relentless bombing and sanctioning of the country, known as ancient Mesopotamia and Sumeria (where writing and scientific innovations were underway over five millennia ago), the world protested and chanted, "No blood for oil!" Little did people realize that the same call could have been made for the obsessive and incessant extermination of whales: "no (whale) blood for oil!" Whales' diving and fecal fumes fertilize the ocean and enable phytoplankton blooms to be formed. The blooms come to the surface and absorb 40 percent of carbon dioxide from the Earth. Further, whale bodies that sink to the bottom serve as a repository for carbon storage—33 tons per whale. Fish also store carbon upon passing away when their bodies sink in the oceans and seas. Commercial whaling and fishing thus result in massive amounts of carbon dioxide being released.[13] Commercialism and capitalism are still the leading causes of climate collapse and excessive Earth heating. This is a crucial ecological, environmental, and sustainability point given that our Earth Mother can only recycle 14–16 gigatons of carbon dioxide while annual emissions are twice that at 36–38 gigatons.[14]

It is the oil-based and oil-centered industrialist "revolution" of the 19th century that sparked the current ecocidal, ecological, environmental, and climate catastrophe crises that we are experiencing the world over today. The U.S. military is the leading customer for U.S. oil production, and the U.S. Navy is the principal conduit for the protection of oil routes and oil to fuel the needs of the U.S. Air Force, Navy, and Army (see image). This sprawling military industrial complex manifest in 172 countries is key to the U.S. empire's ability to police the world, ensuring that its oil access is uninterrupted, while making the military the worst polluter on Earth. Oil production is thus inextricably interwoven with both capitalism and imperialism, signified by the North Atlantic Treaty Organization (NATO) and the Pentagon. Thus, a revolving circle involving the U.S. empire, the Pentagon, oil production control, and protection of the empire persists.

The U.S. has beefed up national oil production to over 12.8 million barrels today largely through the very ecologically and environmentally destructive process of extracting oil from shale all over the country, particularly in Texas and Colorado, and from offshore drilling in the Gulf of Mexico, up from 5.5 million barrels per day in 2010 (see image).[15] While it exported 9.8 million barrels to 180 countries in 2022, it imported 8.3 million barrels of petroleum from 80 countries in the same year, since the latter is cheaper than locally produced extraction costs.[16] This is in stark contrast to the 1970s and prior era, when oil imports were 36 percent of total consumption. The U.S. used 19.78 million barrels of oil per day in 2021, one-fifth of the total daily world consumption of almost 100 million barrels, though it has just five percent of the world's people.[17] The U.S. Department of Defense (DOD) is the "world's largest institutional user of petroleum ... and the single largest institutional producer of greenhouse gases (GHG) in the world," and since 2001, it has "consumed between 77 and 80 percent of all US government energy consumption."[18]

It was access to oil that ensured British colonization of west Asia and parts of west Africa like Nigeria. Oil-producing regimes such as the United Arab Emirates, Saudi Arabia, Iraq, and others serve as client states of the U.S. empire or some European government today. Syria and Iran are the notable exceptions and have thus been targeted by U.S.

Revolving Circle: U.S. Empire, Pentagon Oil Production Empire Protection Graphic (Veronica Rodriguez, 2023).

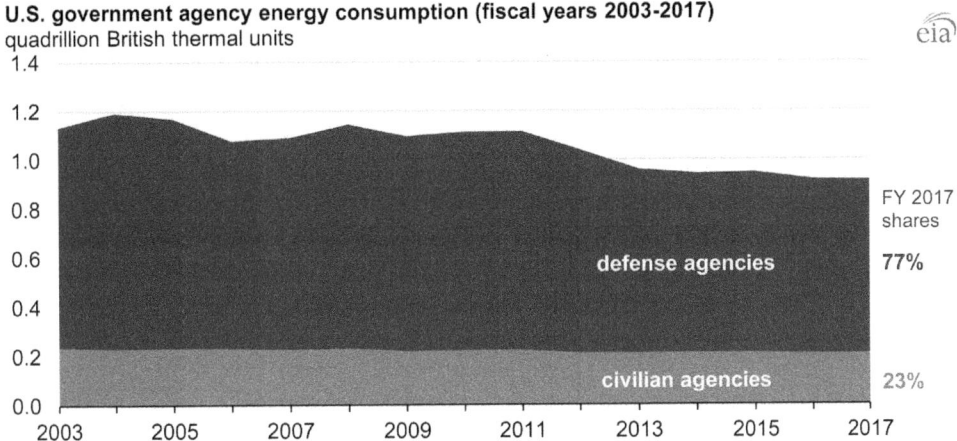

U.S. government energy consumption continues to decline (U.S. Energy Information Administration: Today in Energy, July 25, 2019).

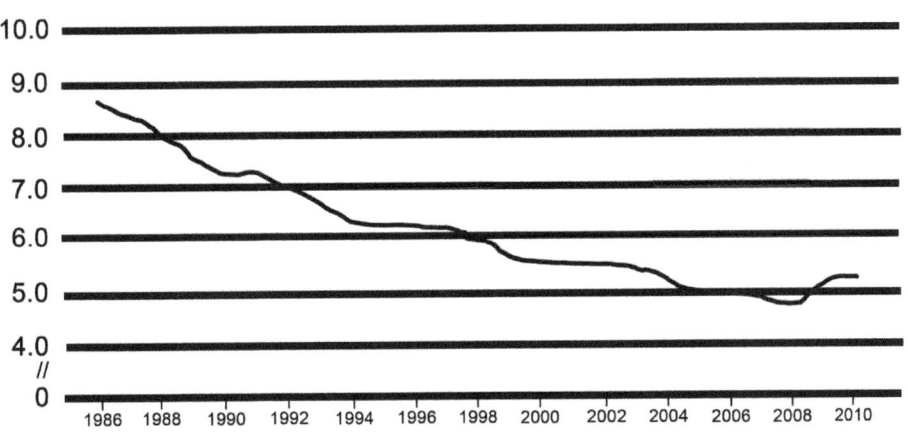

Domestic oil production reversed a decades-long decline in 2009–2010. "Following declines in all but one year from 1986 to 2008, U.S. oil production (crude oil and lease condensate) increased in 2009 and again in 2010. Due in part to Hurricanes Ike and Gustave, average annual production dipped below 5.0 million barrels per day (MMbbl/d) in 2008, then climbed to 5.4 MMbbl/d in 2009 and 5.5 MMbbl/d in 2010, with 2010 volumes representing an 11% increase over 2008" ("Today in Energy: Domestic Oil Production Reversed Decades-Long Decline in 2009 and 2010," U.S. Energy Information Administration, April 27, 2011).

economic embargos and sanctions and have been subjected to military intervention to destabilize their governments. The reason that Iran has been a thorn in the side of the west is that its people were determined to nationalize their vast oil and "natural" gas resources, the world's fourth largest, estimated at 157,530,000,000 barrels of oil reserves (9.5 percent of world reserves in 2016). From 1951 to 1953, Iran worked to nationalize its oil resources, wresting them from British company control and halting the bulk of oil revenues ending up in British commercial coffers. It was for this reason that the U.S. CIA and British MI6 overthrew the democratically elected leader, Mohammed Mossadegh, in 1953, and installed the puppet and tyrannical Shah Mohammad Reza Pahlavi, who ruled with terror and torture until 1979, when he was ejected and exiled and the Council of the Islamic Revolution came into being.[19] This change set the nation into an inevitable collision course with the west, which had been the leading supporter and protector of the imperial Shah. Such historical events serve as reminders that most of the wars in the world today are essentially about capitalist and neo-colonialist access to water, oil, and vital strategic industrial resources, most of which reside in the lands of Indigenous peoples and nations of color or outside the western European axis, including Russia.

It's ironic that the largest oil companies—the "Seven Sisters"—were and are western oil companies, even though the bulk of oil production comes from outside Europe. The "Seven Sisters," Exxon (Esso), Gulf, Texaco, Mobil, Socal (Chevron), British Petroleum (BP), and Shell, became the largest oil companies in the world and remain so as a grouping today, though the oil conglomerate, Saudi Aramco, is the second largest company in the world after Apple, worth $2.2 trillion. Anthony Sampson vividly describes the "Seven Sisters":

> For decades the Companies (with a capital C) seemed possessed of a special mystique, both to the producing and consuming countries. Their supranational expertise was beyond the ability of national governments. Their incomes were greater than those of most countries where they operated, their fleet of tankers had more tonnage than any navy, they owned and administered whole cities in the desert.[20]

Sampson adds that these companies were immune to supply and demand principles as with most other products because of their immense global scope and influence. The Organization of the Petroleum Exporting Countries (OPEC) was formed in Baghdad in

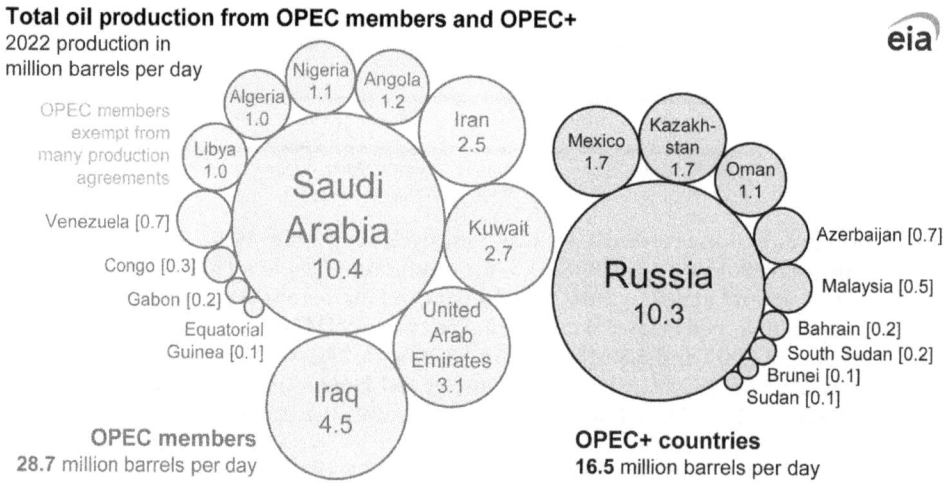

Note: OPEC oil totals are crude oil; OPEC+ oil totals are crude oil and lease condensate. ("What is OPEC+ and How Is It Different from OPEC?" U.S. Energy Information Administration, May 9, 2023).

1960 as a challenge to the Seven Sisters. It was started by Venezuela, Iran, Saudi Arabia, Iraq, and Kuwait.[21] OPEC subsequently included Qatar (1961), Indonesia (1962), Libya (1962), the United Arab Emirates (1967), Algeria (1969), Nigeria (1971), Ecuador (1973), Gabon (1975), Angola (2007), Equatorial Guinea (2017) and the Republic of the Congo (2018). Ecuador withdrew in 2020. Indonesia and Qatar have also since left the organization. Russia, as the world's third largest producer of oil, formed OPEC+ in 2016 due to falling oil prices as the result of the U.S. flooding the market with oil from shale. OPEC+, a close formal partner of OPEC, was later joined by Mexico, Kazakhstan, Oman, Azerbaijan, Malaysia, Bahrain, South Sudan, Brunei, and Sudan. OPEC and OPEC+ cumulatively possess 45.2 million barrels today, some 47 percent of the global daily total. Though the U.S. is the largest world oil producer at 12 million barrels per day, it is not a member of OPEC or OPEC+.

The "Seven Sisters" controlled and dictated the terms of half of all global oil trade and were so markedly influential that even maps of countries in West Asia, where most

Standard Oil's Tentacles Stifling Corporate Competition (J. Campbell Cory, Wikimedia Commons).

oil was located, "showed the region cut up into squares along the coast, marked with initials—IPC, KOC, ARAMCO, AOC—representing the consortia of oil companies, always including some of the Seven Sisters." Kuwait became equated with Gulf and BP and Saudi Arabia with Aramco (today the company is Saudi Aramco). Standard Oil of New Jersey (known as Exxon Mobil today and owned by Vanguard, BlackRock, and thousands of other investors and shareholders) was worth over $276 billion in 2021, after splitting up when its capitalization value was over $1 trillion. It was founded by John D. Rockefeller in 1882, the richest man in the world, who was responsible for building the Rockefeller financial empire based on oil, copper, steel, and other industries, and was firmly influential on the U.S. government.

The Rockefeller family quietly exited the oil business in 2016 to supposedly support environmental causes, yet the bulk of the irreparable ecological, environmental, economic, and political damage over the previous 134 years of Standard Oil's commercial hegemony could never be undone.[22] Shell, as the second leading "Sister," was a British, Dutch, and U.S. corporation, and has since changed its name—first to Royal Dutch Shell and then to Shell PLC in January 2022, rebasing itself in London, with drilling and refining operations in 99 countries, particularly in Africa. It is a dangerous oil company, plundering the Niger Delta region of west Africa for oil and destroying the fragile ecology there, along with its operation in the Gulf of Mexico. Shell is notorious for supplying aviation fuel and 80 percent of the explosive chemical trinitrotoluene (TNT) to the British military in the 1940s and to the apartheid South African military.

The oil supply crisis of 1973 was the first time that OPEC exercised its political and economic muscle as the most powerful financial and energy cartel by demanding a global oil price of five dollars per barrel, two dollars more than the existing price. The "Seven Sisters" were willing to offer a 25 percent increase in the per barrel price as instructed by their respective boards. OPEC was unmoved and thus instituted an oil embargo against the western countries, especially the U.S., because of the latter's military and economic support of the settler-colonial state of Israel and the unbridled dominance of oil price decision-making roles by the "Seven Sisters."[23] This triggered the biggest oil and energy crisis in the western world in the 20th century.

The U.S. military flexed its muscle and its might in the west Asian region in the 1970s and subsequently functioned as a protector of the Wahhābī Saudi Arabian oligarchy and the oilfields there that ensured cheap oil flowing into the U.S. empire. The U.S. economy fully depended (and depends) on the hundreds of billions in petrodollar income of the Saudis, who were the main exporters of oil to the empire. International oil imports were indispensable for half of oil consumption by the U.S. and good for oil prices and profits since oil extraction abroad was cheaper.

During the 1980s, the U.S. armed Iraq with billions of dollars' worth of weapons and military equipment and encouraged the invasion of neighboring Iran, which sparked a horrific and bloody war that saw one million people from both nations annihilated. A huge change occurred in 1990 with the Iraq economy in tatters and in debt for $70 billion to countries surrounding it, even while bordering Kuwait had raised its oil production to two million barrels, twice the OPEC quota. Kuwait's action led to the maintenance of a lower oil price that undercut Iraq and Iran's respective oil revenue based economies, urged by the U.S. It was this economic deterioration that compelled Iraq to invade its tiny neighbor, Kuwait, on August 2, 1990, and that led to the illegal 1991 bombing and invasion by the U.S. and other western "Allied" nations that pulverized Iraq with the raining down of 88,500 tons of bombs in 109,000 aerial sorties. These actions harkened back to the U.S. bombing and occupation carnage in Vietnam, Laos, and Cambodia from 1964 through mid–August 1973. Five million tons of bombs dropped by the U.S. Air Force rained down on Vietnam then, more than all of the total bomb tonnage dropped during the second intra–European war of the 1940s, and another 1.62 million tons on Laos and Cambodia.[24] The U.S. Navy and Marine Corps dropped another 1.5 million tons on the latter. To underscore the living horror of that period in southeast Asia, 80 million of the 260 million bombs dropped on Laos, destroying 37 percent of the agricultural area permanently, did not explode and remain embedded close to the topsoil or on it.[25] On top of this massive military onslaught on a small country, the notorious CIA had invested heavily through the Phoenix program to destroy the Vietcong infrastructure through a concerted "neutralization" program in 1967.[26] Oil and war always go hand-in-hand.

In 1991, the U.S. led the onslaught in Iraq in a 42-day, all-out, relentless bombing crusade, endorsed shamelessly for the first time by the world's leading "peace-brokering" organization, the UN. The genocidal damage and ecological damage suffered by one nation due to such massive bombing within such a short length of time follows in the vein of that of the U.S. atomic bombing of Hiroshima and Nagasaki in 1945, when the equivalent of 15,000 tons and 22,000 tons of TNT were dropped respectively, the annihilation of five million Koreans in 1951, and the extermination of four million Vietnamese over the course of the 1960s and 1970s.[27]

In Iraq, which possessed the best electrical grid in west Asia, the hard fought gains of free and universal education, radical reductions of illiteracy, prenatal and postnatal health care for children (including in rural communities), free health care for all, 42 hospitals constructed from 1982 to 1990, low-interest loans, women's educational and social advancement, allocation of land for those promising to enhance food production that supplemented the cheap availability of food, thousands of miles of highways and dams and hydroelectric flood and irrigation systems, advanced telephone and electrical grid communication systems were mostly erased in 42 days, 1991, and beyond.[28] The U.S. military heavily destroyed eight multipurpose dams, eliminated flood control mechanisms, municipal and industrial water storage facilities, irrigation and hydroelectric power sources, seven of the country's major water pumping stations, 32 sewage facilities, 400,000 of the 900,000 phone lines, 139 rail and automobile bridges, half of the agricultural production centers and food processing plants, two million cattle, and three and a half million sheep herds. Not satisfied with their extermination deeds, again in 1992, U.S. and "Allied" bombs damaged or destroyed 676 schools, 28 civilian hospitals, 52 community health centers, eight universities, and even 31 mosques.[29] Further, the U.S., Israel, and the United Arab Emirates have looted Iraq's famous National Museum with over 120,000 antiquity pieces from ancient Mesopotamia, Babylon, early Jewish history, and other significant West Asian historical eras stolen.[30]

These documented facts serve as a gruesome reminder of the cost of oil as western capitalism demands it: cheap and accessible by any means necessary. No oil producing nation should expect to be immune to the relentless colonializing control of the U.S. empire that still has a military presence in Iraq from over 32 years ago. The oil war legacy in Iraq lingers today as doctors urge women not to have children due to the high propensity for birth malformations in places like Fallujah and Basra in Iraq, where tons of depleted uranium shells were unleashed on the civilians of those places in the 1990s.

The declaration of a no-fly zone in sovereign Iraqi territory and deadly economic sanctions and embargos eventually led to the deaths of 500,000 children. "It's a very hard choice … but we think … the … price is worth it" was the response of then U.S. ambassador to the UN, Madeleine Albright, when asked by Lesley Stahl on CBS's *60 Minutes* about whether the loss of half a million lives was worth it, considering the inhumane and draconian economic embargo against Iraq that included vital medicines, medical equipment, and food supplies.[31] It's the same genocidal slaughter of Indigenous children of Gaza by Israel using U.S.-manufactured bombs and military funding witnessed in October to November 2023, where thousands of little ones and women especially were exterminated in a colonial ethnic cleansing campaign. Oil, water, and energy access even played a role in this latter conflict, where, for instance, 90 percent of the water aquifers within the occupied West Bank were diverted to Israel.

From being the country with the most advanced educational, technological, and electrical grid system in west Asia, Iraq was turned into a dilapidated beggar nation from 1991 onwards, which culminated in the full-blown U.S. invasion of Iraq in 2003 and the establishment of a permanent U.S. military "Green Zone" with U.S. troops and military infrastructure in the country to this day.[32] The U.S. refused to leave even after the Iraqi parliament voted on the expulsion of U.S. troops in January 2020.

U.S. military contractors still occupy parts of Syria in flagrant violation of international law and have coerced Iraq to retain a military presence under the pretext of fighting the terrorist group ISIS. Yet the White House openly acknowledged in 2019 that the military was there to "secure the oil." Most of Syria's oil and oil revenues from Kurdish-controlled areas in Iraq have been expropriated even though they belong to Syria and Iraq by international law. Some of this oil has been shipped via Turkish tankers to Israel.[33]

Oil production and sales have been consistently turbulent since the 1970s because global business trends and revolutionary political events have a direct bearing on the cost of materials for industrial production and the corresponding demand for oil based on vagaries of economic and commercial downturns and upturns.[34] Since 1973, non–OPEC oil producing countries like Mexico, Norway, the U.S., and Canada sought to raise their oil production levels phenomenally so that OPEC's share of oil production dropped sharply from 53 percent in 1973 to 43 percent in 1980 to 28 percent in 1985, but it rebounded to 40 percent in 2020 due to the high cost and depletion of oil wells elsewhere.

In April 1980, following the inception of the Iranian Islamic Revolution in 1979, the price of oil jumped from $14.85 to $40 per barrel for the first time. The 1980s saw the U.S.-orchestrated Iraqi invasion of Iran that caused further disruption of oil production and cost over a million lives. The raising of U.S. interest rates by the U.S. Federal Reserve Bank in 1980 triggered further economic decline and induced a recession in the U.S. that was felt globally due to the hegemonic role of the U.S. dollar in the global capitalist economy, reducing global oil demand and corresponding consumption.

Oil prices rose in the 1990s on account of the U.S. invasion and occupation of Iraq in 2003. The Great Recession of 2008 recalled the 1929 Great Depression, this time precipitated by the U.S. housing subprime mortgage crisis and artificially inflated value of major U.S. banking and financial assets where the largest banks and asset managers listed outstanding debt and loans as assets. The price of oil skyrocketed to $134 per barrel in June 2008, and subsequently to a stunning $147 per barrel, only to skydive to

$39 per barrel in February 2009. Global economic and financial uncertainties resulted in industrial and economic growth indices and projections being significantly reduced. From 2014 through 2016, oil prices experienced a historic 70 percent decline, from $100 a barrel to under $30 in one of the three largest drops since the second intra–European war of the early 1940s. It was the longest in duration, largely due to an oversupply glut and shifting OPEC priorities without a corresponding global economic expansion as anticipated.[35] The U.S. reduced investment in the energy sector and oil-producing economies stagnated. Further, the United States' insertion into oil price influence was facilitated by the plundering of domestic oil-from-shale resources for unearthing oil that boosted local oil production and weighed heavily on the oil price in the global market.

The oil price fluctuation moved into revolutionary downward territory as the global coronavirus pandemic took center stage in spring 2020 *with oil prices decreasing for the first time,* exacerbated by a radically slowing global capitalist economy and the environmental movement's pressure to move away from dependence on fossil fuels toward decarbonized economies. Oil-producing underdeveloped countries, mostly in Africa, that were dependent on oil for 50 percent of governmental revenues and up to 75 percent of export revenue like Nigeria and Angola were particularly badly hit since commodity diversification was slow in these countries.[36]

The negative oil price and affordable global oil was short lived. In 2021, oil prices rebounded and rose to $60+ per barrel in the first half of the year, rising to $79 per barrel in the fourth quarter, though the Energy Information Administration (EIA) wrongfully predicted that oil prices would drop in 2022.[37] This was followed by the skyrocketing price of the fossil fuel on March 6, 2022, of $130 per barrel with the movement

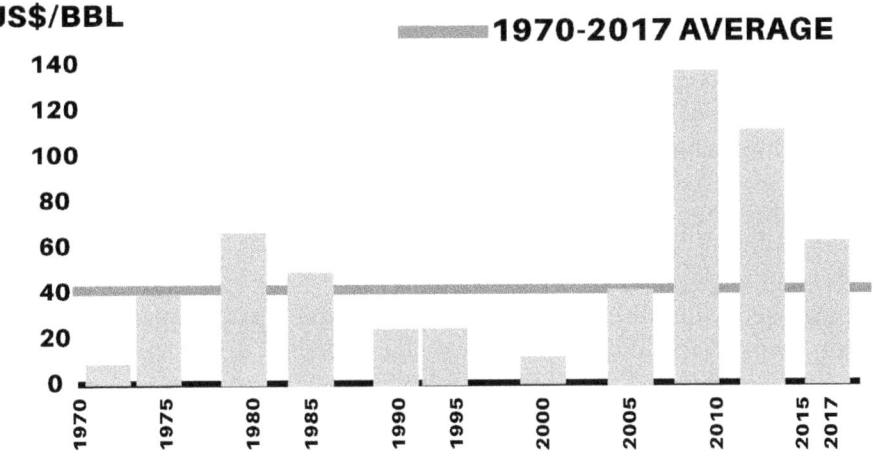

Data sourced from Marc Stocker, John Baffes, and Dana Vorisek, "What Triggered the Oil Price Plunge 2014–2016 and Why It Failed to Deliver an Economic Impetus in Eight Charts," *World Bank,* Blogs, January 18, 2018 (Veronica Rodriguez and Casey Ontiveros).

of Russian troops into southeastern Ukraine over the escalating buildup of North Atlantic Treaty (NATO) forces on Russia's borders. The violation of the 2014 Minsk I and II agreements between the European Union, Ukraine, and Russia that cemented a non-expansion of NATO was a key factor. The agreement stipulated greater autonomy for the southeastern, largely Russian-speaking areas of Donbass in Ukraine that include Donetsk and Luhansk, areas that Ukraine violently suppressed.[38] The saddest feature about this conflict is the tens of thousands of lives lost on both sides and rising prices of fertilizers and food staples like wheat, of which both Russia and Ukraine were leading global producers.

The oil price in March 2022 hovered around $113, topping $120 on March 26, following a series of Houthi–led Yemeni missile strikes on Saudi Arabian oil facilities and refineries, further jeopardizing cheaper oil production and vital access, especially for consumers in the EU and the U.S. This development, accompanied by the Ukraine-Russia conflict, threatened to drag western imperialism into a severe economic recession with escalating oil prices given the heavy dependence of the respective economies on oil, yet prices stabilized in the second half of 2022 to end at $75 per barrel in December that year.

The preoccupation with both oil production and consumption is very much a western capitalist obsession since the U.S. consumes about 19.69 million barrels of oil daily, a fifth of the total world consumption of 99 million barrels, and the European Union is a close second, together using about 40 percent of the total daily global oil consumption. China uses some 11.75 million barrels a day, just over 10 percent of world consumption, even though China has almost one-fifth of the world's people, followed by India at 4.489 million barrels, and Japan at 4.026 million barrels. Russia, with over 150 million people, uses 3.594 million barrels of oil daily. Even though China and India have significant oil consumption, their use is still a fraction of the U.S. and the European Union. China and India each have over 1.4 billion people, constituting about 36 percent of the world, and still use less—15.75 million barrels of oil daily.

However, even with the Green Movement and the organizations subsumed under this umbrella challenging oil production and use for vitiating the environment, one still sees capitalist economics play out, with the largest environmental rights groups receiving the lion's share of grant monies in the U.S.[39] The Sierra Club received over $204 million, Nature Conservancy received 404 grants from 2015 to 2017, joined by other large, well-funded ones like the Environmental Defense Fund, the Resources Legacy Fund, the World Wildlife Fund (WWF), and the Monterey Bay Aquarium Research Institute.[40] These organizations received 64 percent of grants and 80 percent of total grant dollars, some $3.65 billion over the same period. Smaller environmental racial justice organizations concerned with protecting ecologies and environments in communities of color, like Indigenous, Black, and Latina/o people facing lethal toxic waste and mining pollution, for instance—far more pronounced than other communities—received only 10 percent of all grant dollars, even though 56 percent of the environmental foundations were directed toward people of color.

The oil conglomerates have been responsible for making plastic ubiquitous for over a century in every sphere of life, from food packaging (all organically produced, processed, and unprocessed food, too) and fertilizers, to toothbrushes, automobile parts, eyeglasses, pipes, insecticides, perfumes, tapes, CD/DVD players, furniture, tables and chairs, shoes, clothes, and all electronic products including computers, phones, radios, and television sets. Virtually everything we habitually use daily is plastic, though one barrel of oil (42 gallons) only generates 19 gallons of gasoline.[41] Automobiles, too, consist mostly of plastic, hence one of the reasons for California's written health hazard warning in all automobiles.

In scientific studies published in September 2023, the "plastic truth" chicken finally came home to roost. Research by scientists in Japan revealed that the presence of microplastics (airborne microplastics, or AMP) was found in cloud samples and is thus present in rain, deleteriously affecting the soil, plants, microorganisms in the earth, birds, insects, four-leggeds, the sea, and aquatic life. We have also become walking zombies of plastic.[42] Nine different kinds of polymers were found in samples of food and feces alike. In Patna, India, the Indian Institute of Technology found "polyethylene, terephthalate, and polypropylene" as the most common deadly polymers in rainwater that inevitably manifest in drinking water. Ironically, most of the content of handguns is plastic, compounding lethal environmental and health hazards. Many globally routinely eat from plastic containers and drink from plastic bottles, signs of living "plastic lives." The "black gold" Indigenous people cautioned against so that it remained in the ground has now come to haunt us daily even while making life convenient for travel, energy, etc.

Ecocide Caused by a Colonial Chemical-Capitalist and Militarized World

Concretely, pesticides, herbicides, and chemical fertilizers have devastated the ecology in all parts of the world where they've been used. The onset of industrial agricultural food production marked the first chemical-based society since the 1840s or so of what we know of human history, where the natural food that we ate and sustained our health shifted toward employing poisonous and deadly chemicals that have destroyed our bodies, minds, spirits, and all life.[43] Small wonder, then, why mental illness is a pandemic in the U.S. and children and youth especially are struggling to read, pay attention in classrooms and focus, along with adults of all stripes. Modified and chemically processed food and the ubiquity of toxic online technology have everything to do with this transmogrification and deterioration of life. The ideological supremacy of technology that involves deadly chemical fertilizers, pesticides, and herbicides in agricultural production is the leading cause of food toxicity. Indigenous natural food cultivation, like the amazing *maize* (corn) of the ancient Aztec and Mayan cultures of the central part of the western hemisphere along with diverse food crops with multifarious varieties like amaranth, beans, squash, quinoa, chia, avocado, cacao, cassava, papaya, peanut, peppers, pineapple, potato, sunflower, sweet potato, tomato, and tomatillo from 4,000 BCE, has provided the basis for nutritious diets all over the world, especially in the western hemisphere. In Asia, 200,000 varieties of rice were grown and produced in India.

Healthy Indigenous multi-crop production and rotation maintained healthy soils, humans, other life forms, and environments. Agroindustry since the mid–19th century has steadily worsened the plight of natural, chemical-free corn cultivation and production and forced tens of thousands of Indigenous farmers in places like Mexico, Africa, the Philippines, and other parts of the world to adopt GMO seeds and chemical fertilizers in the 20th century. Such lethal seed imposition has accelerated and is being institutionalized globally in the 21st century with funding from billionaire Bill Gates and the former Monsanto corporation now owned by chemicals and pharmaceutical behemoth Bayer. Bayer has been responsible for exporting 54 lethal sickness-causing pesticides banned in the EU to Azania/South Africa and other colonized countries.[44] These pesticides are a leading cause of ecocide. It's impossible to live without our relatives in the insect world especially, since 40 to 90 percent have been decimated and each day hastens further extinction, since 150 to 300 species are vanishing each day. Our wonderful, life-supporting relatives are suffering lethal threats, hastening their and our extinction.

The Paris Agreement (COP 21) and other perfunctory elitist meetings that consume tens of millions of dollars and are directed ostensibly toward reducing Earth heating and global warming are vacuous because the governments of the world are certainly not halting lethal destruction of insects from mining, oil extraction, fracking, etc., now or in the future. All these processes entail violence and poisoning our sacred Earth Mother, substantiating that both capitalist and even socialist forms of industrial development are inevitably headed toward extinction.[45] Economic growth, which still remains a global index to measure the viability of each national economy, in both capitalist and even socialist countries, is driving the world toward oblivion. There's no life, no growth, if the core of life that sustains biodiversity is exterminated. We won't be here much longer given the shocking apathy of those steeped in industrial systems of production to suffering and obliteration.

The lack of water for life since the early 1970s in many parts of the world, particularly in Indigenous communities here in North and Central America, Africa, Asia, the Pacific, and Latin America and the Caribbean, led to water rationing in many places. Billions of humans and trillions of insects, birds, four-leggeds, trees, plants, rivers, lakes, streams, and all natural waterways—indeed all forms of physical life—have struggled for adequate water to live, and many died without gaining enough in time. In this early part of the second decade of the 21st century, water's axiological significance has finally hit home. Eleven western states in Turtle Island have been entrenched in the worst drought in memory, with Lake Mead and Lake Powell dams at less than quarter of their capacity, the lowest level since the 1930s when they were initially filled (picture in preface). The interruption of structural drought by intense bouts of rainfall in 2016–2017 and in 2022–2023 has not radically changed the landscape of lingering "aridification" and dryness.[46]

The megadrought in the western states has been persistent for 22 years, marking the driest era of the Colorado River Basin for 1,250 years, and is so protracted that water experts and authorities have indicated that the term "drought" is inadequate to express the gravity of the environmental decline. They have coined the term "aridification."[47] In desperation, the U.S. federal government will be diverting water from the Flaming Gorge Reservoir on the Green River that runs through Wyoming and Utah to fill the rapidly diminishing Lake Powell, causing disruption to the ecological and environmental integrity of the Green River.[48] The western states are now facing what people in places around the Earth have faced for decades and continue to suffer compounded drought effects: water scarcity. Phoenix, Arizona, received 0.15 inches of rain during the monsoon season of 2023, the lowest ever recorded since 1946.

In Africa, places like Ethiopia and Somalia, where millions of traditional pastoralists have depended on goat, sheep, and cow herds for millennia, devastation by three progressive rain-free seasons resulted in millions of their herds dying from lack of water.[49] So, too, the Sahel region of North Africa, Southern Africa, especially Namibia, Botswana, and Madagascar, the Arab world, particularly Yemen, Jordan, Palestine,

Australia, Latin America, and other regions of the world are experiencing the worst drinking and agricultural water crisis in centuries.

All oil producing areas of the Arabian desert, Siberia, India, and Nepal, where people are forced to pay up to 20 percent of their meager incomes for trucked water delivery,[50] parts of the Philippines like Baguio, and Java Island, Indonesia, where 60 percent of 300 million people of the country lives, are now suffering the most lethal survival crisis in human history. In Java, water scarcity is pervasive, and the annual per capita consumption level has dropped to 1,169 cubic meters from a desperately needed 1,600 cubic meters per capita for decent living. The level is precipitously headed toward a radical drop to 476 cubic meters by 2040 or sooner.[51]

Yet, imposed and mandatory restrictions on water use, especially in the western states, have been few and far between, with certain limits beginning in 2019 and running through the present, mostly in California, where residents in the southern part of the state were urged to voluntarily use a maximum of 80 gallons per day in 2021–2022, with no mandatory requirements.[52] Ironically, too, in southern California, even with urging by authorities of consumption water limits, residents used 19 percent more water in March 2022.[53]

The use of water for consumer, industrial, and commercial purposes will be inevitably curtailed in this century as we have already bypassed the Paris Agreement of COP 21 at the UNFCCC, including machinated car washes, watering lawns, long showers, mining, and industrial production. Most of the world is forced to live on the smallest quantity of water per day, and drinking water scarcity will be the largest crisis everywhere on Earth. Viable and sensible alternatives will need to be found that will entail shorter showers, buckets for bathing, removing lawns in arid areas (now uprooted in many Southern California towns), and developing community pools so that individual household pools using thousands of gallons are avoided. The story of a student who returned to visit family members in Iraq in 2019 and was asked to limit water use in the washing of her long hair provides just one glaring example. The simple, unequivocal, scientific truth is that *water*, the foundation of all life on Earth and in the sea, can't be manufactured no matter how advanced 3D technologies and artificial intelligence may appear to be.[54] The year 2024 was unquestionably the hottest year on record, with the past 10 years being the hottest ever.

Deforestation is proceeding unabated as the demand for both beef and firewood persists. The equivalent of a football field of forest is decimated every six seconds.[55] Both Arctic and Antarctic poles are melting rapidly, and sea level rise is unprecedented, with huge holes in the ozone layer. Sea ice is collapsing so quickly that polar bears, penguins, and other creatures dependent on living in and on the ice are dying in large numbers.[56] Ecocide and violent experiments against the diversity of Earth creatures who have every right to live continue:

> more than 100 million animals—including mice, rats, frogs, dogs, cats, rabbits, hamsters, guinea pigs, monkeys, fish, and birds—are killed in U.S. laboratories for biology lessons, medical training, curiosity-driven experimentation, and chemical, drug, food, and cosmetics testing. Before their deaths, some are forced to inhale toxic fumes, others are immobilized in restraint devices for hours, some have holes drilled into their skulls, and others have their skin burned off or their spinal cords crushed.[57]

Up north in Grandmother's Country, as Indigenous leader Black Elk referred to Canada, 3.9 million animals are subjected to torture and violence in laboratory experiments, followed by Britain. At Porton Down, a military laboratory outside Salisbury, Britain, the U.S. Defense Threat Reduction Agency (DTRA) funds such experiments. There were 121,000 creatures exposed to deadly diseases, chemicals, and pathogens from 2005 through 2016, part of the 3.52 million experimental procedures there.[58]

Over a trillion fish, mostly anchovy and krill, are killed in the oceans and seas each year to satisfy human needs—anywhere between 37 and 120 billion farm-based fish are killed for consumption, and at least 53 million sharks are killed in the fishing business globally.[59] Over 200 million land-based animals and almost three billion "wild-caught" and farmed fish are killed for food each day. In New South Wales and Queensland, Australia, an estimated one billion birds, insects, four-leggeds, reptiles, and other creatures were annihilated in the 27.2 million-acre fires that ravaged the country from late 2019 into 2020, a true genocide and ecocide in the first 20 years of the 21st century.

Meanwhile, the number of people dying from lack of food, malnutrition, and hunger in June 2020 was over 4.2 million, almost half of the 9 million expected to die from such unnatural causes, mostly in Africa, Asia, and Latin America and the Caribbean. Of these, over two-thirds are children. Though one always laments the high number of deaths from diseases like malaria, AIDS, typhoid, cholera, and now, Covid-19, the deaths of those dying from hunger exceed all of these diseases combined. Two and a half billion people in the world still lack clean drinking water. While organizations like the WHO, the FAO, and various agencies associated with the United Nations have consistently documented the data of such shocking realities experienced daily by millions in the world, few agencies have been able to enforce any structural changes in the global political economy due to the capitalist hegemony discussed in chapter 2.

From 1970 to 2010, 68 percent of mammals, birds, amphibians, reptiles, and fish have radically decreased, and there's been a 94 percent reduction in the ability of most known Earth creatures to live sustainably.[60] Today:

- Over a million species of various plants, mammals, birds, reptiles, amphibians, fish, and invertebrates are facing extinction possibly within decades, with the loss of species now hundreds of times more than the past 10 million years, according to UN secretary general António Guterres, speaking at the UN Biodiversity Conference in Kunming, China, in October 2021[61];
- Escalating ecocide sees anywhere between 150 to 300 known species becoming extinct each day, with the International Union for Conservation of Nature and Natural Resources Red List (IUCN) of Threatened Species identifying 98,500 species facing such extinction[62];
- Almost half of insect species are in rapid decline, of which a third are facing threatened extinction, the worst and largest extinction on Earth, with insect biomass dropping 2.5 percent globally, and 60 percent of large insects in Europe and 51 percent of such creatures in North America are in accelerated decline[63];
- Though insects are the leading pollinators of the Earth and provide nutrition to 75 percent of 115 food crops, including coffee, almonds, cherries, cocoa, etc., and fruit like bananas, apples, avocados, blueberries, and apricots, they are now facing mass extinction from oil, gas, air, water, and land industrial pollution, due to deadly herbicides and pesticides like neonicotinoids that have resulted in extermination of almost half of all Indigenous bees in Turtle Island and one-

sixth around the world, with beekeepers suffering up to half of bee community loss[64];
- Almost 40 percent of bees are facing extinction, particularly with the use of chemical fertilizers and neonicotinoids made by the same chemical-war companies, Dow, Dupont, and Germany's Bayer, that have caused destructive impacts on neurological health[65];
- Horrifyingly, 89 percent of the widely seen Turtle Island bumblebees have disappeared, largely due to the undermining of habitat, historical and ongoing use of pesticides in agriculture and gardening, extreme heat and unstable weather conditions, droughts, and erosion of genetic diversity as the result of chemical industrialism.[66] The net result has been the virtual extinction of bumblebees, with no evidence of the creatures in the northeast generally, particularly in New York where a 99 percent annihilating decline is indicated; Illinois, where such creatures have suffered a 74 percent decline; 16 states in the northeast and northwest that have seen the rapid disappearance of bumblebees; 94 percent elimination in the upper northwest, and continued decline in much of the southeast and the Midwest;
- Flying insects across Europe have been radically reduced by 80 percent, resulting in the suffocation of the lifeline of some 400 million birds[67];
- Birds of diverse nature are now facing extinction due to climate change, protracted drought and aridity, and the Earth and sea heating at unprecedented levels, with half of Turtle Island birds, the bald eagle, the three-toed woodpecker, the northern hawk oil, the northern gannet, sparrow, the rufous hummingbird, the common loon, the osprey, the piping plover, the spotted owl, the trumpeter swan, the Baltimore oriole (after which a U.S. baseball team is named, and while the bird is disappearing, the baseball team is thriving), cardinals, hummingbirds, prairie chickens, meadowlarks, and six other species facing extinction, in addition to the three billion that have disappeared since 1970, all due to industrialist and residential destruction of ecological habitat. The Audubon Report of 2014 that reached such conclusions failed to identify industrialist capitalism as the root cause of Earth heating and climate change, while projecting an unreasonably overoptimistic future year of 2080 when such bird disappearances would be visible most drastically given the data noted above[68];
- Three-quarters of globally increasingly scarce fresh drinking water sources have been polluted[69];
- Seventy-five percent of the land and soil has suffered irreparable degradation;
- Seventy-five percent of agricultural biodiversity has been annihilated;
- Forty percent of greenhouse gas emissions are caused by chemical and fossil fuel based agriculture;
- The growing and uncontrollable acidification of oceans due to the Earth's and the oceans' inability to absorb tolerable levels of carbon dioxide from industrial use have destroyed essential elements of the food chain like coral and plankton, particularly in deep cold and warm water reef systems ranging from the Mississippi Canyon, the Florida Keys, the coasts of Hawaii, the mesophotic coral coasts of North Sulawesi, Indonesia, to the Great Barrier Reef in Australia, and the reefs along east Africa and southeast Africa[70];
- The widespread and normative use of chlorine to sanitize drinking water

sources, pools, etc., especially in the U.S., with full knowledge that the roots of such chemicals reside in war industries that annihilated millions in Europe and other regions, continues.

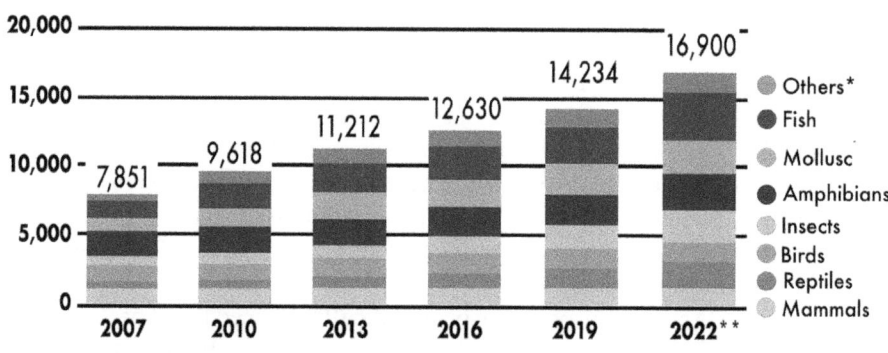

Katharina Buccholz, "Number of Threatened Species is Rising," *Statista,* January 18, 2023 (adapted by Casey Ontiveros).

Climate cycles and circles were severely disrupted with the colonization and occupation of the western hemisphere from the 15th century. Indigenous and African diaspora scholars and educational activists have challenged the false notion that the "Anthropocene" age began around 1963. Educators like Zoe Todd ("on time") and Christina Sharpe (*In the Wake: On Blackness and Being*) have sharply criticized the Eurocentric objectification of creation from its persistent insistence on categorizing, classifying, ossifying, atomizing, dualizing, and hierarchizing all creation and have dated the "Anthropocene" age much earlier in Europe.

Ecocide intensified with the emergence of European industrialism and the heightening of capitalism and colonization through the present. It became particularly acute with the loss caused by ecological violence over the past six hundred years, as Indigenous scholars assert, and not fifty years, as some writers erroneously argue today.[71]

The expanded use of machines and now virtually deified online technologies in recent years that are ubiquitous today like the internet, wireless technologies, and artificial intelligence (AI) just compound these entrenched problems in every sphere and provide no structural solution to the entrenchment of insatiable capitalist thirst for profits. The waste products generated from all these technological processes, like mining for lithium for electric batteries and coltan for cellphones, computers, and other electronic devices, signify the same culture of anthropocentric apotheosis that vitiates the integrity of the natural world. Recycling such waste, plastic, and other toxic materials and eliminating oil use while retaining the capitalist foundation are centuries-old,

3. Oil Production and Its Capitalist History 73

deceptive, ideological subterfuges ensuring that we are hoodwinked into believing that our actions will permanently arrest the ecocidal and pollution crisis enveloping us daily.

According to a team of international climate researchers' analysis of past sea-level records, irruptive sea-level rise was triggered from 1863 onward and has now reached lethal proportions 159 years later.[72] The studies found that the earliest sea level rise was in Turtle Island in the mid to late 19th century and subsequently in Europe, and that the fastest rise occurred from 1940 to 2000, coinciding with expanded industrialist oil drilling, production, and consumption, and the resultant destruction of the fragile marine life ecosystem.[73]

Fracking and Its Effects of Ecological and Environmental Destruction and Earth Destabilization

Fracking entailed expansion from the drilling of shale in north Texas' Barnette region by George Mitchell and gas engineers in 1981 and marked the inception of the so-called "shale revolution."[74] According to the Frontier Group of the Environment

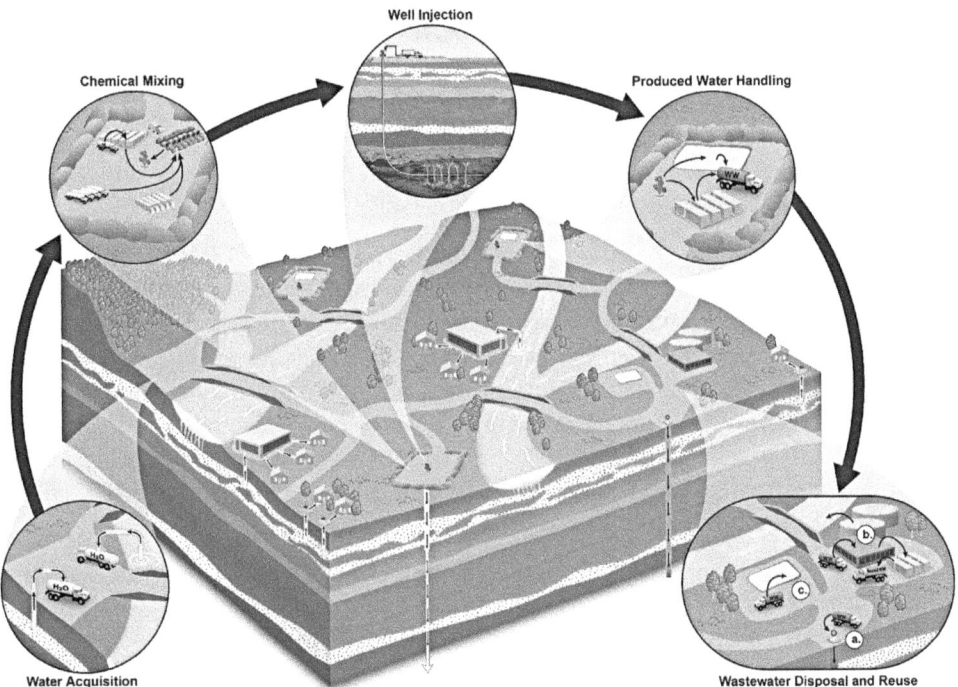

Fig. 1: Figure ES-3. Not to scale. The five stages of the hydraulic fracturing water cycle. The stages (shown in the insets) identify activities involving water that support hydraulic fracturing for oil and gas. Activities may take place in the same watershed or different watersheds and close to or far from drinking water resources. Thin arrows in the insets depict the movement of water and chemicals. Specific activities in the "Wastewater Disposal and Reuse" inset include (a) disposal of wastewater through underground injection, (b) wastewater treatment followed by reuse in other hydraulic fracturing operations or discharge to surface waters, and (c) disposal through evaporation or percolation pits (10) (EPA, https://www.epa.gov/hfstudy/executive-summary-hydraulic-fracturing-study-final-assessment-2016).

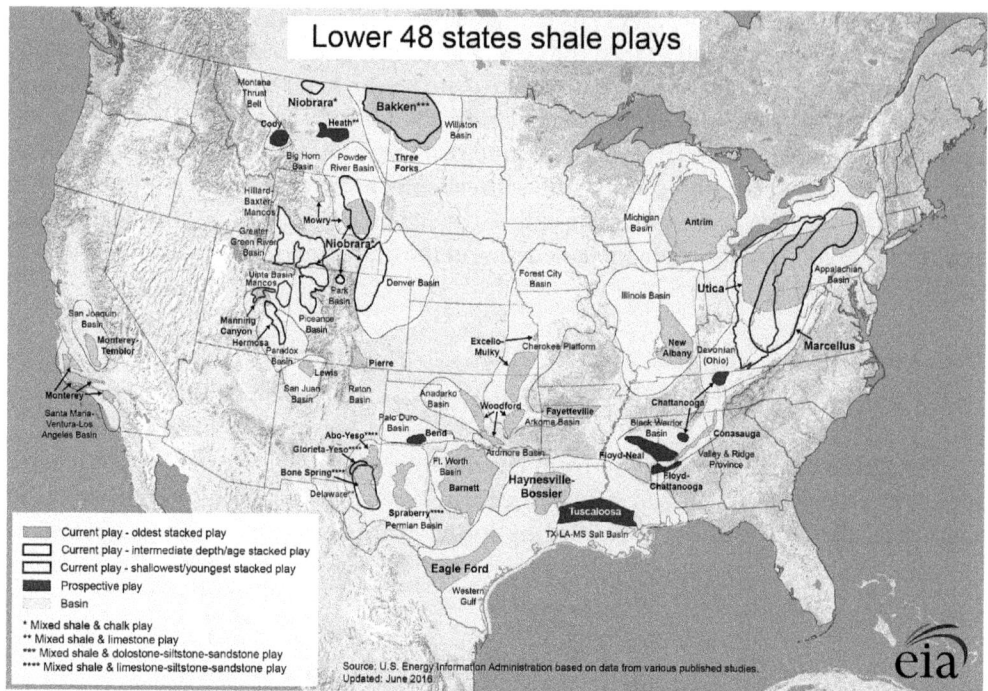

Shale Gas and Oil States (U.S. Energy Information Administration). (Elizabeth Ridlington and John Rumpler, "Fracking by the Numbers: Key Impacts of Dirty Drilling at the State and National Level," Environment America Research & Policy Center, 2013).

America Research & Policy Center's April 2016 document, "Fracking by the Numbers: The Damage to Our Water, Land and Climate from a Decade of Dirty Drilling," fracking involves hydraulic fracturing using over 100,000 gallons of fresh water (see image). In places like California, 116,000 gallons are used in the average fracturing well operation, yet in other locations, up to four million gallons of water (equivalent to what New York City uses every six minutes) are used, during the world's worst drought and water access crises on record for the past 20 years, especially compounded in the past five years.[75] Some U.S. states are reflected in the "shale plays" in the accompanying image.

The fracking reports asserts:

> Since 2005, according to industry and state data, more than 137,000 fracking wells have been drilled or permitted in more than 20 states, but the scale of fracking's impact on our environment can be difficult to grasp.... Fracking uses vast quantities of chemicals known to harm human health. According to the industry-reported data in the FracFocus database, oil and gas wells fracked across the U.S. between 2005 and 2015 used at least:
>
> - 5 billion pounds of hydrochloric acid, a caustic acid (pp. 22–24);
> - 1.2 billion pounds of petroleum distillates, which can irritate the throat, lungs and eyes; cause dizziness and nausea; and can include toxic and cancer-causing agents;
> - 445 million pounds of methanol, which is suspected of causing birth defects.
> - The exact identities of many other chemicals are unknown because they are kept *secret* [italics ours] as proprietary information.[76]

The following table crystallizes the lethal ecological, environmental, health, and climate toll of fracking.[77]

National Environmental and Public Health-Related Impacts of Fracking

Fracking Wells Since 2005	At least 137,000
Water Used Since 2005	At least 239 billion gallons
Toxic Wastewater Produced in 2014, Selected States	At least 14 billion gallons
Land Directly Damaged Since 2005	679,000 acres
Global Warming Pollution from Well Completions in 2014 (methane)	At least 5.3 billion pounds

The environmental and Earth-heating consequences of fracking are well documented. In Pennsylvania, for example, there were 260 instances of contamination of vital drinking water sources from fracking and a 6–7 percent failure rate of wells due to structural inadequacies. Texas, which has over 165,136 oil wells and 86,612 gas wells—with over 12,000 fracked wells in the Fort Worth region since 2011 alone—has precipitated a major environmental and health crisis.[78] These wells are responsible for the bulk of toxic emissions, more so than Texas trucks and other large commercial vehicles and have led to 160 earthquakes from 2008 through 2016 in the Fort Worth-Dallas area.

The fracking industry today is a core polluter of increasingly scarce clean drinking water supplies around the U.S. because chemical-poisoned wastewater from the drilling leaches into rivers, streams, ponds, and the underground water table, even unearthing radioactive materials that come to the surface. One thousand new hydraulic wells were drilled each month from January 2019 to March 2020, and since the Covid-19 pandemic hit, this number is down to 500.[79] The number of fracking wells, chemicals used, water consumed, land disturbed, and wastewater produced in the states below from 2005 through 2015 are as follows[80]:

Estimated Impacts of Fracking, Selected States
(data are cumulative impacts since 2005, except where noted)

State	Wells Fracked	Hydrochloric Acid Used (thousand pounds)	Methanol Used (thousand pounds)	Wastewater Produced in 2014 (million gallons)	Water Consumed (million gallons)	Methane Released from Well Completion in 2014 (million pounds)	Land Disturbed (acres)
Arkansas	6,496	142,406	2,025	unavailable	11,290	144	22,858
California	3,405	1,034	489	1,057	237	140	15,940
Colorado	22,615	68,663	10,042	3,139	19,142	395	105,866
Louisiana	2,883	15,136	2,045	unavailable	4,880	131	16,010
New Mexico	4,318	70,798	4,403	8,592	3,132	125	35,273
North Dakota	8,224	82,198	88,168	unavailable	14,891	517	33,718
Ohio	1,594	105,447	1,942	313	7,771	136	9,118
Oklahoma	7,421	455,225	17,147	unavailable	19,582	546	41,210
Pennsylvania	9,233	1,806,032	5,396	1,821	24,732	295	52,813
Texas	54,958	2,148,789	302,501	unavailable	120,215	2,521	257,272

State	Wells Fracked	Hydrochloric Acid Used (thousand pounds)	Methanol Used (thousand pounds)	Wastewater Produced in 2014 (million gallons)	Water Consumed (million gallons)	Methane Released from Well Completion in 2014 (million pounds)	Land Disturbed (acres)
Utah	4,949	35,926	1,414	unavailable	916	186	35,478
West Virginia	2,670	64,134	1,174	unavailable	7,651	88	15,272
Wyoming	7,277	18,074	5,870	70	2,528	116	29,836
TOTAL	**137,743**	**5,038,953**	**444,786**	**14,993**	**239,166**	**5,340**	**679,148**

"Fraccidents," or "fracking accidents," are so named to refer to the huge devastating impact of fracking of oil and natural gas on chemical seepage into groundwater, contamination of drinking water sources, and other adverse effects on the ecology and environment, with 282 major ones that have occurred in various parts of the U.S. in the Fraccidents Map as depicted below since 2005.

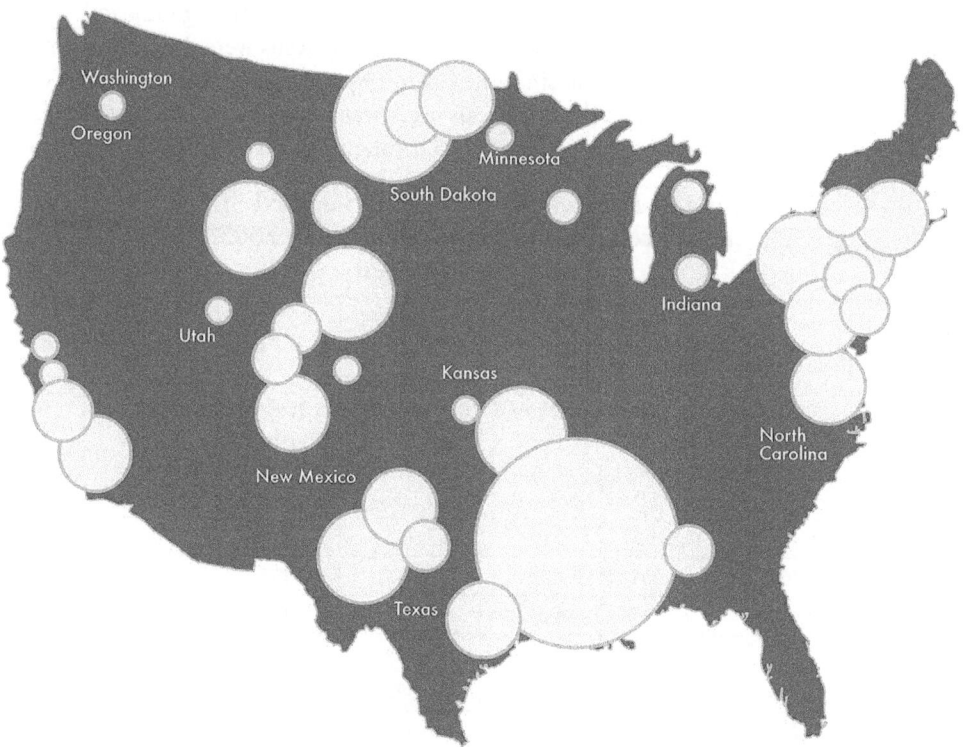

Fraccidents: Fracking Map of Accidents Located

Fraccidents: map of fracking accidents (Casey Ontiveros, https://www.google.com/maps/d/u/0/viewer?mid=1k5UjLbXk4oxCyfCM7xwUSgj8aDg&hl=en_US&ll=39.17836426815686%2C-98.48509310000003&z=3, accessed on April 22, 2022, CC-BY-SA 3.0).

For the past four decades, the oil and gas industry has secured special pollution exemptions that protect it against U.S. environmental laws that require violations of drinking water and air quality to be reported. The U.S. Congress and government are a true friend of oil and gas, as specified here[81]:

EXEMPTIONS
Federal safeguards for oil and gas production are missing. Many of the fundamental environmental statutes have exemptions for oil and gas production, leaving aspects of those activities largely ungoverned at the federal level.

Clean Air Act
The oil and gas industry is exempt from critical requirements to assess, monitor, and control hazardous air pollutants.

Safe Drinking Water Act
Fracking is exempted from the Safe Drinking Water Act pollution control measures unless diesel is used in the fracking process.

Clean Water Act
Oil and gas operations are exempt from important permitting and pollution control requirements of the Clean Water Act, including the stormwater runoff permit requirement. In addition there is a loophole that allows certain wastewater produced by oil and gas wells to be discharged into surface waters in the western United States.

Resource Conservation and Recovery Act
Oil and gas waste is exempt from the testing, treatment, and disposal provisions that govern the assessment control and clean-up of hazardous waste under this law, and, by extension, from the Comprehensive Environmental Response Compensation and Liability Act (aka "Superfund"), which adopts the same definition of hazardous waste.

National Environmental Policy Act
When oil and gas companies lease federal lands, they are exempted from some requirements for environmental impact reviews.

The grave threat by fracking to drinking water sources continues the criminal culpability of oil, gas, mining, and power companies that resulted in the release of 1,100 tons of uranium waste and 94 million gallons of radioactive water into the Rio Puerco River in New Mexico in 1979. This was the aftermath of an earthen dam break at the United Nuclear Corporation's mill near Church Rock and occurred four months after the lethal gas leakage from a cooling malfunction that melted part of the #2 nuclear reactor at the Three Mile Island plant in Pennsylvania in 1971. In the case of the Rio Puerco spill, the Diné community complained about illness and livestock being born deformed or without limbs. People expressed their grievances to the U.S. government about the lethal health exposure effects on Diné miners from protracted uranium and coal mining operations in Arizona and New Mexico in the late 1940s and the 1950s. Yet there was a consistent deaf ear from the EPA and the government. Concerns about short- and long-term impacts were dismissed by the New Mexico Environmental division as "quite limited."[82] The Centers for Disease Control reinforced the same racist defense, stating that the risk of eating radioactive-contaminated sheep "appeared small."[83] These cases, along with the disastrous Gold King mine spill on the Animus river outside Durango, Colorado, on August 5, 2015 (when the EPA attempted to obscure its own care-less responsibility by ignoring the potential for such a catastrophic spill), substantiate the charge by Indigenous and other people of color of "environmental racism."[84] To date, the core problem of poisoned water by mining in Colorado and other states lingers, and though there have been environmental clean-ups worth tens of millions of dollars, the toxicity of drinking water following the 2015 spill on the Animus river persists, making clean drinking water progressively elusive.[85] The $32 million settlement offered by the U.S. government to the New Mexico state government does not mitigate the effects of three million gallons of highly lethal poisonous arsenic waste that flowed through the Diné Nation in the region, including Utah.[86] Both the Animus River and the San Juan Rivers are sacred to the Diné people, and the poisoning of essential water supplies for drinking, agriculture, bathing, etc., underscores the gravity of the deadly toxicity for all life from mining and extractionist technologies.

Fifty thousand Diné people are deprived of running water in Arizona and Utah, while non–Indigenous people generally have always taken access to water for granted. Colonialist environmental racism lives. Further, plans to transport 350,000 barrels of crude oil daily from the Uinta Basin, Utah, part of the vast Colorado plateau, via a newly constructed, multi-billion-dollar, 88-mile rail line through mountains, forested areas, and lands of the Ute Indian Nation, to refineries near the Gulf of Mexico, demonstrate such assertions.[87] Notwithstanding the Ute Tribal Council's unusual approval

Gold King Mine Rupture on August 5, 2015 (Environmental Protection Agency).

of the plan for economic reasons in contrast with most other Indigenous nations, the long-term destructive effects and the high potential for a massive hazardous oil spill into the tributaries of the Colorado River from such a plan are being conveniently obscured by a White House that duplicitously claims that its policies are environmentally protective. Indigenous people are held hostage by colonialism here and everywhere that this system operates: either support oil drilling, mining, fracking, etc., or face starvation. Such is akin to casinos that have become the sole source of support and only means for survival for so many desperate Indigenous peoples in Turtle Island, even though the casino is underpinned by capitalist monetary principles.

The U.S. government subsidizes such ecological, environmental, and land destruction with subsidies to oil and gas companies from taxpayer monies, fueling capitalist profits for the corporate energy industrial complex and disguised as "providing cheap energy" to the public. An analytical research study by the Stockholm Environmental Institute in June 2021 revealed that:

> two tax incentives alone—the expensing of intangible drilling costs and percentage depletion provisions—increased the expected value of new oil and gas projects by billions of dollars in most years and by over $20 billion in certain high-price years. This translates to a median increase in expected value of $4 per barrel of oil equivalent or more for projects in those high-price years (2008 and 2010–2014).[88]

The boom in U.S. shale production over the past couple of decades to undercut oil production prices as determined by OPEC, for example, raised the value of oil and gas projects lucratively in the "Bakken Formation in 2005–2006, the Appalachian and Haynesville regions in 2008, the Eagle Ford play in 2009–2010 and the Permian basin in 2011–2015."[89]

It's a dramatic irony that when oil and shale gas prices were low, the U.S. government stepped in to provide these tax-funded subsidies to the numerous gas projects in Appalachia from 2010 to 2012—and even to marginal oil projects in 2016—that provided a boon to these destructive corporations which otherwise would have been put out of business. Three subsidies were and are key in this scheme, demonstrative of how the U.S. government enhances corporate profits at the expense of the working class and often impoverished majority in the country who receive very little in any serious tax subsidies. The gas subsidies are "the accelerated amortization period for geological and geophysical expenses … because it allows these expenses to be deducted faster than they otherwise would be under standard amortization schedules"; "the expensing of intangible drilling costs (IDC) subsidy" that "acts in a similar manner as the geological and geophysical (G&G) subsidy, but for a much larger class of capital expenses: well drilling and construction"; and "the percentage depletion subsidy" that "allows independent firms to deduct on their taxes 15% of the gross value of oil and gas production (up to 1,000 barrels per day of crude oil or the equivalent amount of natural gas), rather than make deductions based on the actual capital invested."[90] Such governmental handouts often inflate and elevate these deductions that are heavily disproportionate to the original money invested and provide surplus income to the oil well producer during the well's life.[91] These actions continue the underhanded activity of the government in subsidizing oil and gas "exploration" for over a century under the pretext of "cheap oil." Yet the "consumer" is totally unaware that their hard earned taxes that line government coffers are subsidizing some of the largest energy corporations in the country, the bulk of subsidies directed toward obscenely high profits so that very little goes to investment in civilian infrastructure and job creation.[92] The government functions essentially as an energy capitalist protector and insurer.

In 2017, when the U.S. Congress reduced the corporate tax rate from a marginal tax of 35 percent to a flat 21 percent, large corporations, including oil- and gas-drilling companies, saw their profits rise, particularly with reduced costs of fracking sand and fracking materials, thus reducing the need for continued subsidies even when gas and oil prices dropped.[93] Regardless of the contextual and policy changes, these companies' profit margins are assured at the highest level by the government, and CEOs are laughing all the way to the bank.

Further, the huge extraction of water through the fracking process undermines farmers' agricultural production needs of access to water for irrigation since many of the oil and gas operations occur in many of the drought-stricken states of the west. Energy companies are able to pay much more than farmers for acre-feet of water. In 2012, oil and gas companies paid $3,000 for an acre-foot of water in Colorado, 100 times what farmers paid.[94] The destruction of the land, ecology, and the annihilation of threatened species of birds, insects, four-leggeds, fish, etc., is an inevitable consequence of this dangerous oil and gas plundering of the Earth. All mining operations have caused ecological and environmental collapse over two centuries globally and persist today.

The lethal and catastrophic side effects of oil and gas drilling, mining, and fracking on the natural cycles of the Earth Mother are incalculable. Earthquakes, tremors, tsunamis, tornadoes, floods, land and coastline collapse, and other Earth-destabilizing processes are the inevitable result. Again, the fracking impact report from 2005 to 2015 notes that in 2014, bringing new fracking wells into existence caused 5.3 billions of methane

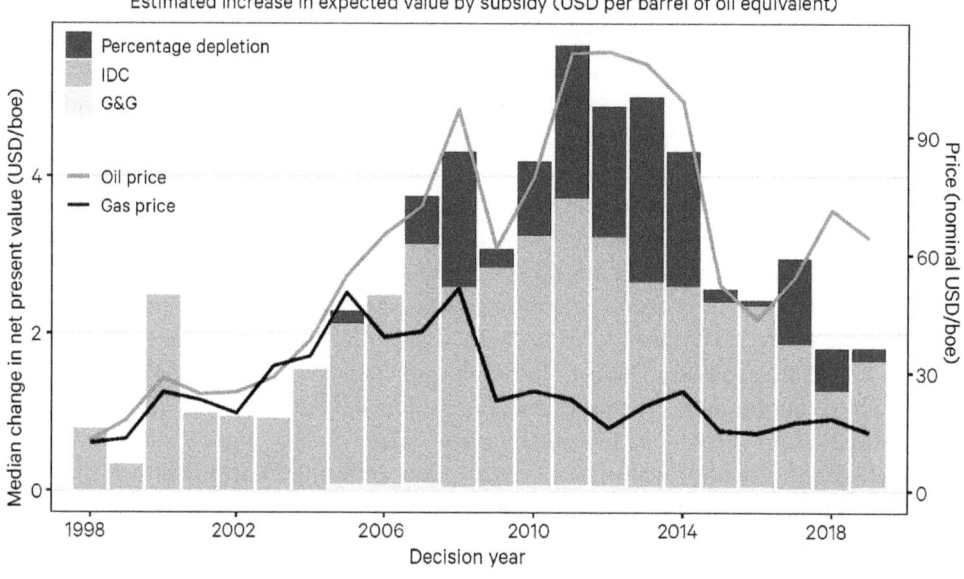

Fig. a. How government subsidies aided the U.S. shale oil and gas boom, 1998–2018 (Swedish Environmental Institute-SEI, P. Erickson and P. Achakulwisut, "How Subsidies Aided the U.S. Shale Oil and Gas Boom," Report, Stockholm, Sweden: Stockholm Environmental Institute, June 23, 2021).

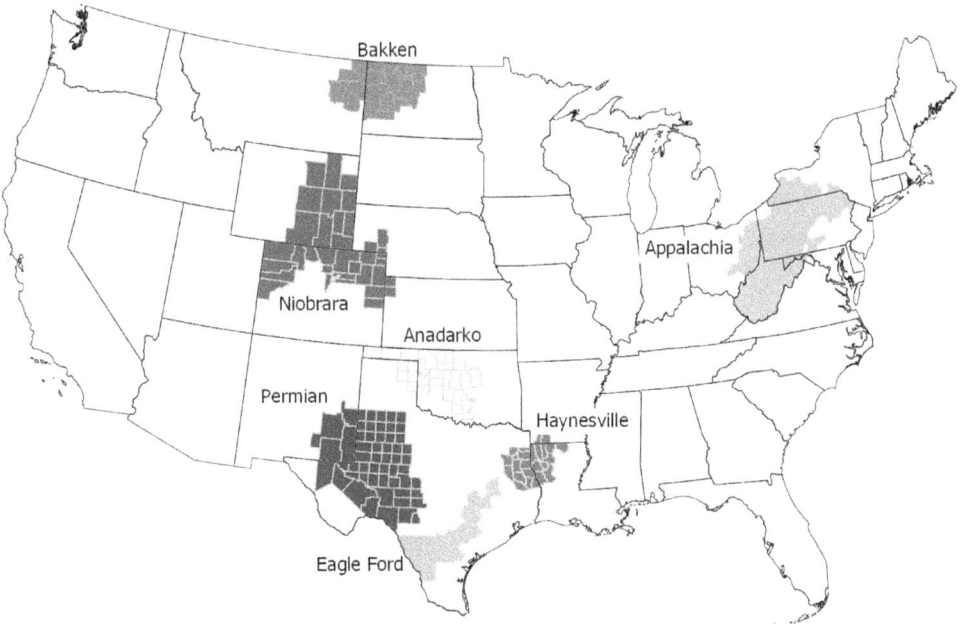

Fig. b. U.S. map of major unconventional oil- and gas-producing regions (U.S. EIA, https://www.eia.gov/petroleum/drilling/).

being emitted, equivalent to 22 coal-fired plants, even while the fracking industry is fully aware that methane is 86 times more lethal in causing Earth heating over a 20-year period.[95] Fracking has been a leading factor in the eruption of 659 earthquakes in the central and southeastern part of Turtle Island, particularly in places like Oklahoma, where a 5.7 magnitude earthquake occurred in the town of Prague in 2011, causing extensive damage to 14 homes nearby.[96] *Inside Science,* a publication underwritten by organizations like the American Physics Society, which supports fracking and hydraulic fracturing because it makes energy more cheaply available, dismissed the serious role of fracking in Turtle Island, notably from the 2011 earthquake in Prague, Oklahoma, and subsequent ones in that state especially. The publication contradictorily noted that fracking does not cause earthquakes and that the Prague event was caused by oil and gas developers, not fracking![97]

Fracking is at the root of these earthquakes because it's responsible for pumping wastewater into the Class II wells, into which gas and oil wastewater from the fracking process are injected to withstand the pore pressure in the Earth's crust. If there were no fracking, there would be no need to inject wastewater into these wells, thus the rapid explosion of earthquakes in fracking states like Oklahoma would not have skyrocketed from the sporadic ones from 1972 to 2008. There were 50 in 2009 and over a thousand in the state in 2010.[98] Oklahoma and Kansas had a very negligible history of tremors. Both have been shaken by the explosive number of tremors and earthquakes following the fracking "boom" in the 2000s, with Oklahoma experiencing 564 quakes with 3+ magnitude in 2013 and a swarm of such quakes of similar level magnitude in the fall of that year, including a 3.8 magnitude one on December 16.[99] There have been over 300 such quakes in the central and eastern states from 2010 to 2012, in contrast to 21 annually from 1967 to 2000.[100] Seventy-seven earthquakes were felt outside Youngstown, Ohio, in

January 2014, an area unaccustomed generally to such activity. Fracking waste injection back into wells in the Earth has been the leading cause of such tremors, as many geologists and seismologists note.[101]

There are over 850,000 waste wells in the U.S. today—a lethal destruction of the structural and balanced integrity of the Earth—which have been responsible for injecting over 30 trillion gallons of toxic liquid into the far depths of the Earth, with no container at the base![102] Over 180,000 of these wells are Class II wells that operated in the country in 2021, most of which continue injecting fluids from gas and oil fracking operations, over four times up from 40,000 wells in 2013.[103] The EPA notes that two billion gallons of such toxic fluids are injected into the Earth in the U.S. each day, with most wells operating in Texas, Oklahoma, California, and Kansas.[104] Again, while "science" and the EPA often speciously claim that these wells are safe because they test drinking water samples in nearby areas and rigorously monitor the environmental toxicity of injection wells, they really have no idea whether there are leaks underground and whether seepages eventually creep into the drinking water supplies in communities where such wells exist. Nor do scientists and the EPA acknowledge the lethal effects on terrestrial and marine life because they routinely assume that they can control earth and sea processes through technological and scientific knowhow. Injection wells penetrate many thousands of feet deep into the Earth, and it's impossible to monitor the underground leakage of toxic chemicals and waste for each well. The leakage inevitably seeps into aquifers around Turtle Island that provide drinking water to the residents of the nation. Again, the claim of "total knowledge" and "assurance of safety" and "minimal damage" is the assumption of capitalist industrialism and its scientific defenders, myopic and greedy for lucrative short-term industrial gain and profits, all at the cost of the lives of all on the Earth. Assurances of "minimal damage are baseless":

> In 2010, contaminants from such a well bubbled up in a west Los Angeles dog park. Within the past three years, similar fountains of oil and gas drilling waste have appeared in Oklahoma and Louisiana. In South Florida, 20 of the nation's most stringently regulated disposal wells failed in the early 1990s, releasing partly treated sewage into aquifers that may one day be needed to supply Miami's drinking Water.[105]

Over 150,000 of the 680,000 underground waste and injection wells across the country inject lethal industrial fluids deep below Earth's surface.

Earthworks has documented the deadly Earth-disrupting consequences of fracking in Turtle Island:

Earthquakes caused by fracking:
- **British Columbia Oil & Gas Commission**: *Investigation of Observed Seismicity in the Horn River Basin*, August 2012.
- **CBC**: Fracking causes minor earthquakes, B.C. regulator says.
- **British Geological Survey:** Fracking and Earthquake Hazard.
- **Cuadrilla Resources:** *Geomechanical Study of Bowland Shale Seismicity*, 2011.
- **Schlumberger:** Seismicity in the Oil Field.
- **Kansas Agland:** What's known—and suspected—about induced earthquakes.
- **Washington County (PA) Observer Reporter:** DEP links quakes to fracking in 2016 Lawrence Co. event.

Earthquakes caused by fracking wastewater injection:
- *Science:* Injection-Induced Earthquakes 2013.
- **National Academy of Sciences**: *Induced Seismicity Potential in Energy Technologies*, 2012.

- **The Earth Institute, Columbia University:** Wastewater Injection Spurred Biggest Earthquake Yet, Says Study.
- **Journal of Geology:** Potentially induced earthquakes in Oklahoma, USA: Links between wastewater injection and the 2011 Mw 5.7 earthquake.
- **Capitol Confidential:** Tkaczyk, Avella question DEC's seismic work.
- **Mother Jones:** Fracking's Latest Scandal? Earthquake Swarms.
- **Southern Methodist University:** Causal factors for seismicity near Azle, Texas.[106]

Fracking also enhances "ozone smog," which is a death recipe, especially for elderly people and those with compromised respiratory and immune systems. Los Angeles issued serious health warnings to elderly residents to remain indoors during the summer of 2020 when smog levels were astronomically high since such smog exposure shortens life spans. The evidence of serious ecological, environmental, and health hazards from fracking is indisputable, stemming from the use of over a thousand chemicals injected into the Earth during the fracking process (plus many other chemicals that are undisclosed by the fracking and "natural" gas and oil industry).

Chemical pollution of air, water, and Earth has deadly effects. In a study of 1,021 chemicals by Yale University's School of Public Health, researchers found that of the 240 substances analyzed, 157 were very dangerous chemicals including arsenic, benzene, cadmium, lead, formaldehyde, chlorine, and mercury, all known for causing serious impacts on peoples' health, particularly cancer.[107] These volatile organic compounds have resulted in heart and lung disease and in reacting with nitrogen oxides produce ozone, a highly reactive and lethal gas that compounds breathing difficulties for people suffering with chronic asthma and respiratory ailments, particularly in pollution-infested cities and environments like Manhattan, New York (toxic dumps, public utility transportation, sanitation trucks, etc.); Houston, Texas (oil refineries and gas flames burning 24/7); and Los Angeles, California. In the latter city, where the South Coast Air Basin that includes LA, Orange, Riverside, and Bernardino counties sees 12 million vehicles in daily operations, tailpipe pollution rose by over five percent since 2013 and caused PM2.5, a very toxic air pollutant that has shortened life spans and increased levels of cancer, asthma, cardiovascular disease, and even diabetes.[108] The aforementioned LA counties are in violation of ozone and particulate matter emissions standards, and coupled with carbon monoxide, nitrogen dioxide, sulfur dioxide, and lead discharge, have caused serious health problems for the elderly and for many susceptible communities of color in the region, including premature births in pregnant women.[109] A 2012 comprehensive study on sky-rocketing pollution levels in LA County by the University of Southern California found that "air pollution not only makes things worse for people with asthma but can actually cause asthma to develop in healthy children."[110] In the same year, over 45 million people in the country lived within 300 feet of a highway, many of whom were children.

In Julie Sze's study of race and class in New York City, *Noxious New York,* she demonstrated how Black, Brown, and Asian working-class communities in places like Manhattan, the Bronx, and Long Island were most adversely affected by the city's toxic waste sites, truck emissions, sludge and waste leaking into rivers and waterways, and incinerator dumps, corroborating the charge of U.S. environmental racism. Consequently, asthma has become a pandemic among these communities.[111] It was fierce and protracted environmental justice demonstrations and campaigns by a coalition of activist groups against the Brooklyn Navy Yard Incinerator, the Bronx-Lebanon medical waste incinerator, the Sunset Park sludge treatment plant, and the North River sewage treatment plant that made a significant difference in slowing or terminating such death-dealing, racist city council practices of environmental injustice against vulnerable communities. These actions remind us that it is only through incessant struggle against and resistance to environmental injustices that the world that will see the demise of ongoing industrialist ecological and environmental violence against the earth, air, and water and the accompanying destruction of the fragile web of life.

In a 2015 study by the University of Pennsylvania and Columbia University, researchers revealed that hospitalization rates for people living in Pennsylvania areas where fracking and hydraulic fracturing were prevalent were much higher than for those living in fracking-free zones. Sifting through 90,000 hospitalization records in Bradford, Susquehanna, and Wayne counties in the state showed that the rate of people hospitalized with cardiac problems was 27 percent higher than in non-drilling areas. The rate of those experiencing dermatology problems to the point of requiring medical treatment was 45 percent higher in drilling areas than in non-drilling ones, with 79 wells per 100 square kilometers as the yardstick for such comparisons.[112] Fumes from the trafficking of fracking materials in trucks and toxins emitted from fracking zones were the leading factors in cardiovascular problems, and 75 percent of the chemicals deployed in "natural" gas production were very harmful to the skin, eyes, and other sensory organs. The fracking industry's environmental and health standards violations from expanded air pollution and drinking water contamination demonstrate that the U.S. government and the energy industry have always been in cahoots with each other in the frantic obsession to lead the world in "natural" gas and cheap oil production. Further, the impact on climate change and Earth heating is exacerbated through the escalating release of methane, a very dangerous Earth-heating gas, that "traps more than 84 times more heat in the atmosphere than carbon dioxide" and reduces lifespans. Methane and other climate pollutants contribute toward a third of the intensification of Earth heating.[113]

In 2012, gas and oil production was the second largest energy extraction cause of

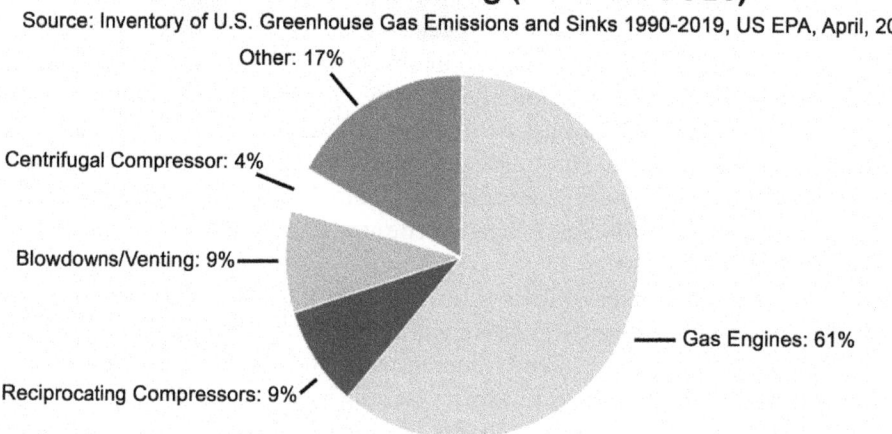

Breakdown of Sources of U.S. Greenhouse Gas Emissions and Sinks from 1990–2019 (EPA, https://www.epa.gov/).

methane emissions in the U.S.[114] In 2019, 8,000 oil and gas operations in the U.S. emitted almost a third of methane, of which Texas constituted 20 percent with over 1,180 operations, including 350 operations in and around Houston alone. These cumulatively caused 700 million metric tons of carbon dioxide equivalents (MMTCO2e) of methane gas, compounding the climate crisis.[115]

While the world focuses on carbon dioxide and is obsessed with the chimerical "green energy" syndrome, the fracking and fossil fuel and gas conglomerates uncaringly persist in their deadly enterprise, with the U.S. government regulation agencies endorsing such activities to manipulate the price of such energies on the world market to U.S. capitalist advantage. The U.S. exceeded Russian oil and gas production in 2009. Its industrial and energy activities are among the leading causes of Earth heating.[116]

During a 2016 summer research visit to the tar sands area around Fort McMurray in Alberta, Canada, we observed toxic smoke billows in the sky 24/7 that can be seen for miles. Similar to fracking, the violence and contamination by extraction of tar sands for oil has poisoned vital fresh water in the boreal forest in Alberta, some 140,000 square kilometers, equivalent to the size of England. Through relentless plundering and excavation of the water and the Earth by the four largest extraction companies, Cenovus Energy, Canadian Natural Resources Limited (CNRL), Imperial Oil Limited, and Suncorp Energy, explosive rates of cancer and deaths, especially among Indigenous First Nations people, particularly elders, in the area are pervasive.

The Fundamental Ineffectiveness of the U.S. Oil Pollution Act of 1990

The Oil Pollution Act of 1990, while seemingly directed toward preventing ecological destruction and serious environmental damage, is impuissant, perfunctory, and ineffective since it does not address the root cause of such pollution. Notwithstanding this fact, it is nevertheless important to highlight the major spills since the Act was legislated by the U.S. Congress in 1990 principally because each of these toxic events has had lasting impacts on our Earth Mother and all life in the specific location of such spills and surrounding areas. We are indivisibly connected regardless of where we are.

The U.S. National Oceanic and Atmospheric Administration (NOAA) saw fit to document the lethal ecological and environmental cost of spills in the aftermath of the Oil Pollution Act of 1990. This law was promulgated in the aftermath of the devastating Exxon Valdes spill in Alaska on March 24, 1989, one of the worst in world history, when 11 million gallons of oil spilled into Prince William Sound. It was congressionally passed so that the EPA's ability to "prevent and respond to catastrophic oil spills" would be "streamlined and strengthened."[117] The EPA evidently has been unsuccessful in its ability to supposedly "prevent" catastrophic spills, just as the Clean Air Act and the Clean Water Act of 1973 have failed to deliver anywhere close to such, as noted below[118]:

The EPA was unable to prevent major spills for almost every year from 1990 through 2020, as noted below. For each of the past thirty years, oil spills have been substantial, continuous, deadly, disastrous, and extremely destructive by polluting precious water on Earth and in the sea, decimating plant and marine life particularly, and causing untold damage to river systems, lakes, and other fresh water sources. These need to be detailed here to underscore the ecocidal and environmental deterioration caused by

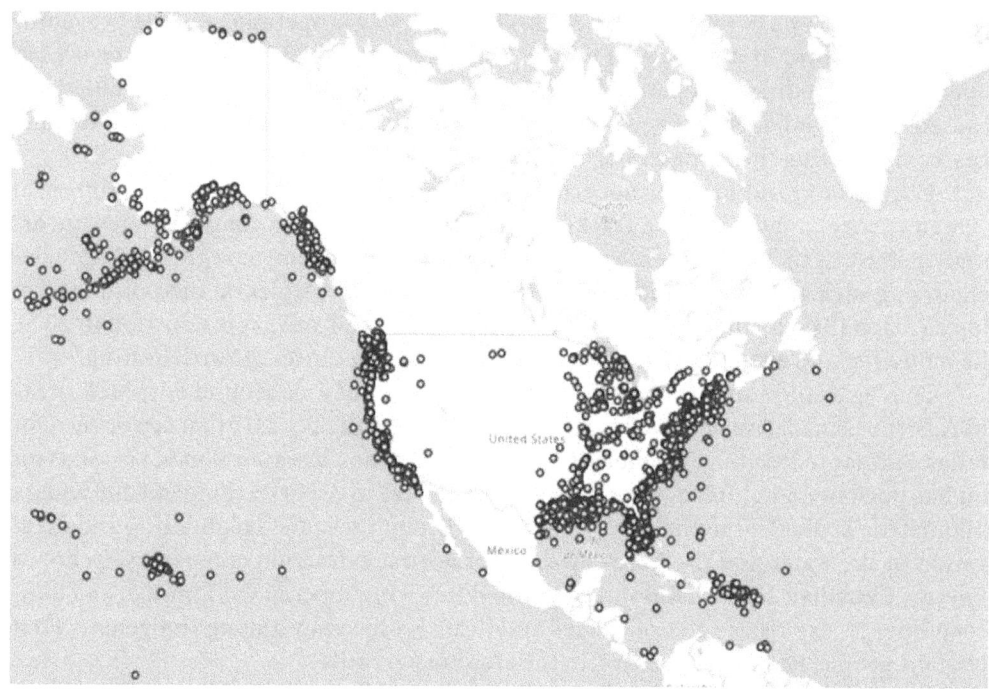

Location of oil spills where NOAA's Office of Response and Restoration provided technical support (NOAA, Megan Ewald, "The Oil Pollution Act of 1990: 30 Years of Oil Spills," National Oceanic and Atmospheric Administration, U.S. Department of Commerce, August 18, 2020).

normative oil drilling, production, transportation, and consumption in industrial capitalist societies, particularly in the west, but also in other leading industrialist nations in western Europe and China, Japan, Russia, India, Korea, Singapore, Brazil, Malaysia, and virtually all the countries of the world forced to depend on oil for living.

Date	Source	Location/Region	Damage/Result
1990	Exxon Bayway, NOAA	New York	killing at least 700 birds and destroying fish, crabs, and clams, as well as depleting significant sections of salt marsh
February 7, 1990	California Dept. of Fish and Wildlife	Huntington Beach, Orange County, California	release of 416,598 gallons of crude oil, annihilating thousands of birds, fish, and other marine life, and severely damaging the beach's coastline
1991	Ramsey Clark, *The Fire This Time: US War Crimes in the Gulf* (International Action Center, 2002)	U.S. bombing of Iraqi oil tankers	7.5 million barrels
1991	NOAA	Lake Salvador, Louisiana	1,300 gallons of oil in the lake and killing at least 1,300 birds

3. Oil Production and Its Capitalist History

Date	Source	Location/Region	Damage/Result
1992	NOAA	Chiltipin Creek, Texas	ruptured pipeline spilled 124,000 gallons of light crude oil and destroyed a 38-acre marsh, seriously affecting birds, insects, reptiles, trees, and plants
August 10, 1993	NOAA	Tampa, Florida	poisoning 13 beaches and destroying sea turtles, manatees, and mangroves
1994	NOAA	San Juan, Puerto Rico	emission of 800,000 gallons of oil off the island's coastline, destroying significant marine creatures and much of the ecology
1995	NOAA	Dixon Bay, Louisiana	oil and "natural" gas were released, lethally affecting migratory birds and sediment deposits
1996	NOAA	Rhode Island	827,000 gallons of household heating oil spilled. More than 9 million lobster; 150 million surf clam; 4.2 million fish; 2,100 birds; and over one million pounds of worms, crabs and mussels killed and affected
1997	NOAA	Puerto Rico	coral reef damage
August 24, 1998	NOAA	Honolulu, Hawaii	marine life, dead birds in the immediate area and 100 miles away
1999	NOAA	Beaver Creek, Oregon	5,388 gallons of oil, polluting spawning areas of salmon; steelheads were decimated, and indigenous ceremonies were disrupted
2000	NOAA	Patuxent River in Maryland	80 acres of wetlands and marsh areas were badly polluted and over 1,000 birds and other creatures were killed, permanently affecting the life span of these animals
2000	Oil Spill Response	Cape Town, South Africa	oil spill
2001	NOAA	10 miles south of the Waikiki Beach in Oahu, Hawaii	nine deaths, oil spill and leaked radiation
2002	NOAA	Cooper River, South Carolina	30 miles of shoreline after spilling 12,500 gallons of oil
November 2002	Euronews	Galicia coast of Spain	15 million gallons of oil that contaminated 600 beaches, killing 300,000 birds
2003	NOAA	Rhode Island and Massachusetts	fish, shellfish, birds, and other organisms within the fragile marine habitat for 100 miles

Date	Source	Location/Region	Damage/Result
2004	NOAA	waters off Delaware	oil spill that caused environmental destruction in Pennsylvania, New Jersey, and Delaware
2005	NOAA	Gulf of Mexico	1.9 million gallons of a heavy oil mixture to sink deep onto the ocean floor, decimation of sea life
2006	NOAA	Puerto Rico	reef damage
2007	NOAA	San Francisco–Oakland Bay Bridge	53,000 gallons of oil released
2009	NOAA	San Francisco Bay	400 gallons of fuel, killing hundreds of birds, affecting shoreline and Alameda Island
2010	NOAA	Deepwater Horizon Oil Platform, Gulf of Mexico	205 million gallons of oil over three months; 11 people dead; 16,000 miles of coastline; countless species of fish, birds, mammals, coral, and the ocean floor destroyed; 83,927 square miles of fishing closed off
2011	NOAA	Washington, D.C.	17,000 gallons of mineral oil, bottom invertebrates, migratory birds, insects
2012	NOAA	New Jersey, New York	255,180 gallons of diesel fuel, 158 people, $70 billion in environmental and structural damage
2013	NOAA	Sandusky Bay/Lake Erie	ran aground, no loss of coal or fuel
March 22, 2014	NOAA	Galveston Bay, Texas	168,000 gallons of oil; 160 miles of Texas coastline; bottlenose dolphins, turtles, birds, and diverse marine life destroyed
2015	NOAA	Refugio Beach, California	100,000 gallons of oil; massive destruction of fish, birds, seals, and other marine and aquatic life and the coastline
2016	NOAA	Gulf of Mexico, seafloor	80,792 gallons of oil leaked and oil slicks were visible at the surface, destroying fish larva, invertebrates, dolphins, fish, and birds
2017	NOAA	southwest Juneau, Alaska	oil sheens that killed herring eggs and migrating smolt salmon
2018	NOAA	Lake Washington/Rattlesnake Bayou, Louisiana	13,280–1 million gallons of oil, sea life, and birds
2019	NOAA	St Simons Sound, Golden Ray, Georgia	350 vehicles overturned from a carrier, causing a fire and releasing oil into the shoreline and marshes

Date	Source	Location/Region	Damage/Result
August 1, 2020	WSOC-TV	Huntersville, North Carolina	1.2 million gallons of oil
October 2, 2021	NOAA	Huntington Beach and Newport Beach, California	Coastal ecological devastation, Southwest California
December 27, 2021	*USA Today*, January 13, 2022	St. Bernard Parish, New Orleans	fish and wildlife devastation
February 2022	NOAA	Marshfield, MA	sunken fishing vessel leaking diesel off coast
	NOAA	Ilwaco Marina, WA	fishing vessel fire
	NOAA	Key West, FL	KEYS energy pipeline
	NOAA	Deerfield Beach, FL	Tug Sea Eagle aground
	NOAA	Mud Lake, LA	discharge of crude oil at liquid oxygen plant
	NOAA	Southwest Pass Area, LA	discharge of crude oil from transfer pipeline
	NOAA	Westlake, LA	ongoing air discharge from chemical plant due to power outage
	NOAA	55 miles outside of Louisiana	satellite imagery detects oil anomaly at inactive well platform
	NOAA	East Cameron, Louisiana	abandoned well discharging gas and condensate
	NOAA	Long Island, New York	gasoline tank truck spill
	NOAA	Prince of Wales Island, Alaska	commercial fishing tender sinks in Sumner Strait
March 22, 2022	NOAA	Neva Strait, Alaska	breeding herring and other fish and sea creatures
April 19, 2022	NOAA	Plaquemine Parish, Louisiana	chlorine leak; residents told to stay indoors and turn off AC
January 3, 2023	NOAA	Port of Milwaukee, Wisconsin	14,000 gallons of discharged fuel
January 6, 2023	NOAA	Summerland Beach, California	oil sheen from inactive oil field
February 10, 2023	NOAA	Newport, Oregon	2,000 gallons of red dye diesel on soil and tracks and deposited into the Yaquina River
February 22, 2023	NOAA	Lopez Island, Washington	visible sheen of bilge residue
March 9, 2023	NOAA	Westport Marina, Washington	ship caught fire
March 13, 2023	NOAA	Bolivar Peninsula, Texas	leaking pipe was found leaking natural gas from an old well

Date	Source	Location/Region	Damage/Result
March 16, 2023	NOAA	Anacortes, Washington	derailed train spilled 7,000 gallons of diesel and 2,000 of fuel on soil
March 29, 2023	Inside Climate News	Midland, Texas	402,486 gallons of crude oil gushed out of an EnLink Midstream pipeline
April 7, 2023	NOAA	Cameron, Louisiana	1,050–42,000 gallons of fuel into surrounding marshlands; caught fire on April 10
April 8, 2023	NOAA	Hylebos Waterway, Tacoma, Washington	fishing vessel caught fire; shelter in place ordered for the surrounding area
April 26, 2023	NOAA	Hackberry, Louisiana	1,260 gallons of crude oil was spilled in the surrounding marsh area on the Intracoastal Waterway
May 5, 2023	NOAA	Deer Park, Texas	explosion from refinery; harmful levels of chemicals polluted the surrounding area
May 17, 2023	NOAA	Corpus Christi, Texas	unknown amount of hydrogen sulfur and sulfur dioxide were released into the air
May 20, 2023	NOAA	35 miles west of Pacifica, California	aircraft crash; 1,000 gallons of Jet A fuel into Pacific Ocean
June 11, 2023	NOAA	Ashland, LA	3,402 gallons of high-temperature oil
June 24, 2023	NOAA	Reed Point, Montana	17 derailed train cars containing sodium hydrosulfide, asphalt liquified petroleum, molten sulfur, and scrap metal

In actuality, spilled oil is never recovered, especially from oceans, rivers, lakes, and waterways as western environmentalists and government officials regularly claim. The damage is permanent because the earth, the ocean, and the air are not disparate and compartmentalized entities; they are all part of the foundational fabric of life in constant movement and interaction with each other, of which all life is inextricably interwoven. These spirit powers are *not* distinct and isolated from each other. In *Hakai Magazine* on July 12, 2016, Andrew Nikiforuk penned an article with an incisive title, "The Oil Spill Cleanup Illusion: Why do we pretend to clean up oil spills in the ocean?" In the article, he underscores how the oil spill cleanup movement was actually promoted by the oil industry and designed to appease the consciences of liberal environmental activists who need to feel that they are doing something positive to address the plethora of oil spills by oil tankers serving the mammoth oil industry. He highlights research that was conducted by German biologist Sylvia Gaus, who had worked and studied in the tidal flats of the Wadden Sea, in the North Sea, a vast ecological habitat especially for birds, that had suffered from a lethal oil spill of over 26,417 gallons in 1998, resulting in 13,000 birds being killed.[119] What Gaus learned is both shocking and revealing but, for

3. Oil Production and Its Capitalist History 91

The Deepwater Horizon mobile offshore drilling unit (MODU) operations on the Macondo Well at Mississippi Canyon Block 252, April 20, 2010, and resultant explosion and fire that tragically took 11 lives and injured 16 others (U.S. Coast Guard, Department of Homeland Security).

Indigenous peoples, ancient veritable knowledge: that the damage done to the ecology, in this instance, the birds especially, is irreparable, and cleaning the oil off the feathers of the fouled birds is, in fact, harmful, because the cleaning destroys part of liver and kidney function and increases mortality of the cleaned birds because they are unable to reproduce. Such findings were reconfirmed during a 1996 study of cleaning up oil residues off birds in California, and in 1997, when brown pelicans were cleaned after being poisoned by an oil spill and died much sooner than normal. Similarly, in November 2002, the oil tanker M/V *Prestige* ripped in two and released over 15 million gallons of oil that contaminated 600 beaches off the Galicia coast of Spain, killing 300,000 birds.

Gaus reminded environmental cleanup authorities in Europe, Turtle Island, and elsewhere that the survivability rate for cleaned-up, oil-soaked birds is 1 percent, and that a dime-sized oil spill can kill at least one bird. The rare occasion of a "successful" cleanup following a toxic oil spill, like the oil tanker M/V *Treasure* spill off the South African coast near Cape Town in 2000, is rare indeed. Gaus emphasizes that cleanup efforts deploying oil dispersants like Corexit, the leading ostensible industrial cleanup chemical, causes "more harm than good," and essentially affects the viability of marine life because such chemicals are toxic and designed to deceive the public with the *illusion* of cleaning up oiled beaches and marine waters. Nikiforuk stresses that such deceptive cleanup practices are what many environmental scientists outside the oil industry

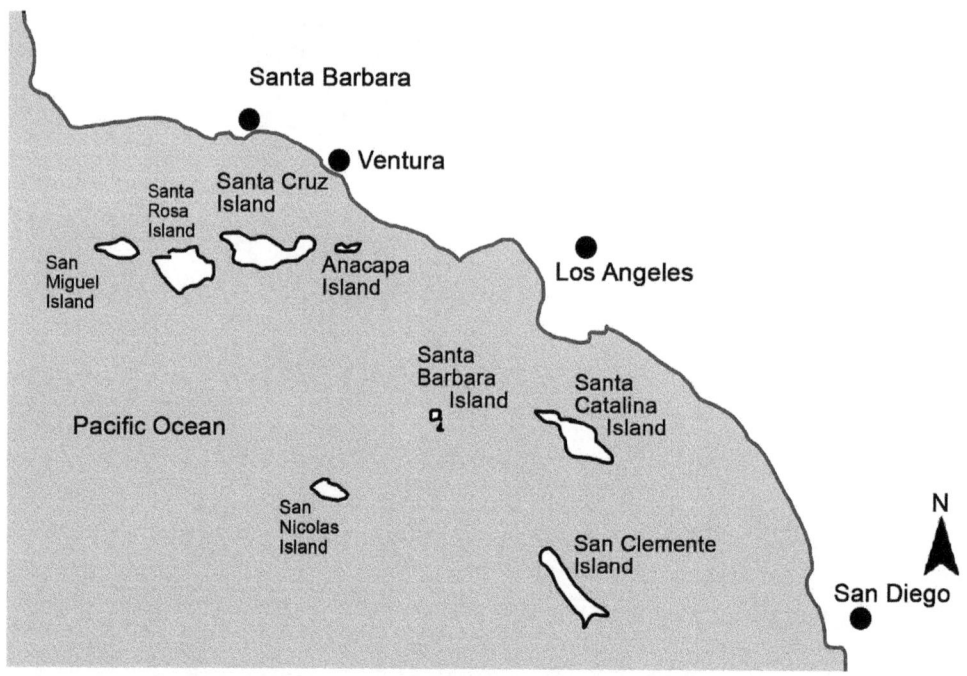

Off-shore oil drilling locations off the southern California coast (John B. Smith, Offshore Technology Conference, April 30–May 3, Texas, "California Offshore Oil and Gas Decommissioning Outlook and Challenges," https://www.slc.ca.gov/wp-content/uploads/2018/10/Offshore-California.pdf).

describe as "prime theater" for gullible public consumption that obscures the lethal effects of oil on the oceans and waterways of the world.

The billions of dollars spent on cleanup efforts, like those in the aftermath of the BP Deepwater Horizon spill in the Gulf of Mexico, is futile in the final analysis because it is an ideological ploy to portray Big (and profit-obsessed) Oil as "moral" responders to such deadly crises. This subterfuge refuses to accept that oil production, refining, and transportation causes massive annihilation, especially of terrestrial and marine life, and permanently destroys the teeming endless microbial and organic life in the oceans, seas, earth, and air. Greed is the basis of the deceptive game of "cleanup," a capitalist fabrication that conditions us into falsely believing that oil-based living is indispensable to life since there is no alternative (the TINA syndrome). Only Indigenous ancestral wisdom that has preserved 80 percent of the world's biodiversity can infuse common sense in global cultures. Only then can the scientific truth of the spirituality, complexity, fragility, and indivisibility of life in and on the Earth and in the air and sea prevail and compel a change of course for us. All else is oblivion.

Indigenous Cultural Perspectives on Oil and Energy Extraction

Oil extraction from the dear Earth Mother is bloodshed of all life. All creation and creating of the Spiritual Universe possess inalienable rights to live in the manner that the Earth Mother and the Eternal Beginningless and Endless Universe have constructed life.

3. Oil Production and Its Capitalist History

Oil's history is gruesome principally because its production and imperialist foisting on the vast majority of the world's people caused wars, ecocide, climate instability and collapse, refugees, bloody internecine conflict, and permanent contamination of air, water, and earth wherever it has operated. It is at the core of the capitalist system built on accruing gargantuan profits for transnational corporations and billionaire and millionaire elites at the cost of lives of billions the world over. It is the basis of the destruction of the world in which we live, and through its indispensability to the automobile and to industrialism, oil is the leading enemy of life itself since it is progressively undoing our existence. Oil drilling, fracking, and mining are the leading causes of Earth's heating in unprecedented ways that have never been experienced in the known history of the Earth, pronounced more than ever in 2023 when average temperature rise reached over the tipping point of 1.5°C (34.7°F).[120]

Natural law is the law for Indigenous nations, not colonial regimented law. Indigenous Indians of the Amazon, for example, like the Waorani, Kichwa, Achuar, Tagaeri and Taromenane Indians of the Yasuní preserve in Ecuador, the most bio-diverse and Indigenous self-sustaining region on the Earth, have resisted toxic oil companies like Texaco, which drilled for oil in Ecuador from 1964 to 1990. In 2001, Texaco changed ownership due to litigation by Indigenous Amazonian Indians, represented by New York environmental lawyer Steven Donziger and other attorneys. Donziger did a spell in prison in New York under a fabricated "contempt of court charge" brought by a judge who has interests in Chevron. Texaco, and subsequently Chevron, were responsible for deliberately dumping 18 billion gallons of oil residue in the Indigenous Amazon forests and river, around 140 billion gallons, on sensitive forest, fresh drinking, and bathing water sources, covering 4,400 square kilometers (1,700 square miles).[121]

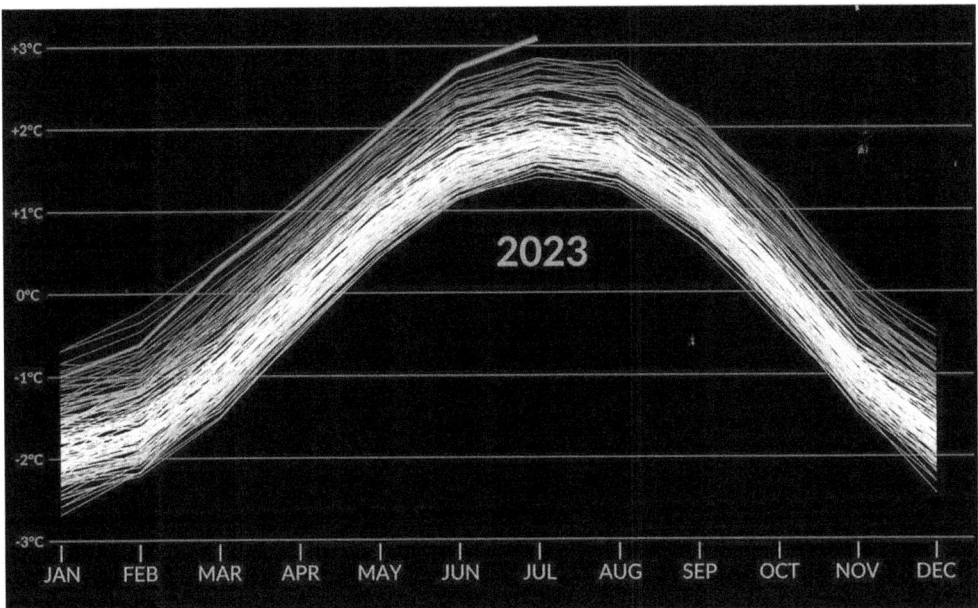

Increases of Earth's temperatures in Celsius from 1880 through July 2023 with bottom-most curves starting in 1880 and a baseline period of 1951–1980 (NASA's Scientific Visualization Studio, https://svs.gsfc.nasa.gov/5137/).

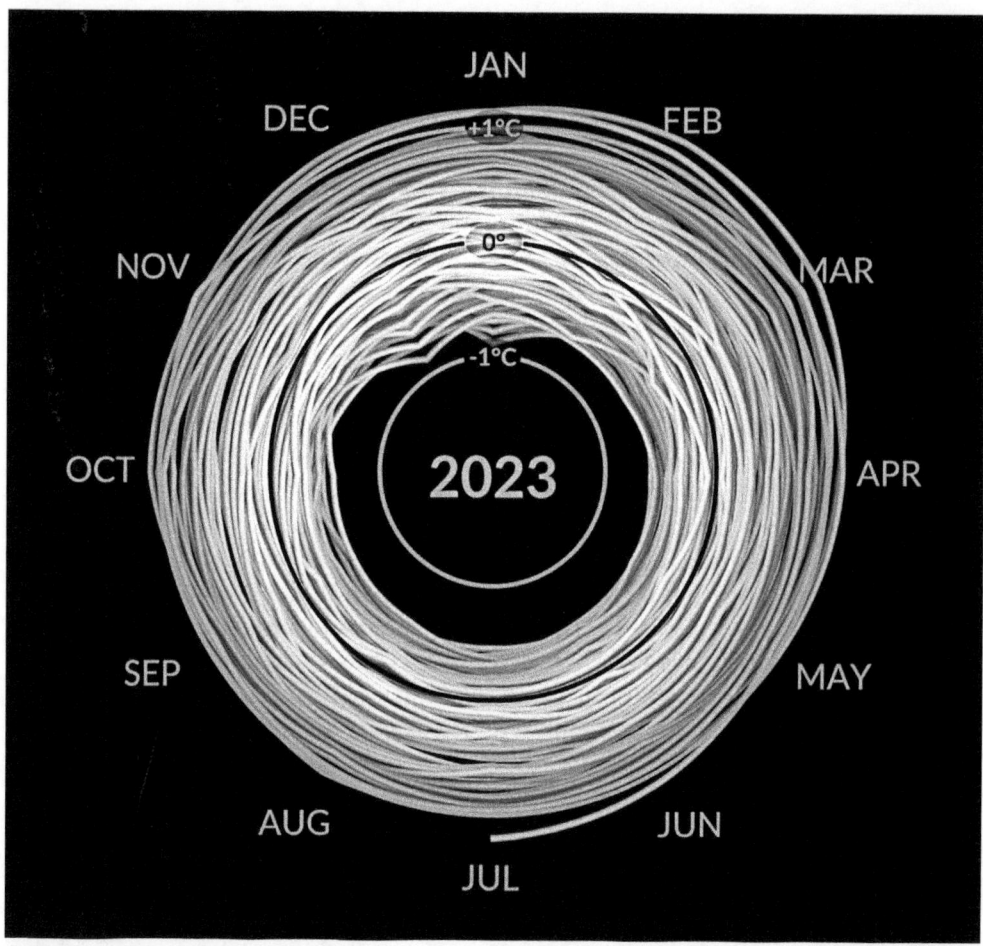

Climate Spiral showing monthly temperature anomalies 1880 through July 2023 (over the baseline period 1951 through 1980) (NASA's Scientific Visualization Studio, https://svs.gsfc.nasa.gov/5137/).

Similarly, the Ogoni people of the Niger Delta in Nigeria have consistently rejected and fiercely opposed (even with weapons) Shell's invasion and destruction of the vital life source of the people there since the river has become fully contaminated, destroying drinking water and fishing with ongoing oil sludge residues and 24/7 gas flares that poison the ecology and environment. On April 22, 2022, the explosion at an illegal crude oil refinery in Abaezi forest in the Local Government Area of Ohaji Egbema in Imo State, Nigeria, killed 100 people, burning their faces beyond recognition.[122] "Mainstream" media and news outlets in Nigeria and the west heaped blame on "illegal oil refining," eliding any coverage of the root cause of the grinding impoverishment and irreparable damage to the Niger Delta and the adjoining region: oil drilling and production. Nigeria's capitalist economy feeds directly into western capitalism, fueling colossal profits for oil conglomerates. *Vanguard News Nigeria* was one of the few local Nigerian news outlets that reported on a joint statement made by 11 Civil Society Organizations (CSOs) in Nigeria that revealed that the forest around the refinery was also destroyed and much of the life there was burnt to death. The statement stressed that the root causes of impoverishment, unemployment, ecological destruction of traditional Indigenous fishing, and depleting food sources needed addressing in such catastrophic incidents. The statement underscored that the cycle of death from oil production and the inevitable derivative explosions from the horrific pipeline fire at Atiegwo (near Jesse), Nigeria, on October 18, 1998, killing 1,082 people, the artisanal refinery in the Rumuekpe community of Rivers State, and the refined petroleum storage fire on April 11, 2022, at Bonny-Bille-Nembe jetty in Port Harcourt, which killed five people, including a pregnant woman and a two-month-old baby, requires radical solutions.

Impoverishment has been the net product of 70 years of relentless oil drilling, disregarding Indigenous Ogoni opposition in the Delta region. The continued exploitation of oil, destruction of the forests, and the poisoning of the vital fishing source Niger Delta has rendered people homeless and without viable forms of revenue from productive employment. The 11 CSOs noted that the previous three years had seen both ecological annihilation and impoverishment compounded with no respite from oil plunder or state attempts to halt such destructive operations.[123]

The neo-colonial, oil-conglomerate-protecting regime of Nigeria, as with many of the regimes ruling African countries that possess oil resources like Angola, Gabon, Algeria, Libya, Chad, Sudan, Equatorial Guinea, the Democratic Republic of the Congo, and the like, has suppressed the Indigenous peoples' resistance. These regimes depend on oil export revenues for their national economies, which they have refused to diversify since the monolithic oil-based economies serve the elites and their clients in the west well. This dependence was bloodily marked with the suppression and eventual hanging execution of playwright and internationally renowned human rights activist Ken Saro-Wiwa and eight other Ogoni leaders—Barinem Kiobel, John Kpunien, Baribor Bera, Saturday Dobee, Felix Nwate, Nordu Eawo, Paul Levura, and Daniel Gbokoo—in Nigeria on November 10, 1995.[124] Today, the resistance continues under the Movement for the Survival of the Ogoni People (MOSOP), a coalition of various Indigenous groups representing over a million people, and organizations like the Ogoni Solidarity Forum with Celestine Akpobari as national coordinator. The Forum is part of the Peoples' Advancement Center, which complements the work of MOSOP, the Chikoko Movement, and the resistance of the Ijaw people under the Ijaw Youth Council (IYC).[125] The Ogoni Solidarity Forum is affiliated with the International Indigenous Peoples' Movement for Self Determination and Liberation (IPMSDL) based in the Philippines, a global Indigenous peoples' organization active on all continents, with which we are active. IPMSDL works with major Indigenous organizations and groups resisting colonialist and corporate invasion and occupation of ancestral lands for oil, minerals, and energy resources.[126] The number of targeted assassinations of activists defending land and environmental rights globally against oil, mining, deforestation, and commercial companies in 2020 was 227, of which one-third were Indigenous people, and of this number, 10 percent were women.[127] Nineteen percent of all attacks were in Asia.

From the era of nineteenth-century industrialism through the present, the violent destruction of the Earth and life on land and in the air and sea has become progressively worse. Reckless capitalist industrialist-obsessed oil drilling on land and offshore for oil, coupled with extractive commercial mining, has unleashed lethal consequences that no corporation or government can withstand. Yet, the oil conglomerates and their political supporters uncaringly persist in deceptive practices of advertising as part of their collective claim of "concern for the environment" and "green energy" in the first two decades of the 21st century. Is it surprising to learn, then, that the leading climate-polluting and Earth-warming causes are not the majority of the world's eight billion people, but the ruling one percent billionaire class that emits half of all carbon pollution on Earth, twice that of the most impoverished 3.1 billion half of humanity?[128]

Indigenous peoples locally and globally have refused to unearth, plunder, and extract oil, gas, and coal that belong to the Earth Mother anyway because such actions constitute violations of natural law and the sanctity of life. These destructive acts have caused untold suffering and destruction of life, now even the unfolding of life, especially of humans. Unprecedented heating of the Earth, protracted droughts, shocking cold and freezing temperatures never heard of in many places, annihilating floods and storms, and perpetual climate and ecological collapse all attest to this assertion. Over the past two centuries, Indigenous peoples have urged, begged, and demanded in local, national, and international forums that this incessant insanity and perpetual myopia on the part of industrialization and globalized capitalism end immediately to preserve life. Arrogant regimes and capitalism consistently repeat the same fatal errors and fail to acknowledge their fundamental epistemological flaw: human technologies can neither "save" the Earth nor ensure the present and future of our collective existence.

Rejection of ancient Indigenous wisdom, knowledge, and teachings propelled us into the irretrievable abyss of an unavoidable, deadly collision course with our universal Mother, Earth. The outcome stares us in the face daily. The purpose of this persistent writing, with all of its limitations of language and accessibility, is to accept reality and act decisively in making a break with globalized capitalism and ecological violence, even at the cost of global capitalist and economic collapse so that we simply preserve and sustain our lives and those of our relatives. There is no other path to life.

Summary and Conclusion

This chapter has demonstrated the myopia of capitalism that has always perceived the sacred Earth Mother as a football to be kicked around, assuming falsely and absurdly that the vast, countless, myriad forms of life on Earth, in the Earth, and in the sea and air are expendable for "our" progress. Ultimately, no form of life is indivisible; the destruction of one leads inextricably to the annihilation of the whole.

The next chapter will delve into yet another important correlative subject of climate change and the devastating effects of the chemical geoengineering of the weather as a supposed solution to global warming and associated corporate and militaristic intervention.

4

Chemical Geoengineering of the Weather

Corporate and Militarist Intervention and Effective Climate Collapse

> From space, the masters of infinity would have the power to control the earth's weather, to cause drought and flood, to change the tides and raise the levels of the sea, to divert the Gulf stream and change temperate climates to frigid.
>
> ...we now have on record the appraisal of leaders in the field of science, respected men of unquestioned competence, whose valuation of what control of outer space means renders irrelevant the bookkeeping concerns of fiscal officers.
>
> It lays the predicate and foundation for the development of a weather satellite that will permit man to determine the world's cloud layer and ultimately to control the weather; and he who controls the weather will control the world.
>
> —Vice President Lyndon B. Johnson, 36th president of the United States, at Southwest Texas State University (1962)[1]

Introduction

This chapter will examine the widely discussed subject of Earth heating and climate chaos and instability—even collapse—by illuminating the history of industrial chemical use, including gas weapons production, U.S. weather monitoring and chemical engineering, and climate modification. It will highlight the collective impact of these facets on ongoing climate disasters plaguing daily existence, unprecedented sea level rise, prolonged and entrenched drought, unheard of extremes in heat and cold with constantly spiking temperatures high and low, unimaginable floods and consequential destruction, and devastation of plant life and soil fertility that has exacerbated impoverishment and "climate refugees" from the rapid disintegration of island countries around the world, especially in the Pacific. This chapter will illuminate Indigenous cultural and cosmological perspectives on recent historical and current crises and explain how Indigenous knowledge can function as the basis for Earth-centered and Earth-friendly solutions to this escalating and unfolding life collapse, in concert with all environmentally and ecologically concerned communities.

The History of Chemical Use

Chemical composition and medieval alchemy were pioneered by the Muslim, Persian, and Arabic-speaking Jăbir Ibn Hayyăn, who combined concepts of Earth, Wind, Air, and Fire, with that of combustible sulfur and metallic properties of mercury. Robert Boyle, the seventeenth-century Irish-Anglo chemist, is generally credited with separating chemistry from alchemy, and in the subsequent centuries, others like eighteenth-century Scot Joseph Black began isolating carbon dioxide. Nineteenth-century Frenchman Louis Claude Cadet de Gassicourt isolated arsenic compounds and formulated cacodyl oxide, possibly one of the earliest "discoverers" of the synthetic organometallic compound.[2]

Chemical use during the European Renaissance became the basis for the emergence of the chemical industrial society in which we live (or as Indigenous peoples, are forced to live) today, resulting in deep and irreparable harm to all life on the Earth, Sea, and Air. The technologies employed in the chemical-military industrial complex were responsible for the deaths of tens of millions in the wars in western and eastern Europe in the second decade of the 20th century. Today, these are supplemented by geoengineering and weather modification by government and military industrial corporations. It is critical to understand the historical roots of weather modification and precisely how this ideology came to assume such decisive proportions, especially in this era of the third decade of the 21st century to which we are subjected daily in all quarters of the Earth.

The roots of geoengineering, like the genetic modification of seeds, mark the era of the first chemical-grounded society of human cultures. Carlos Petrini, the Italian traditionalist and innovator of the *Slow Food* movement, which marks a return to traditional Indigenous ways of living and cultivating food naturally in balance and reciprocity with the Earth and the rest of creation, recalls the destructive shift in the 19th century. He notes that from the 1840s, chemical fertilizers were deployed for food production for the first time ever, introducing poisonous inorganic compounds into food cultivation that eventually became overused and pervasively normative in the western world especially. From 1987 through 2007, agriculture has deployed more than twice as many chemical fertilizers as we had ever previously produced! Petrini laments that the "agroindustry of food production has become the model of development in a world in which technology reigns," destroying fragile ecosystems and environments that Indigenous people had carefully nourished and nurtured in preserving biodiversity from ancient times to the present.[3]

Chemical weapons were first deployed on April 22, 1915, outside the Belgian city of Ypres, when German soldiers ignited chlorine gas canisters from 5,000 chlorine barrels, releasing thick trails of yellow smoke that reached 6,000 French troops and soon annihilated them. In 1918, German chemist Fritz Haber, an enthusiastic World War I advocate, received the Nobel Prize in chemistry for his work on poison gas. Alfred Nobel, the Norwegian industrialist chemist inventor of dynamite and explosives used in war like ballistite, formulated the Nobel "prize" in various disciplines to assuage his guilty conscience from making a fortune from dynamite and explosives production that were responsible for much decimation of life in wars, mining, and industry. The U.S. military industrial complex became the center of chemical research in the escalation of war from the early 20th century, following in Germany's footsteps and preparing for challenges for the Cold War with the Soviet Union and the eastern bloc.

The U.S. Bureau of Mines; leading universities like Yale, the Massachusetts Institute of Technology (MIT), Johns Hopkins University, the University of Michigan, American University; and others—with almost 2,000 academics—all played a key role in the advancement of war that deployed poisonous gases. The U.S. Bureau of Mines thus became the principal "civilian organization" engaged in poison gas research, paving the way for the lethal and expanding "military industrial complex" decades later. These eventually led to the Manhattan Project and the invention and subsequent detonation of the atomic bombs on Hiroshima and Nagasaki in August 1945.[4] Such diabolical designs persist, particularly in the U.S. proliferation of cluster munitions that were used with intensified lethal effect in Vietnam, Laos, Cambodia, Iraq, Afghanistan, the former Yugoslavia (now Serbia, Montenegro, and Kosovo), and these particularly torturous weapons were shipped to Ukraine and bunkbuster munitions to Israel's military in 2023. They remain unexploded and lethal in many places in southeast Asia, Iraq, and Afghanistan.[5]

When the intra–European war of 1914–1918 ended, the U.S. was producing gas weapons four times that of Germany previously, and 1,700 chemists were recruited from universities and corporations to join the newly formed Chemical Warfare Service of the U.S. military under Woodrow Wilson, initially as volunteers, but soon as solid participants and war advocates.[6] Over 200,000 people were killed in Europe from chemical gas warfare, of which close to 70,000 were from the U.S. Chlorine is used ubiquitously for water cleansing and treating wastewater, and other toxic chemicals like arsenic, lead, and cadmium are used for industrial agricultural fertilizers that kill vital organisms in the Earth.

Atomic testing involved massive chemical explosions, as O'Connor, a U.S. military service member who was present at the initial atomic test explosions at the Nevada Test Site in 1951, explained. He resultantly suffered from persistent nausea, repeated blood clots, and polymyositis, where his muscles began to slowly collapse. Howard Hinkie, another U.S. military serviceperson who was present at the Atomic Energy Commission (AEC) test series codenamed "Teapot," fathered three children, two of whom were born deformed after he had spent ten years at the Nevada Test Site.[7] O'Connor charged that he had been irradiated twice, though many senior military leaders denied his accounts as "hallucinations." Following the declassification of classified military documents describing the nuclear tests, it came to light that:

> Staff memorandums from the army's Training Methods Division which worked with the "human-research units" of the various service branches, reveal a disturbing pattern of systematic manipulation of soldiers' attitudes towards the bomb and atomic testing. These reports *fairly rejoice* [italics ours] at the success with which the average GI was misled and misinformed about the dangers of radiation and its biological effects.[8]

O'Connor had been telling the truth after all. In fact, "nearly 100,000 servicemen, mostly soldiers and marines, were brought to the test site to witness one or more atomic blasts."[9] Many of these persons died prematurely or were dying from a range of cancers following exposure to some 83 atomic tests that were five to six times more lethal than the bombs dropped on Hiroshima and Nagasaki.[10]

The health and mortality toll on people's health given the history of chemical warfare and atomic testing in Nevada and other places in Turtle Island and in the Pacific, especially on women and young people, is incalculable, and most deleterious for children and youth. Michael Uhl and Tod Ensign warned of the lingering toxic and lethal effects of such chemicalization of society and peoples' consciousness of this deadly adversity growing progressively:

> Ultimately came the dawning realization that each of us was being zapped by a growing barrage of gamma rays, microwaves, and x rays; pesticides; preservatives; and ten thousand other chemicals. Even the basic staples of life are no longer safe—neither the air we breathe, the Water we drink, nor the food we eat. Indeed, the morning dew, age-old symbol of freshness and purity, is slowly turning to acid as a thousand pollutants infect the atmosphere.[11]

The reference to "ten thousand chemicals" is patently no exaggeration. Our bodies are now saturated with chemicals from exposure to radiation, wireless technologies, radiation of food, genetically modified (GMO) seeds, chlorinated water, x-rays for medical treatment, and a host of other processes.

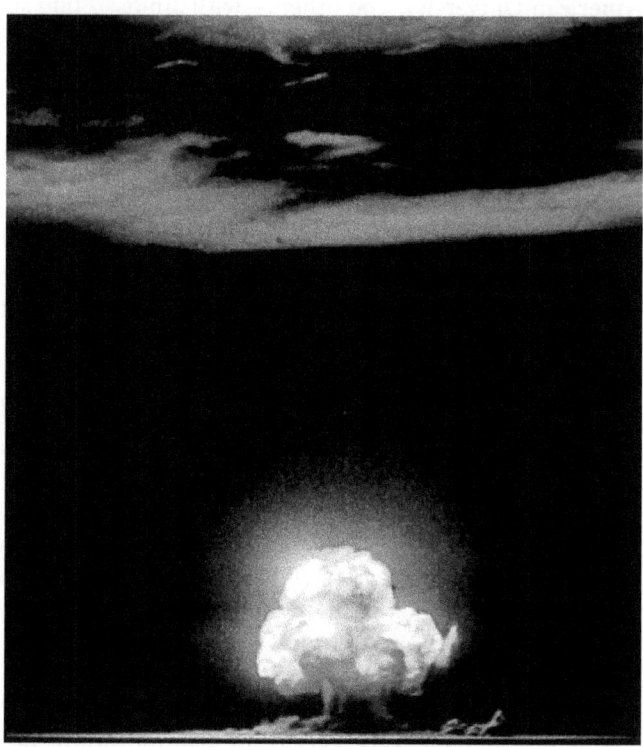

Trinity explosion, first nuclear test, *Jornada del Muerto* desert, southeast of Socorro, New Mexico, July 16, 1945 (National Nuclear Security Administration/Nevada Field Office, Wikimedia Commons).

The History of U.S. Geoengineering, Particularly for Weather Modification and Military Hegemony

Geoengineering's history has far-reaching roots. The U.S. Congress charged the War Department with being responsible for meteorological observations and strategic meteorological monitoring for war interventions in the early 1900s. The war of 1914–1918 served as the springboard for the launch of the Weather Service of the U.S. Army. The function of the Weather Service was "to provide the Army at home and American Expeditionary Forces (AEF) in Europe with all the meteorological information needed; to supply the aviation fields, the coast artillery stations, the ordinance proving grounds, and *the gas warfare service* [italics ours] with such meteorological and areological data as might be useful to them; and to undertake special investigations in military meteorology and related problems."[12]

The entire field of weather monitoring and prediction, as known in civilian circles, has been the province of the U.S. military from its inception, a legacy that persists in weather forecasts today, where though the National Oceanic and Atmospheric Administration (NOAA) falls under the Department of Commerce, it's an interwoven partner with the Department of Defense (DOD), particularly with the Navy. The National Defense Authorization Act of 2021 that became Public Law 116–283 on January 1, 2021, includes NOAA and its Ocean Policy Committee that formally established the Ocean Resource Management Subcommittee in August 2018. The original and primary focus of weather monitoring, while engaged in weather prediction and preparedness, was heavily focused on global war preparedness and intervention.

Shortly after the launching of the intra–European war in 1941, the War Department formed the Defense Meteorological Committee on January 21, with the mission to "coordinate wartime and civilian and military weather activities" and then became the Meteorological Committee, the Joint Chiefs of Staff (JCS) in 1942. In 1941, too, the Air Corps became the Army Air Forces (AAF), constituted by the Air Corps and the Air Force Combat Command. It subsequently became the Joint Meteorological Group on June 1, 1967. In 1942, the Air Corps Weather Service changed to the AAF Weather Service, with an expanded function of providing weather monitoring support to the Army Ground and Service Forces, thus becoming directly involved in combat operations. With the end of the war in 1945, the U.S. had nine Weather Service squadrons with 900 weather stations in Turtle Island and 700 of these stations abroad, particularly preparing for invasions of southeast Asia and the Pacific Islands. It was these weather stations in the Pacific, like the island of Hawaii (colonized by the U.S. since 1893 when

Chem trails Absegami (Atlantic City) (Phoebe Farris, 2023).

colonial sugar planters led by Sanford Dole, the progenitor of the U.S. Dole corporation, overthrew the Hawaiian monarchy under Queen Liliuokalani), that served military purposes. The subsequent invasion by 300 U.S. Marines paved the way for the illegal colonization of Hawaii in 1898 that was fiercely opposed by Indigenous Hawaiians and even found to be illegal by the U.S. government-appointed Blount Commission in 1894.

Other Pacific islands like Guam, Samoa, the Northern Mariana Islands (CNMI), Palau, the Federated States of Micronesia (FSM), and the Marshall Islands, were all colonized by the U.S. government in violation of international law, in which the UN was a willful participant providing political cover to such colonization and occupation.[13] The principal reason was the need for the U.S. empire to dominate the Pacific militarily so that the large nations of east Asia like Japan and China would be challenged globally in the context of the Cold War. Dean Acheson, former U.S. undersecretary of state and secretary of state under then President Harry Truman, served as the lead foreign policy advisor to Truman from 1945 to 1947, and again from 1949 to 1953, along with the "Wise Men"—Charles Bohlen, Averell Harriman, George Kennan, Robert Lovett, and John McCloy—who became "architects of the Truman Doctrine, the Marshall Plan, and Cold War containment policy."[14] Under this policy, National Security Council document 48/2 charted the industrialization of Japan as the economic powerhouse for U.S. companies, and southeast Asian countries as providing cheap labor and a market for U.S. goods. Dean Acheson was key in elevating the centrality of the U.S. "militarized national security state," inspired by the armaments industry boon to economic growth during the second intra–European war of the early 1940s.[15] As a close associate of English economist John Maynard Keynes, Acheson championed military spending as providing economic stimulus and profits for the U.S. capitalist establishment, a legacy firmly entrenched in the current U.S. political economy with military industrial contractors being one of the largest recipients of U.S. governmental spending, as noted below.

Fifty-seven cents of every discretionary dollar in tax revenues in the U.S. is earmarked for military and war expenditures, spending $1.2 million per minute.[16] While the official fiscal Pentagon budget for 2023–2024 was a monumental $885 billion, especially considering economic suffering, impoverishment, and hunger in the U.S. and the world, this was and is not the complete defense and national security budget. The sprawling military industrial and national security complex budget exceeds $1.25 trillion, including various slush funds and funding of a myriad of national security agencies. The breakdown for fiscal year 2020, for instance, was as follows:

- Pentagon "Base" Budget: $554.1 billion
- War Budget total, which includes funding for the Overseas Contingency Operations (OCO) as in Syria, Iraq, Afghanistan, west and east Asia, Europe, the Pacific, and Africa, for 22 bases operated by AFRICOM in 39 countries, including 11 in Somalia: $173.8 billion
- Department of Energy/Nuclear Budget: $24.8 billion
- Defense-Related Activities Budget (most for the FBI and homeland security issues): $9 billion
- The Veterans Affairs Budget, which covers the total cost of providing medical and health services support for hundreds of thousands of veterans, many

suffering from post-traumatic stress syndrome (PTSD) and brain injuries or permanent disability following return from overseas wars (projections of lifelong care for those returning from Iraq and Afghanistan alone were $1 trillion): $216 billion
- The Homeland Security Budget, which covers the cost of "the Coast Guard, the Federal Emergency Management Agency (FEMA), Customs and Border Protection, Immigration and Customs Enforcement (appropriately called ICE), Citizenship and Immigration Services, the Secret Service, the Federal Law Enforcement Training Center, the Domestic Nuclear Detection Office, and the Office of Intelligence and Analysis": $69.2 billion[17]
- The International Affairs Budget, which includes the U.S. State Department and the USAID, of which over 10 percent is allocated to the Foreign Military Financing Program, most for "military aid" for Israel and Egypt, but also for "Jordan, Lebanon, Djibouti, Tunisia, Estonia, Latvia, Lithuania, Ukraine, Georgia, the Philippines, and Vietnam": $51 billion[18]
- Intelligence Budget, which funds 17 separate intelligence agencies, including "the CIA; the National Security Agency; the Defense Intelligence Agency; the State Department's Bureau of Intelligence and Research; the Drug Enforcement Agency's Office of National Security Intelligence; the Treasury Department's Office of Intelligence and Analysis; the Department of Energy's Office of Intelligence and Counterintelligence; the National Reconnaissance Office; the National Geospatial-Intelligence Agency; Air Force Intelligence, Surveillance and Reconnaissance; the Army's Intelligence and Security Command; the Office of Naval Intelligence; Marine Corps Intelligence; and Coast Guard Intelligence and Office of the Director of National Intelligence": $80 billion
- Defense Share of Interest on the National Debt: $156 billion (of the $500 billion national annual interest on the debt)[19]

The arms expenditure in the world totals $1.9 trillion, with the U.S. budget constituting some 63 percent of global military expenditures.[20] Mandy Smithberger and William Hartung from the Center for Defense Information at the Project for Government Oversight satirically, but correctly, note:

> our final annual tally for war, preparations for war, and the impact of war comes to more than $1.25 trillion—more than double the Pentagon's base budget. If the average taxpayer were aware that this amount was being spent in the name of national defense—with much of it wasted, misguided, or simply counterproductive—it might be far harder for the national security state to consume ever-growing sums with minimal public pushback. For now, however, the gravy train is running full speed ahead and its main beneficiaries—Lockheed Martin, Boeing, Northrop Grumman, and their cohorts—are laughing all the way to the bank.[21]

These writers note in a more recent article that in 2020–2021, the Pentagon spent "nearly $204 billion on various service contracts—more than the budgets for the Departments of Health and Human Services, State, or Homeland Security."[22]

Following Lyndon B. Johnson's speech in 1962, the U.S. military began the process of engineering the weather over Vietnam and over neighboring Laos and Cambodia, which the U.S. military bombed relentlessly over the next 12 years, twice the tonnage of bombs used in all of the 1940s intra–European war, resulting in millions of deaths for which no reparations have been substantially paid.

U.S. military bases in Africa today and planned construction through 2025: 39 in 22 countries (Casey Ontiveros, based on data from Nick Turse's article, "Exclusive: The U.S. Military's Plan to Cement its Bases in Africa," *Mail & Guardian,* South Africa, May 1, 2020).

The evidence of geoengineering has been particularly clear over the past decade, along with the genetic engineering of seeds and medicines as supposed ways of addressing global hunger and curing serious diseases like cancer. Recent effects of climate instability are reflected in the aggravation of Earth heating, drought, unprecedented destructive storms, earthquakes, floods, and tsunamis, like the one that caused the Fukushima Daiichi nuclear power plant incident in 2011 and the disaster on the island of Tonga in January 2022. The massive loss of life has been exacerbated by the engineering of the weather especially by the Pentagon in the U.S., Europe, China, Saudi Arabia, UAE, Israel, and other places where cloud seeding to generate rain occurs routinely in violation of the *UN Convention on the Prohibition of Military or Any Other Hostile Use of Environmental Modification Techniques,* which the ENMOD Convention promulgated in 1978 after the U.S. war against Vietnam. Though ENMOD "has been ratified by 78 countries, including Russia, the United States, Britain, China and Germany," many of the signatories refuse to abide by the convention.[23]

After 1945, the U.S. detonated 66 nuclear bombs in the Bikini and Enewetak atolls of the Pacific, a flagrant violation not only of the lands of Indigenous peoples of these islands and the region, but also a direct contravention of international laws that protect the right to land and life of all sovereign peoples. On June 30, 1946, the military detonated the first nuclear bomb on Bikini Atoll, forcibly removing people from the island for advancement of "Operation Crossroads." Similarly, in 1947, the people of the Enewetak islands were forcibly relocated to prepare for the execution of Operation Sandstone

4. Chemical Geoengineering of the Weather

Flamethrower tank in Vietnam spraying napalm (Donn A. Starry, Wikimedia Commons, National Archives: https://catalog.archives.gov/id/532440).

when three atomic tests were conducted. In March 1948, impoverished Bikini islanders were taken against their will and moved to Kwajalein and then to Kili, where no enclosed harbor existed and no fishing could occur, depriving the Indigenous people of their lands and livelihoods.

Between 1951 and 1952, more atomic bomb tests were conducted, and for the first time in history, the hydrogen bomb was detonated, with a shocking "10.4 megatons, or 750 times larger than the bomb dropped on Hiroshima" in a "test known simply as 'Mike.'"[24] This resulted in an entire island being erased from the map through vaporization in October 1952. On November 4, 1962, the Air Force Weather Agency (AFWA) furnished ten EB-50 reconnaissance aircraft and strategically placed the 6th WS mobile rawinsonde (a method of observation of upper air streams, wind speed and direction, temperature, pressure, and relative humidity using balloon-borne radiosondes monitored by radar or radio communication) in support of Operation Fishbowl. The latter involved the detonation of "a 1.59 megaton yield nuclear warhead at 69,000 feet altitude near Johnston Island, 717 miles west-south west of Hawaii ... part of a bigger operation called Dominic I."[25] These "operations," as the military glibly uses the term, involve massive death and destruction of life, and permanent poisoning of Water, Earth, and Air, and leave indelible lethal footprints manifest in deformed baby births, malfunctioning sea life reproduction, and other malformations in the terrestrial, aquatic, and celestial spheres.

Beginning in January 1996, the U.S. Air Force launched a focused emphasis on weather control when ranking senior major William Tasso formulated a position paper entitled, "Incorporating 'Own the Weather' into PME [Professional Military Education] Curriculums." The training curriculum adopted a three-tier approach that stressed

"'Know the Weather,' 'Apply the Weather,' and 'Own the Weather.'"[26] In August of that year, Colonel Tamzy House, along with other Air Force personnel, published "Weather as Force Multiplier: Owning the Weather in 2025," which was geared to "examine the concepts, capabilities, and technologies the United States would require to remain the dominant air and space force in the future," with a specific anticipation that "in 2025, US aerospace forces can 'own the weather' by capitalizing on emerging technologies and focusing development of those technologies to war-fighting applications."[27] It was this publication that formally advocated the systematic military "use of a future weather-modification system to achieve military objectives." Thus, a significant shift was marked in which "own the weather" meant "modifying the weather" and not simply "knowing the weather, applying the weather, and owning the weather," an expanded development of the Air Force's Windall's 1995 memorandum that formalized "the Combat Weather Facility (CWF) as a reinvention laboratory."[28] In essence, the U.S. Air Force was then implementing an integrated weather modification program in line with hegemonic military ambitions and elevating the role that Air Force bases play around the country. Davis Monthan Air Force Base in Tucson, Arizona, for example, in conjunction with military industrial contractor RTX Corporation (formerly Raytheon) plays a major role in repeated chemical sprays over the city (especially over the summers, but throughout the year), resulting in toxic chemical cloud formation that can eventuate in rain harmful to vegetation. Such measures are part of the desperate effort to dim the light of the sun to "stave off the effects of global warming" and are now the cause of ill health of all lives, especially those with respiratory illnesses, and diminishing plant growth and productivity.

The obsessive hegemonic control of resources of the Earth is a foundational factor in the phenomenon of unprecedented Earth heating and climate and ecological instability and collapse. For instance, the U.S. military is *the* largest polluter on Earth. H. Patricia Hynes, a former professor of environmental health at Boston University School of Public Health, elaborates in her 2011 article, "The Military Assault on Global Climate":

> By every measure, the Pentagon is the largest institutional user of petroleum products and energy.... Yet, the Pentagon has a blanket exemption in all international climate agreements.... It's a loophole [in the Kyoto Convention on Climate Change] big enough to drive a tank through, according to the report, "A Climate of War."
> In 1940, the US military consumed one percent of the country's total energy usage; by the end of World War II, the military's share rose to 29 percent. (1) Oil is indispensable for war.[29]

The U.S. Air Force (USAF) bases locally and the over 800 around the world are leading sources of pollution, generating 1,200 Superfund toxic sites because of military waste depositions. Hynes highlights the significant levels of toxic emissions from U.S. fighter aircraft on a single flight, daily occurrences on U.S. bases and regular in war events:

> The US Air Force (USAF) is the single largest consumer of jet fuel in the world. Fathom, if you can, the astronomical fuel usage of USAF fighter planes: the F-4 Phantom Fighter burns more than 1,600 gallons of jet fuel per hour and peaks at 14,400 gallons per hour at supersonic speeds. The B-52 Stratocruiser, with eight jet engines, guzzles 500 gallons per minute; ten minutes of flight uses as much fuel as the average driver does in one year of driving! A quarter of the world's jet fuel feeds the USAF fleet of flying killing machines; in 2006, they consumed as much fuel as US planes did during the Second World War (1941–1945)—an astounding 2.6 billion gallons. (3)[30]

Hynes adds that "militarism is the most oil-exhaustive activity on the planet," with expanded production and use of gas-guzzling planes, tanks, armored cars, and naval vessels, crystallized, for example, with 2,000 M-I Abrams tanks (now being used by the Ukrainian military after being supplied by the U.S. in 2022–2023) in Iraq in 2003 using 250 gallons of fuel every hour.[31] The more gas consumed, the higher the rate of carbon emissions, consequently becoming a leading factor in Earth heating and climate chaos as we experience it today.

The terminology of "energy security" is a misnomer for hegemonic domination that, at its core, is directed toward controlling and expropriating the vast oil, water, "natural gas," and mineral resources globally. According to the International Energy Agency (IAE), which consists of mainly industrialized capitalist western and eastern European countries and includes Japan, Turkey, and Korea as members and China, India, Brazil, Indonesia, South Africa, Thailand, Morocco, and Singapore as associate members, "energy security" implies "ensuring the uninterrupted availability of energy sources at an affordable price."[32] The IAE claims that its mission is its commitment "to shaping a secure and sustainable energy future for all." While the globalized, industrialized, capitalist mirage insists on using such cheap rhetoric including a "secure and sustainable energy future for all," it glibly ignores and tramples upon the fundamental rights of the majority of the world's people struggling for sustainable water supplies in the present.[33]

Vijay Prashad, director of the TriWorld Institute, accurately challenged attendees at the 2021 COP 26 Summit in Glasgow and asked why the west was obsessed with the future when over two billion people lack basic food and water and are mired in deep poverty today, with another two billion living on the edge earning a pittance to survive on $1 or $2 per day in the informal sector.[34] Such ostensible emanating concerns of "energy sustainability" and "renewable energy" serve as the smokescreen for the invasion and occupation of lands and countries that have been subjected to coups, military occupation, "regime change," and support and maintenance of client surrogate states. Countries with vast energy and mineral resources have become consistent targets of such invasions and "regime change." Places like Saudi Arabia and many of the neighboring Gulf states, Iran, Chile, the Congo, Iraq, Afghanistan, Sudan, Libya, Somalia, Venezuela, Syria, Haiti, Guyana, Mali, Niger, Kosovo, the Baltic countries, and other resource-rich areas of the underdeveloped world in the Pacific, Latin America, and the Caribbean are examples. Hynes' assertions about climate capitalism, militarism, and imperialism are absolutely correct when she contends that the nearly 1,000 (officially around 800) U.S. military bases globally are about "sweeping over all major oil resources—all related, in part, to projecting force for the sake of energy security."[35] She adds that the greenhouse gas emissions are accentuated owing to the production of military equipment, vehicles, and munitions used to protect U.S. oil routes, and that 20 percent of the U.S. defense budget is for "oil security," compounding the impact on climate and Earth heating. Such facts should be factored in when it comes to environmental assessment and addressing climate change issues.[36] The "energy security" hype is about the control of oil, and the conduit for such control is through military aggression and, in the 21st century, cyber and electronic wars using drone and robotic technologies.

U.S. military consumption of oil jumped from 1 percent of U.S. oil use in 1940 to an astounding 39 percent by the second intra–European war of the late 1940s, yet the U.S. still insisted on an unjustified exemption for its military cutting greenhouse gas emissions at the 1997 Kyoto Convention on Climate Change.[37] The U.S. government thus has a huge loophole in its commitment to reduce greenhouse gases and adhere to the one degree centigrade requirement of all nations and entities: the Pentagon, demanding exceptionalism for "national security" reasons. NATO, the core military alliance of western colonial nations, accounted for 55.8 percent of world military spending in 2021, along with allied nations that have "mutual military pacts" with the U.S. that spent another 6.3 percent, totaling 60 percent of all global military expenditures. This unequivocally signifies a "carbon bootprint!"[38]

Costs of War, published by the Watson Institute of International and Public Affairs at Brown University in December 2019, patently explains the principal role of the U.S. military in war against other nations and peoples, "counter-terrorism operations" (often concealing other geopolitical and economic designs), and the focus on "energy security." The Pentagon is the leading world user of fossil fuels, particularly since the events of 9/11 in New York City and the intensification of the war machine from thereon, necessitating the revealing details highlighted below:

> With an armed force of more than two million people, 11 nuclear aircraft carriers, and the world's most advanced military aircraft, the US is more than capable of projecting power anywhere in the globe, and with "Space Command," into outerspace. Further, the US has been continuously at war since late 2001, with the US military and State Department currently engaged in more than 80 countries in counterterror operations (3). Indeed, the DOD is the world's largest institutional user of petroleum and correspondingly, the single largest institutional producer of greenhouse gases (GHG) in the world (5).[39]

In 2017, total U.S. military greenhouse gas emissions exceeded that of Sweden, Denmark and Portugal and were even higher than all CO_2 emissions from the production of iron and steel domestically.[40] From FY1975 to FY2018, over 3,685 million metric tons of CO2 equivalent were emitted.

How does one celebrate "Earth Day" each April 22 (Earth Day is every day) when such toxicity from Big Oil and Big Military nonchalantly intensifies their contamination, pollution, Earth heating, climate deteriorating, and ecological collapse? Leaving toxic chemicals in the Earth and poisoning vegetation so that growth is radically diminished has a direct bearing on the Earth becoming hotter. Trees, forests, and vegetation everywhere protect the Earth against the intense rays of the Sun. When hazardous waste is produced, as the military does, it exacerbates heating by poisoning the ground. The DOD's dumping of depleted uranium, 2,4-D and 2,4,5-T (Agent Orange), lead, herbicides, defoliants, and other toxic chemicals produced more hazardous waste than the combined waste generated by the five largest U.S. chemical companies. Nine hundred of the 1,200 Superfund sites requiring such cleanup are abandoned U.S. military bases and operational centers, with 39,000 locations spanning 19 million acres in Turtle Island being heavily contaminated with poisonous chemicals fully leached into the soil.[41]

Research by Adam Liska and Richard Perrin from the University of Nebraska in Lincoln demonstrates that the "US military operations to protect oil imports coming from the Middle East are creating larger amounts of greenhouse gas emissions than once thought," from 8 percent originally assumed to actually 18 percent.[42] U.S. supertankers engaged in military security in the Persian Gulf emitted 34.4 million tons of carbon dioxide each year, and the wars in and around Iraq generated another 43.3 million tons of such toxic gases in the U.S. occupation of the country. Liska and Perrin called for regulation measures by the EPA to intervene in halting the astronomical proportion of greenhouse gases emitted by the U.S. military security complex in line with the U.S. 2017 Energy and Independence Security Act where "biofuels have to meet specific reductions of greenhouse emissions—from 20 to 60 percent—under gasoline to qualify for substitution" and radical reductions in "direct emissions and indirect emissions ... must include what is being put into the air from burning fuels" and "additional emissions" generated from fuel production.[43] The U.S. government spends almost $100 billion in the protection of maritime oil transit routes, causing warming of the Earth and compounding problems rather than solving climate change crises issues.[44]

Even Europe is a military toxic disaster.[45] The German military is responsible for 58 percent of the contamination of Germany's airspace.[46] In 1992, what the U.S. military consumed by the way of oil annually could have maintained U.S. public transportation for over two decades.[47] In 2017, the U.S. military purchased almost $8.7 billion worth of oil, generating 25,000 kilotons of carbon dioxide, making the agency the single largest contributor to Earth heating within the U.S.[48]

The Role of Geoengineering in Climate Chaos and Collapse in Recent Years

Dane Wigington, the coordinator of www.geoengineeringwatch.com, has invested all his time and energy into investigating the role of geoengineering. The point of advancing this very dangerous climate-life changing system is for the purpose of global control by a handful of elites—illusory and surely misguided, but real today. Effects of solar management programs designed to ostensibly reduce the warming effects of the Sun, known as Grandfather and Father to many Indigenous nations, have the following consequences:

- Completely disrupting the global hydrological cycle, causing record droughts and deluges;
- Destroying the ozone layer, which is allowing lethal levels of UV radiation to reach the surface of the planet, including ultraviolet C radiation;
- Completely contaminating the biosphere (from the clouds to the ground) with highly toxic heavy metals, polymers and chemicals. Peer-reviewed studies prove that some of these elements (like aluminum and barium) have negative effects on root systems, causing trees/crops to reduce or stop nutrient intake;
- Contributing to extreme global tree, foliage, and crop die-off, a result of factors already cited;
- Greatly contributing to fire intensity and volatility due to climate engineering (like aluminum and barium, which are incendiaries). This incendiary dust settles down through the atmosphere, coating virtually everything on Earth's surface.[49]

The solar panel revolution that began in the early 1980s and has become a leading industry since the early 2000s in California as a solution to shifting from dependence on fossil fuels and "natural gas" for consumer energy is also problematic. Not only do solar farms encroach on Indigenous sacred ancestral lands of the Chemehuevi and Yoeme (Yaqui) peoples in the California desert, but they are now found to leave toxic waste dumps since chemicals like cadmium, lead, antimony, and plastics impurities even in glass panels are projected to reach 78 million metric tons in 2050.[50] Further, huge tornados and severe storms can break solar panels and scatter toxic debris across large land spaces, such as the 2015 tornado in southern California that shattered 200,000 solar modules at the Desert Farm solar farm and Hurricane Maria in 2017 that destroyed most of Puerto Rico's panels on its second largest solar farm, which was responsible for 40 percent of the nation's solar power generation.

No other issue is as urgent and pressing today as the current crisis of ecological and insect collapse and associated Earth heating and climate chaos and instability. Yet, during a research visit with Indigenous communities in Bama Way, through Guugu and Yimithirr Yalanji country, Australia, in the summer of 2014, a Guugu Yimithirr elder, Willie Gordon, shockingly remarked that "global warming was a hoax." Elder Gordon explained that the heating of the Earth and the intensive climate chaos and instability all life has experienced is not just the unfolding of what people generally call "Nature," but is, in fact, the result of the geoengineering climate measures. The underlying cause of Earth and Sea heating and entrenched droughts and aridification is from continued fossil fuels, mineral mining, and Earth predation since the western industrial era in the 19th century and intensification of oil-based and mineral extraction water-depleting capitalist production and economies.

The events of the past decade remind us that the world has gone awry and life becomes more precarious daily. Floods of the worst lethal proportions, tornadoes of such ferocity that many structures are turned to dust and debris within minutes, freezing conditions including icy snow and sleet in places generally unaccustomed to such weather, and extremes in heat and heat waves never experienced before occur daily in all regions of the Earth. The net effect has been unusually high premature and disaster-driven death rates and unparalleled environmental and ecological destruction, particularly in underdeveloped countries of the world. Yet, privileged industrialized capitalist nations who often feel "protected" due to monetary and materialistic security are not immune to the unfolding climate and weather catastrophes.

In 2016, all 50 U.S. states saw temperatures of 100°F or higher. Yet the worst was still to come. The worst heat wave in recorded history occurred in 2023. The hottest day since 1936 with 49.6°C (121°F) on June 29, 2021, in Lytton, home of the Lytton First Nation of the Nlaka'pamux band, followed the previous two-day records of 46.6°C and 47.9°C in this northern Turtle Island location.[51] The temperature at Quillayute, Washington, located near the Hoh Rain Forest on the Olympia peninsula a few miles from the Pacific Ocean, saw 110°F on the same day, 11°F higher than the previous record set on August 9, 1981.[52] In July 2023, Death Valley in California saw regular temperatures of high 110s and even into 120s for days in the second week, extending its July 9, 2021 record of 129.9°F (54.4°C) and 130°F (54.444°C) and following the 134°F highest temperature on Earth ever recorded there indisputably in 2013. But such deadly weather cycles didn't stop there. In 2023, China, the southwestern region of Turtle Island, and southern Europe experienced protracted heat over extended weeks never occurring before.[53] Phoenix saw temperatures of 100°F (almost 38°C) for numerous weeks at a time, including beyond the normally hottest July month. Maximiliano Herrera, who studies global weather records, noted, "What we are seeing now is totally unprecedented worldwide.... It's an endless waterfall of records being smashed."[54] The monsoon summer season, which has generally been the saving rainy period, saw 0.15 inches of rain, the lowest since records have been kept for this era.

4. Chemical Geoengineering of the Weather

The incredible eruption of massive, widespread incinerating fires in the world in 2020 and 2021 are seared into our memories. The ecocidal fires in Australia spread over 11.46 million hectares (28,318,277 acres, equivalent to the size of the British Isles and larger than the U.S. state of West Virginia and the 2018 California fires) from September 2019 through January 2020 that killed over three billion of our precious relatives stand out as the worst ecological catastrophe in our known history. One hundred forty-three million four-leggeds, 2.46 billion reptiles, 180 million birds, 51 million frogs, countless birds, and the teeming life within the Earth that can't be documented, particularly in the southeast of Australia, were erased in the span of a few months.[55]

The persistent fires in California and the remaining 11 states of the west in Turtle Island, which saw millions of acres of fires incinerating millions of four-leggeds, birds, reptiles, insects, and other life forms and razing forests, fields, and buildings over the past three years in particular, lingered even into the heat of summer 2023. In 2020, more than 4.3 million acres of land were burnt leaving 100 million trees decaying into "dead wood." In 2021, almost 1.9 million acres of land was burned in the state.[56]

The role of "science" funded by large corporations engaged in the process of "dimming the Sun" and "cooling the Earth" has ironically swung in the opposite direction: uncontrollable heat. Forest and vegetation fires have thus become much more widespread, dangerous, and intense in every form and favor combustion that generates such huge columns of smoke, obscuring the Sun's light significantly.[57] An October 4, 2021, research report by the U.S. National Academy of Science, "Global Urban Exposure to Extreme Heat," which measured extreme heat exposure in 13,115 cities from 1983 to 2016,

Kangaroo and her joey who survived the forest fires in Mallacoota, Australia, in 2020 (Jo-Anne McArthur, Unsplash).

found that such exposure grew by 200 percent and almost tripled over that period, with "Total urban warming elevated exposure rates 52% above population growth alone," and direly affecting 1.7 billion people around the world.[58]

In the U.S., there were over 31 catastrophic mega-disasters requiring $20 billion+ in reconstruction and partial restoration costs from 1980 to 2021.

On July 20, 2021, while the world struggled to protect her citizens against the Covid-19 pandemic and its deadly effects, China experienced its worst ever flood catastrophe in Zhengzhou with 644.5 mm (25.38 inches) of rain within a 24-hour period, equivalent to a year of rainfall in the area. The floods were responsible for the deaths of 347 people and $30 billion in damage, a repeat of the 1998 floods that caused $48 billion in damage. These costs were similar to those suffered by Thailand following flooding from tropical storm Nok-Ten in 2011, $47 billion, which together with the aftermath of the Tohoku earthquake and the flooding along with the tsunami-triggered Fukushima nuclear catastrophe totaled $116 billion, with 815 Thai deaths and over 20,000 deaths in Japan.[59] In the case of Japan, with the forced evacuation of 500,000 people due to the tsunami and environmental and structural damage exacerbated by the nuclear plant collapse, the country incurred $360 billion in rebuilding, evacuation, and housing replacement costs. From July 27 through September 1, 2021, Cox's Bazaar in southeastern Bangladesh, a major fishing port and tourist area, saw destructive floods, with 1300 mm of rain (52 inches of rain) and 463 monsoon-triggered incidents. Bangladesh is experiencing annual receding of its original land area since its inception as

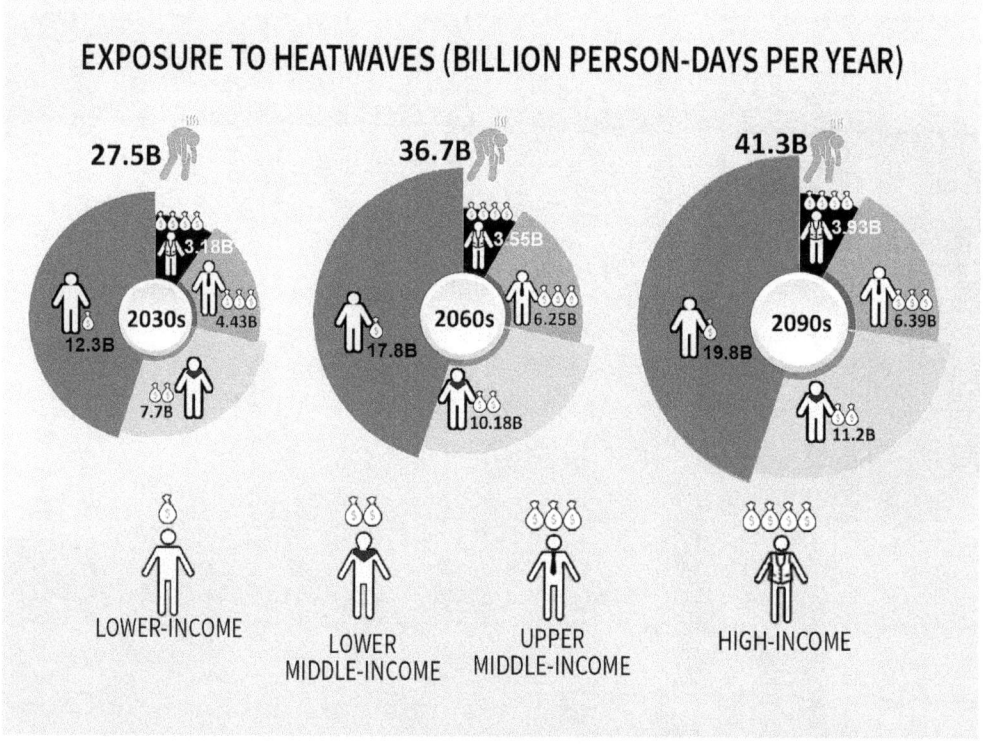

Increasing heat stress inequality in a warming climate (Mohammed Reza Alizadeh, *The Conversation*, February 10, 2022, CC BY-ND. Adapted and edited by Veronica Rodriguez).

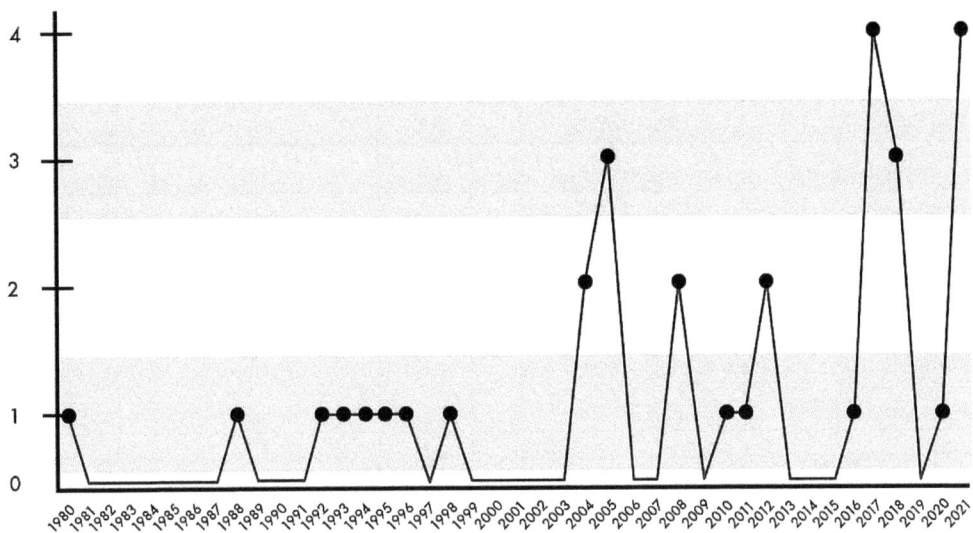

Global Weather Mega-Disasters Costing $20+ Billion, 1980 - 2021

Catastrophic weather events requiring over $20 billion in reconstruction: 1980–2021 (Casey Ontiveros, from data from NOAA and EMDAT, international disaster database, EMDAT, at https://yaleclimateconnections.org/2022D/01/the-top-10-global-weather-and-climate-change-events-of-2021/).

a country in 1971 by one quarter each year, similar to the capital of Indonesia, Jakarta, where one-third of the city could become submerged under water in the next couple of decades.[60]

Global Weather Mega-Disasters Costing $20+ Billion, 1980–2021[61]

Disaster	Location	Year	Damages (in Billions USD)	Deaths
Hurricane Katrina	U.S. LA/MS/AL/FL	2005	182	1,085
Hurricane Harvey	U.S. TX/LA	2017	141	89
Hurricane Maria	U.S. PR/VI	2017	102	2,981
Hurricane Sandy	U.S. NY/NJ/CT	2012	80	159
Hurricane Ida	U.S. LA/MS/NJ/NY/CT	2021	75	96
Hurricane Irma	U.S. FL/GA/SC/PR	2017	56	97
Hurricane Andrew	U.S. FL/LA	1992	54	62
Flooding	China	1998	51	3,656
Flooding	Thailand	2011	49	813
Drought/Heat Wave	U.S. Midwest/East	1988	48	454

Disaster	Location	Year	Damages (in Billions USD)	Deaths
Flooding	U.S. Mississippi River	1993	41	48
Hurricane Ike	U.S. TX/LA/MS	2008	40	112
Drought/Heat Wave	U.S. Midwest/East	2012	37	123
Hurricane Wilma	U.S. FL, Mexico, Cuba	2005	36	35
Drought/Heat Wave	U.S. Midwest/East	1980	36	1,260
Hurricane Ivan	U.S. AL/FL	2004	31	57
Flooding	China	2021	30	347
Hurricane Michael	U.S. FL/GA	2018	28	49
Winter Weather	China	2008	27	145
Flooding	North Korea	1995	27	68
Hurricane Rita	U.S. LA/TX	2005	27	119
Hurricane Florence	U.S. NC/SC	2018	26	53
Wildfires	U.S. Western	2018	26	106
Drought/Heat Wave	China	1994	26	104
Flooding	China	2016	26	475
Winter Weather	U.S. South & Central	2021	24	226
Hurricane Charley	U.S. FL	2004	24	35
Flooding	China	2010	23	1,691
Flooding	Germany/Belgium	2021	22	240
Flooding	China	1996	22	2,775
Hurricane Laura	U.S. LA/TX/MS/AR	2020	20	54

Though the idea of a wall around northern Jakarta (which is losing the most land due to rising oceans) has been proposed, it was rejected by residents in 2017, the only option thus being natural mangroves serving as a bulwark against catastrophic flooding. Yet, 93 percent of the mangroves in that region of Jakarta have been destroyed due to industrial development and accumulation of toxic waste, a wake-up call to the permanent collapse of natural ecosystems due to capitalist industrialism.[62]

By continent, Africa experienced 15 percent of such disasters from 1980 to 2021, with severe and prolonged droughts responsible for 95 percent of lives that perished over that period. Yet floods from the effects of Earth and Oceans heating were 60 percent of the disasters, with one of the most recent being in the southeastern tip of Africa, the Kwazulu coastal areas and eastern Cape border region of Azania/South Africa in April 2022. Over 440 lives were lost and over 40,000 homes destroyed, particularly those in informal dwelling areas, a legacy of the apartheid-colonial era where Black people were forced to build shacks and make shelter dwellings following migration from impoverished rural areas to the cities, like eThekwini (Durban).[63]

The downpour of 400 mm (16 inches) of rain over a four-day period was equivalent to three-quarters of the average rainfall in the country, 300mm within a 24-hour period, the worst experienced in its known history. The effects of the constant rain over a four-day period from April 11–14 on the region that was caused by movements of currents from the lower equator washed away thousands of dilapidated dwellings in impoverished communities and sustained $1.3 billion in damages to homes and infrastructure. The South African Weather Service analyzed the unprecedented phenomenon and indicated that the cause was not a tropical cyclone so typical of such deluges of rain, but a tropical system:

Mangroves under threat in Indonesia (Aldino Hartan Putra, Unsplash).

> a cut-off low in the upper reaches of the troposphere ... moving seawards, off the eastern coast of South Africa. Cut-off lows are associated with widespread instability in the atmosphere, which can promote periods of prolonged rainfall.... For KwaZulu-Natal ... the effect of the cut-off low system has been markedly enhanced by the presence of sustained low-level maritime air which has been fed in from the southern Indian Ocean, thus driving the system to produce more rainfall. The original source of the maritime air was from warmer, sub-tropical parts of the ocean.[64]

However, it needs to be noted that the "instability" within the atmosphere is from the combination of warming of Air and Sea from industrialist fossil fuel and coal use and weather geoengineering impacts in Africa that are western in origin, not African.[65] Weather systems around the Earth are all indivisible, as *Water* is, and what occurs in the west has ramifications for what happens in the east and vice versa, as with the north and the south. Climate instability on the African continent and failed agricultural production are being further compounded by the warmer Indian Ocean and prolonged

Landslide caused by catastrophic flood in Kwazulu, Azania/South Africa, in April 2022 (Copernicus Programme of the EU: Emergency Management Service, https://www.copernicus.eu/en/access-data/copyright-and-licences).

Road destroyed during severe Kwazulu flooding, Azania/South Africa, April 2022 (GCIS, Government of South Africa. Licensed under Creative Commons, CC BY-ND 2.0).

drought as in the western hemisphere, especially in Turtle Island, and other parts of the world. Though Africa contributes just 4 percent to global warming and heating, the continent has experienced the brunt of these industrialized capitalist-caused problems and has suffered protracted drought and temperature rise twice the world average. Racism, too, in areas of science has made African climate scientists invisible in the entire discourse on global warming, with only 3 percent of climate scientists being African.[66]

Europe experienced its highest mortality rate due to flood in 2021, with Germany and Belgium in July seeing 240 deaths and $43 billion in damages, the worst since 1985. The pattern of intensive and excessive flooding continues a course of climate instability and unprecedented destruction and deaths from 1993 in the Netherlands and Belgium, many parts of Europe in 2002, Bulgaria in 2015, France, Italy, Spain, Slovenia, Bosnia, Spain, and Malta in fall 2021 and winter 2022 (particularly Spain), and England and Wales in February 2022.[67] Turkey and Tunisia, too, suffered from major Mediterranean storms and floods in October 2021, leaving five dead. Valencia on the eastern Mediterranean in Spain was drenched and flooded with "201.1 liters of rain per square meter over a span of 24 hours, the largest rainfall registered for May, 2022 since records began to be kept in 1871," and causing the metro system and travel tunnels to close.[68] In the summer of 2023, it was Greece making world climate news, with some of the world's longest lasting fires that destroyed 150,000 hectares (370,600 acres) of lands and forests from July through August. Making matters worse, following the controlling of the fires at the end of August there, a downpour of almost 30 inches (20 mm) of rain (the same amount that normally falls in a year) fell within a 24-hour period in early September. At least 43 people were killed as a result of fires and floods, as well as other life on the land. Libya experienced catastrophic damage from the same Storm Daniel that overwhelmed Greece. Over 5,000 were killed and 10,000 were missing, along with thousands of other creatures, after two dams collapsed in Derna on September 12, 2023. Morocco suffered one of its worst disasters on September 8 when a magnitude 6.8–6.9 earthquake hit Marrakesh at the foothills of the Atlas Mountains, killing 2,100 people. The weather apparently is no longer "cooperating" with forecasts and predictions. The big question is why there's been such a radical shift even with the Earth heating in unprecedented ways principally due to fossil fuels, gas, and mining extraction. Are the Earth, the Sun, the Water, the Wind, and other weather spirits simply acting up, or is something else at play?

As witnessed over the past thirteen years particularly, the toll of capitalist-industrialist-caused Earth heating and shocking extremes in protracted drought and flood deluges is all-encompassing. The nations of the South that normally suffer these catastrophes—like Haiti's deadly earthquake in 2010, Syria's equally horrific 7.8 magnitude quake, and Turkey's quake in February 2023—are not exclusively undergoing cataclysmic weather conditions. Europe and the Americas are now all in the same boat—as the Indigenous Indians have said, "We are all Indian now!" No nation or system is immune from this devolution and radical imbalance in weather conditions with extremes of every kind.

Texas experienced its worst temperature freeze in recent history in February 2021, never witnessed before except for the last major freeze in 1899. The freeze was caused by a cold Arctic front that swept across Texas, Mexico, and parts of Georgia during winter storm Uri. Temperatures dropped to below and just above zero and into the teens and low thirties from February 11–20, 2021, in virtually all 254 state counties, resulting in 246 people dying from hypothermia, fires, carbon monoxide poisoning, auto collisions from freezing roads, and other triggered conditions that saw power outages for several days for at least three million people. The unprecedented freezing of gas and energy pipes essentially shut down ERCOT, the state's electricity grid, for days, since the network of electricity and gas pipes was not adequately winterized and insulated to withstand such freezing temperatures.[69] It's critical to note that one of the catastrophic results of weather geoengineering is the shocking imbalance and production of cold and hot fronts at various times of the year, sending temperatures skyrocketing in winter, like 70°F in Alaska in the extreme north of Turtle Island.

The WMO report notes that from 1970 to 2019, weather instability, climate change, floods, and water hazards were responsible for "50% of all disasters, 45% of all reported deaths and 74% of all reported economic losses."[70] The report notes that the toll of these disasters has been catastrophic, "killing 115 people and causing US $202 million in losses daily." Ninety-one percent of these disasters were in underdeveloped countries, and droughts caused 650,000 deaths, especially in Africa and Asia. Destructive storms with tornados, tsunamis, and other climate instability resulted in 577,232 deaths, floods caused 58,700 deaths, and unparalleled spikes in temperature caused 55,736 lost lives. Eleven thousand disasters cumulatively caused $3.4 trillion in damages all over the world. It is also revealing to realize that the severest hurricanes from 2012 through 2017, were deadliest in 2017. In that year, Hurricane Harvey caused 88 deaths and almost $100 billion in damages in the Texas region; and Hurricane Maria caused 2,975 deaths in Puerto Rico, 65 in Dominica, 5 in the Dominican Republic, 4 in Guadeloupe, 4 in Turtle Island, and 3 in the U.S.-controlled Virgin Islands, plus $90 billion in damages. Hurricane Irma resulted in 143 deaths and $58.2 billion in damages—over a third of all economic losses from the 10 major climate catastrophe disasters from 1970 to 2019.

In 2019, India experienced the worst flooding in 25 years between June and October with monsoon deluges, killing at least 1,600 people.[71] While there have been

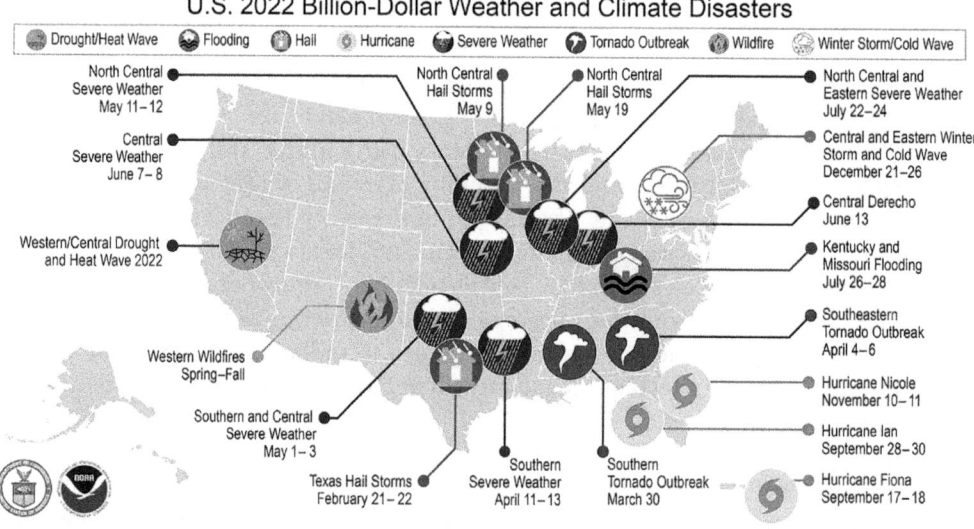

U.S.: billion-dollar and above weather and climate disasters in 2022 (U.S. Department of Commerce, NOAA).

4. Chemical Geoengineering of the Weather 119

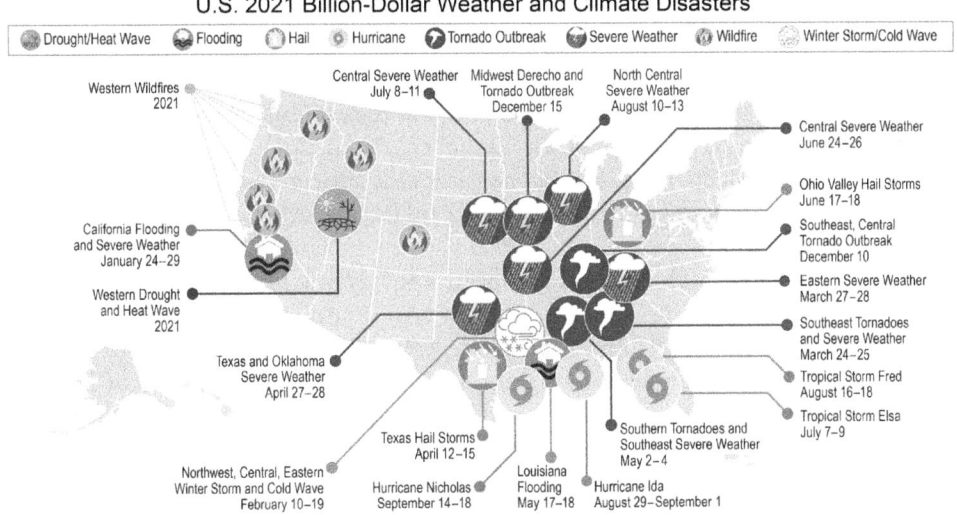

U.S.: billion-dollar and above weather and climate disasters in 2021 (U.S. Department of Commerce, NOAA).

climate-caused or related disasters globally, the largest and costliest climate change disasters and shocking spikes in high and low temperatures or low or torrential rain downpours have been in Turtle Island.

In Russia, 100 fires raged across the Republic of Sakha (Yakutia) that burned 125,600 hectares (310,364 acres) in August 2023. Seven hundred fires in the same region since January this year burned 1.2 million hectares of forest, part of the annual fires experienced in this extremely cold region each summer.[72] In May 2022, widespread brush fires in the Krasnoyarsk region of Siberia caused by speedy winds reaching 70 miles per hour destroyed forest and the ecology in parts, with 7 people dead, 14 hospitalized, numerous people injured, and 480 buildings in the province receiving severe damage, of which 350 were razed to the ground and turned to ash.[73] On May 21, 2021, devastating floods caused hundreds of deaths, leaving 2 million stranded in northeast Bangladesh, including many Rohingya refugees, with 57 dead after the Barak River flooded. In neighboring India, 47 people perished in floods and landslides, of which 14 deaths were in northeastern Assam, where 3,200 villages with 850,000 people were seriously affected.[74]

The climate meltdown persists in 2023 and will continue to escalate given all recent and current indicators as delineated in this section of this chapter. It's good to see over 1,000 scientists from mainly Europe and the U.S. engage in civil disobedience and chain themselves to the doors of corporate office and government buildings, including the White House, on April 6, 2022, to demand immediate action and addressing of cataclysmic climate change.[75] These protests will certainly need to be intensified and expanded all over our Earth Mother and are overdue. Resistance is needed on every level by children; youth; women; workers; and individuals from the environmental, social justice, scientific, and academic communities, led by the Indigenous guardians of our Earth Mother, through constant prayer and ceremony with Nature Spirit Forces determining our lives.

Very Lethal Dangers of Weather Geoengineering and the Accelerated Undoing of Life Itself

The world needs to awaken to the very lethal effects of weather geoengineering and its accelerated pace of destruction of the ozone and the accompanying poisoning of the Earth with injection of highly toxic and lethal chemicals high in the sky illuminated earlier, which leave deep, long-lasting chemical cloud formations as witnessed in the summer and throughout the year, particularly in Tucson, Arizona, the western states, and across the country.

Multiple unpredictable, unprecedented apocalyptic weather conditions have irrupted around the world. The explosion of a pyrocumulonimbus cloud that was triggered from intense heat from the monumental wildfires around Savon in so-called British Columbia on June 29, 2021, is attributable to the exacerbation of heat and unstable climatic conditions, with weather geoengineering being a major factor. These kinds of clouds have been occurring with increasing frequency, especially over areas with raging summer fires in Turtle Island and Australia.

The role of capitalist industrialism grounded in oil drilling, mining plunder, and extraction of what belongs to the Earth, coupled with military corporate weather geoengineering are key factors in the constant eroding of the protective ozone layer and have been documented by a former NASA contract engineer. The NASA engineer, who insisted on anonymity, explained on www.geoengineeringwatch.org that other climate scientists and engineers are so attached to their materialistic and monetary way of life that they are afraid to expose the truth of the serious ozone depletion and hole in the northern hemisphere Arctic. Amazingly, this hole has become as large as the ozone hole

Chemical spray trails in Tucson, Arizona, January 2022 (author photo).

4. Chemical Geoengineering of the Weather

Pyrocumulonimbus cloud from fires over the Alpine National Park, Australia, during the 2019–2020 fires (Merrin Macleod, Wikimedia Commons).

in the southern Antarctic, making "ozone recovery" fictional. He indicated that this evasion of truth is part of a pervasive "scientific" ideology to suppress the fact that the ozone layer is disappearing, rather than regenerating from the heavy depletion caused by the use of chlorofluorocarbon (the product of fluorine and chlorine, two very toxic chemicals combined with carbon) aerosols and sprays over the past century. The public and the world at large are being deceived into falsely believing that things "are not as bad as they seem" and "are recovering" from atmospheric degrading and deterioration. The arrogance of scientists and other "climate experts" from leading U.S. universities like Stanford, Rutgers, Cornell, Harvard, and Calgary, claiming that weather geoengineering is viable though there *may be* some serious "side-effects" and that this is a price "well worth paying" contradicts science's supposed function of enhancing life.

Jeopardizing food production globally is one of the results of geoengineering since chemical sprays affect and destroy food crops and trees, along with other sectors of the ecological food chain. Impoverishment is endemic in many of the nations of Africa, Asia, and Latin America and the Caribbean today given protracted drought and rainfall scarcity, reduced soil productivity due to increased aridity, high ocean acidification due to warming seas and ocean currents, and now the NATO-Russia conflict over Ukraine that has caused energy prices to rise astronomically. The *Congressional Hearing on Geoengineering: Parts I, II, and III* before the Committee on Science and Technology House of Representatives on February 5, 2009, February 4, 2010, and March 18, 2010, indicated such, *inter alia*:

Chemical spray trails showing dried oleander tree leaves witnessed for the first time, Tucson, Arizona, June 8, 2022 (author photo).

> Food and Water Security—A large-scale initiative impacting weather patterns could greatly modify the precipitation patterns in particular geographic areas, jeopardizing local food and fresh water supplies for local populations. For example, drought incurred by unforeseen impacts of artificial cloud formation could suppress crop growth. Poor [impoverished, to be accurate] and developing [underdeveloped, to be accurate] countries may be particularly susceptible to such impacts.[76]

What's clear from this detailed congressional hearing report is precisely what the NASA contract engineer Ray (his last name is omitted for confidentiality), who has conducted research on ozone depletion and compiled UV data of radiation of 250 to 300 nanometers, for example, on April 11, 2020, confirmed. Official government sources like the U.S. Food and Drug Administration (FDA), whose purpose is being "responsible for protecting the public health by assuring the safety, efficacy, and security of human and veterinary drugs, biological products, medical devices, our nation's food supply, cosmetics, and products that emit radiation," claim that "UVC radiation from the sun does not reach the earth's surface because it is blocked by the ozone layer in the atmosphere."[77] It's a flagrant lie that there's no ultraviolet radiation reaching the Earth with wavelengths of between 250 and 300 nanometers since NASA engineer Ray has publicly documented such UV radiation of 250 to 300 nanometers reaching Earth that penetrate the skin by at least 2mm and have lethal implications for all life.[78] Not only is the rapid and imminent depletion and diminution of the ozone layer affecting life on the Earth land mass, but its most severe and lethal effects have destroyed much of the sea phytoplankton that are responsible for 66 percent of all oxygen consumed, an indispensable link in the global food chain. The result of the radical erosion of the ozone layer is what has been called the "Canfield Ocean" model that will see the radical dissipation of oxygen, erasing many lives in the ocean and on land dependent on oxygen in the early 2030s. The ionizing and "microwaving" of the atmosphere is the result of U.S. military and industrialist "science," like the Lucy Alamo Projects on Hydroxyl Generation and Atmospheric Methane Reduction, which involves "a radio/laser system for destroying the first hydrogen bond in atmospheric methane when it forms dangerously thick global warming clouds over the Arctic."[79] Such technology employs "Radio frequencies" that "are used in generating nano-diamonds from methane gas in commercial applications over the entire pressure range of the atmosphere up to 50 km altitude." Nanodiamonds are made by "by detonating an explosive in a reactor vessel to provide heat and pressure."[80] Hundreds of meters below the Arctic Ocean floor lie massive reserves of frozen methane that's intrinsic to the fragile balance of sea and land ecosystems, the undoing of which could inevitably occur with increased sea warming and destabilization of the Earth and Sea naturally balancing both carbon and methane. This indispensable balance would now be totally disrupted with capitalist industrialist emissions of both carbon dioxide and methane since such carbon emissions (36–38 gigatons annually) are over twice what the Earth can comfortably recycle (16–18 gigatons).[81]

Methane emitted from coal, gas, and petroleum extraction—along with nitrous oxide from agricultural production fertilizers from the U.S. and the capitalist industrialized countries especially—generates an atmospheric heating 100–1,000 times more lethal than excessive carbon dioxide released. In this recent period, "the whole northern hemisphere is now covered by a thickening atmospheric methane global warming veil from Arctic methane emissions at the level of the jet streams, which is spreading southwards at about 1 km a day and already totally envelopes the United States," with no indication of methane emissions declining.[82]

The following diagram highlights sources of 2020 U.S. methane emissions.[83]

These scientific endeavors and experiments focused on weather engineering by projects such as the High Frequency Active Auroral Research Program (HAARP) are funded by the U.S. Air Force, Navy, and the DARPA, the latter of which also funds U.S. bio-weapons labs in 25 countries around the world, particularly in eastern Europe and Africa.[84] HAARP is a University of Alaska, Fairbanks project and, according to the university's website, involves no U.S. Navy or military agencies. This is highly unlikely considering the original funding sources and U.S. military and aerospace defense demands especially in 2023. HAARP is extremely dangerous in that the ionosphere consisting of atoms converting to positive ions, a very fragile umbrella over the atmosphere, is being disrupted and destroyed by perturbing energy rays blasted at the sky with the 180 antennas over a 30-40 acre (12-16) hectare rectangle under the rubric of research, followed by measurements of the effects.[85] This perturbation contributes toward the heating of the ionosphere because it produces a microwave effect. The ionosphere, connected to the atmosphere, together with the ozone layer, blocks the ultraviolet rays of the sun and other dangerous rays from outer space, preserving life in our Earth sphere. Eroding this very fragile balance of the ionosphere and atmosphere inevitably results in emission of such energy rays like extremely low frequency (ELF) electric waves back to the Earth, accelerating warming and devastating life on our Earth Mother. Such corrupting intervention in the natural cycles, streams, and currents of the Air, the Earth, and Sun radically disrespects these ever powerful Spirit Forces of Nature. Resultantly, these processes destroy insect, bird, sea life, and other movements of creatures dependent on natural air streams, including humans,

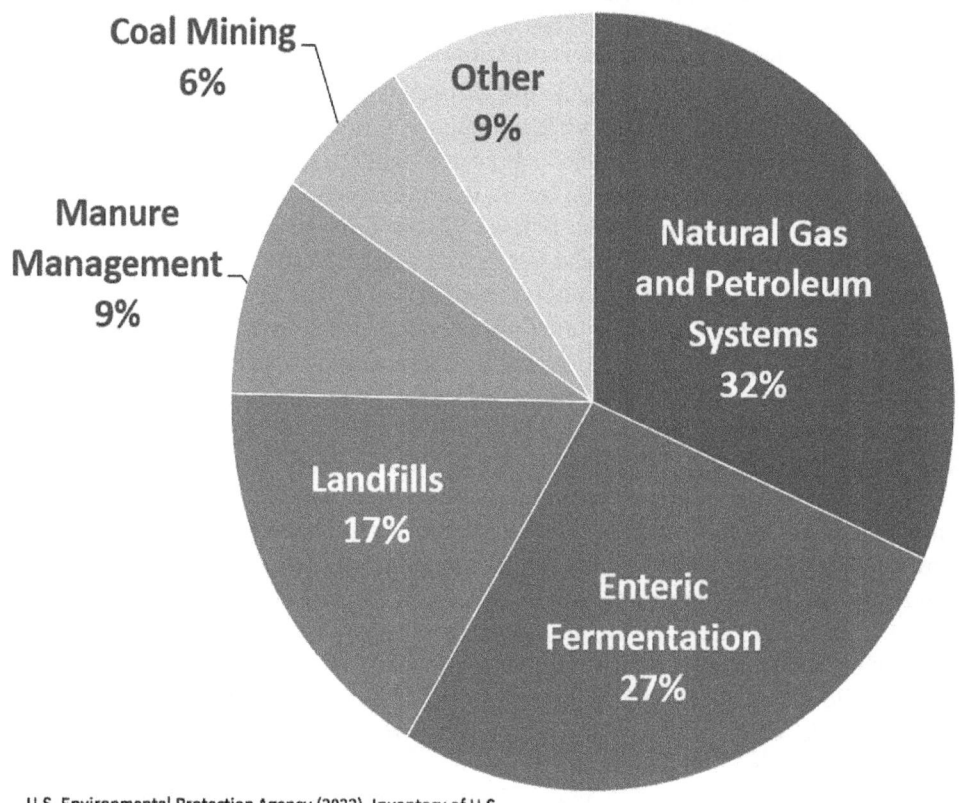

U.S. Environmental Protection Agency (2022). Inventory of U.S. Greenhouse Gas Emissions and Sinks: 1990-2020

Sources of 2020 U.S. methane emissions (EPA, "Overview of Greenhouse Gases," https://www.epa.gov/ghgemissions/overview-greenhouse-gases).

through radiation and radio beams and signals. They only sow more destruction and disharmony and constitute lethal threats to the integrity and preservation of the fragile, harmonious, and intricate web of life.

For instance, an anti-hail cannon, "a cannon device that fires shock waves up into thunderclouds to prevent hailstones from forming," patented by the U.S. under number 5,445,321; laser "rainmakers," which shoot laser waves into clouds to supposedly function as "cloud seeding" chemicals, where tiny droplets can be transformed into rain; "making clouds cry," using small airplanes to spray silver iodide and similar chemicals to imitate ice nuclei so that what's considered "disruptive weather" is prevented; "flying suicidal" robotic planes into low altitudes of a tornado for "stopping tornadoes cold" to mitigate the power and thrust of hurricanes, patented by the U.S. under number 7,810,420; and "halting hurricanes" by using "fleets of ships to mix warm surface Water with colder Water from the ocean's depths" to "help prevent billions in storm damage and save lives," under U.S. patent number 20,090,173,386, are some of the methods of

Electromagnetic Radiation and Hubs Harmful to Ecology and Humans Alike

The lethal effects of chemicals in industrialism as indicated in the first section of this chapter linger in more ominous ways globally today. The threads of radiation and chemical warfare have come full swing with the total saturation of Turtle Island and the world with "wi-fi," including the latest high-speed, highly toxic C-band "5G" run by billionaire wireless corporations Verizon and AT&T. These corporations that "Warn Shareholders of Risk but Not Consumers or Neighbors" have inserted liability waivers in their phone contracts so that they are shielded from claims of "pollution" from these wireless devices that may have caused cancer and other ailments.[87] In 2010, Lloyd's Insurance of London, a leading global insurance conglomerate, issued a report that "the danger with EMF [electromagnetic fields] is that, like asbestos, the exposure insurers face is underestimated and could grow exponentially and be with us for many years."[88] This truth has been conveniently obscured and suppressed in the capitalist owned and controlled media establishment which still insists that the effects of EMF and EMR (electromagnetic radiation) studied are inconclusive. We are all scapegoats at the altar of greed and technological expedience.

5G has undermined aircraft altimeter low visibility landings by the FAA, even though the private FCC approved such safety-threatening technologies.[89] While the U.S. public and the world generally gullibly swallows the propaganda that these electromagnetic radiation emitting technologies are benign and are signs of evolutionary industrialized "progress" in the 21st century, they are also unwittingly subjected to daily lethal radiating emitting mediums. Expanded and intensified Sun-dimming sky chemical sprays by military industrial corporations like RTX Corporation compound this lethal, industrial, technological chemical exposure. The imminent dangers of electromagnetic radiation documented in the poignant 2017 documentary *Generation Zapped* (produced by Sabine El Gemayel) substantiate this serious environmental, health, and social concern of now growing numbers of people.[90] The explosion of high-speed 5G wi-fi technologies is unquestionably destructive for us all, especially for children, because children's physiology (particularly the brain) is still in formative stages. Wi-fi and cell phone towers and hubs disrupt this growth because of the heating impact on and damage to blood cells and other body parts, and triggering of serious health impairments like "breast, cervix, lung, and esophagus cancers" and "Type 1 diabetes mellitus." These dangers are now evident from consistent scientific studies published in many peer-reviewed journals in the United States and in Europe, especially over the past 13 years.[91] Electromagnetic radiation (EMR) generated from cellphone towers and phones using wireless technologies have also found that "school-aged adolescents exposed to higher levels of RFR exposure had delayed fine and gross motor skills, spatial working memory, and attention in comparison to those exposed to lower RFR levels."[92] Equally deadly, electromagnetic radiation from cellphone towers destroys birds, bees, flying insects, trees, and other forms of life, as documented in research on the effects of non-ionizing electromagnetic fields (EMF) on flora and fauna and rising ambient EMF levels in the environment. These forces are major contributors to ecocide and toxicity and reduced life in the Air, Earth, and Water. Unsurprisingly, though the WHO has ruled that EMR is carcinogenic, it ambiguously states that the long-term effects of such exposure haven't been fully established, demonstrating the WHO's obeisance to large pharmaceutical conglomerates and capitalist foundations like the Gates Foundation. Again, as with the commodification of life, including the organisms and vital living systems that are interwoven with and indispensable to us being on the Earth, court rulings in many industrialized countries, including Aotearoa/New Zealand, have declared that internet technologies are more valuable than trees, substantiating just how absurd the capitalist world has become.[93] Trees are key in the need to both address the historical effects of forest degradation globally and in ecological and environmental preservation and in absorbing carbon dioxide from excessive industrial emissions.

Lena Pu's research on the deadly health dangers of massive wi-fi hubs and routers at U.S. schools with 60ghz technologies during the height of the Covid-19 pandemic demonstrated this fact.[94] Wi-fi computer access for learning and instruction is virtually standard, thanks to high-tech companies providing free computers in many U.S. school districts that are dependent on supplementary educational funding often from private enterprise. Little investment for grounded ethernet connections exists in K-12 education.

Further, the growing deficiency in basic literacy and comprehension skills with the expansion of wireless technologies and the explosion of the internet (which has made reading printed books virtually non-existent in many places) need to be redressed so that such technologies are heavily monitored for health and social reasons particularly in children's educational circles. For healthy living, we were all meant—children especially—to be outdoors most of the time, unlike capitalist industrialized culture, which requires the opposite in employment and educational venues. Massive office complexes generally have no open windows and depend on artificial air conditioning and electrical lights, consuming unnecessary and excessively high levels of energy.

Indigenous Cultural Perspectives on the Crisis of Ecocide, Earth Heating, and a Chemically Based World

The pre-colonial and pre-western industrialist delicate balance of the temperatures of the Sea, Earth, and Air, which has preserved life in these various sectors for billions of years, is now being destroyed in escalating proportions. The acidification of the oceans and the destruction of biodiverse coral reefs like the Great Barrier Reef off the east coast of Australia (where cancerous lesions were found on 15 percent of coral trout) and other areas of the world are just two example effects.[95] Australia sits under the largest hole in the ozone layer, and just like fish contracting cancer from the now penetrating rays of the Sun on Earth and Sea, two of every three adults will suffer from skin cancer before age 70 in that colonized Indigenous people's land.

Indigenous cultures have always maintained and insisted on the reverence of all life and the ultimacy of the Earth, Sun, Moon, Sky, and Sea—all intrinsic to the teetering balance needed for all life. *An injury to one, is an injury to all,* Indigenous cultures hold, and a destruction of a tiny element of the interwoven mosaic that industrialist capitalism considers "just one of those unavoidable side-effects" causes the destruction of the whole and results in the undoing of life itself. Yet, Indigenous peoples also consistently explain that *the Earth as the Mother of Life* does not abandon the most vulnerable ones who are Her/Their children, the four-leggeds, birds, insects, creatures of the Sea, plants, and the rest of Nature. These creatures simply live according to their nature—an example that we humans desperately need to follow, too. Unlike the grounding principle of Indigenous cultures that cherishes and embodies humility and the simplicity and naturalness of life, unbridled technological industrialism strives foolishly and constantly to "make Nature better." The net result is destruction and extinction facing us all. Things only get worse from the persistent anthropocentric efforts to improve on what the Creator/Creation/Creating has bestowed upon us all. The Earth is Ultimate, and She/They owns/own us, Hataali Jones Benally repeats in his instructive teaching about these Spirit Powers from which we originate and to which we long and inevitably return. In essence, we are nothing but specks of dust from which we were all made, returning to dust at the end of our physical existence to form new life in other forms, albeit sacred and significant. The underlying problem is that the industrialist capitalist system has no ability to openly acknowledge its core and foundational limitations. Our arrogance, hubris, egotism, and self-illusion are core obstacles to the solution to unending the irreversible climate collapse and ecological annihilation we all suffer today. While industrialist technology may provide palliatives in providing temporary relief akin to a pill furnishing relief for a headache, it overlooks the root cancerous tumor: deification of the human and our intellect. The ideological solutions of capitalist, high-tech interventions of chemical geoengineering; poisonous aerosol and pesticide spraying on life; GMO seeds, plants, insects, vegetation, and four-leggeds; chemical infusions; robotic mimicking of the human through artificial intelligence (AI); and endless bloody wars as antidotes for peace and stability will only end in oblivion, as the ancient Indigenous seers have prophesied.

Indigenous people have constantly warned that incessant wars on the Sun, Earth, Air, and Sea cultures and the accompanying destruction and deterioration of life are indications of a deeply troubled and lost humanity that requires spiritual cleansing and returning to traditional ancestral ways of living simply so that others may simply live. Indigenous cultures understand our very deep roots in the Earth and that chemicals, minerals, and other industrial metals that entail drilling, fracking, excavating, and uprooting for capitalist greed and materialism are core violations of the spiritual integrity of the Earth. The result of almost three centuries of such consistent plunder in the name of "human progress" and "civilization," particularly from the early 20th century and through the first two decades of the 21st century, is stark for the world to behold: growing ecocide; poisonous contamination of land and Water and accompanying wars; uncontrollable Earth heating and forest fire eruptions; prolonged and protracted drought; drying up of rivers, streams, and waterways; melting, fragile icecaps; accelerating ocean level rise; depletion of the protective ozone layer; catastrophic tornados, hurricanes, and tsunamis whose intensity is unprecedented; proliferating disease and pandemics; GMO food and organisms; pervasive legal and illegal drug consumption; violence against people of diverse religions, cultures, languages, classes, gender, and places of origin, especially climate, economic, and political refugees; prison torture and repression; homelessness of the destitute, unemployed, and underemployed compounded in the U.S.; exploitation of all workers; and rampant depression, especially among youth. The list grows with each passing day.

Hataali Jones Benally and his daughter, Jeneda, who are traditional Diné/Navajo healers, revealed after a spiritual analysis following the historic solar eclipse in Turtle Island on August 21, 2017, that the Sun, Moon, and Earth, who were all aligned harmoniously in direct conversation with each other in the sky unanimously agreed that they were *against* humans who have become so uncontrollably destructive of all forms of life in their arrogance. The result they foresaw is catastrophe everywhere, with many more lives taken. Shortly thereafter, on August 25, Hurricane Harvey struck the southeast with a fury not seen in years, causing significant death and destruction. The *Air,* which is the basis of all physical life, the *Water,* and the *Earth,* which are life-giving, have become vitiated by chemical industrial injection and pervasive use. In all twelve western states of Turtle Island, drought and lack of moisture in the Air are norms. Rain is scarce and people suffering from dryness allergies experience aggravated health problems. Hataali Jones also stresses that this sacred land, Turtle Island, "neither speaks nor understands … English." What he is referring to is the traditional way of learning, living, and knowing, principally mediated through speaking, communicating, and praying in one's ancient Indigenous ancestral language, the core conduit for retention of ancestral cultures rooted in the Earth and the Spiritual Universe. Reclamation of our ancestral Indigenous languages is indispensable for being rooted and re-rooted in our Mother Earth. Moreover, Indigenous ancestral cultural knowledge and nature ceremonies are key in environmental and life preservation. For example, outside *Skukchon* (Tucson), Arizona, is *Babad Dawag* (Frog Mountain in English), or Mount Lemmon, sacred to the ancient Tohono O'odham people from early times to the present and so named for the spiritual frog that lives 9,000 feet high in the mountain who determines the weather over the region. The violation of these spirit beings is deadly. The spiritual uprooting from and rejection of our roots in the Earth Mother has had very lethal consequences, especially with weather systems where the desperate resort to chemical engineering and other high-tech measures have only made living conditions on Earth much worse. Indigenous people have consistently and unflinchingly echoed the clarion call globally: "Leave our Earth Mother alone.… Let Her/Them be!" This Indigenous consciousness and reverence for the Earth led Bolivia under former president Evo Morales from the Aymara nation to insert "the rights of the Earth" into the country's constitution, as did Venezuela and Nicaragua.

Environmental and ecological activists are critical in the struggle to address the crisis of the Earth heating and both climate and ecological collapse and are called to work closely with Indigenous people as allies since we all have a common cause of love and reverence for the Earth and all life. It will require all our energies coalescing to undo the extensive and even irreparable damage done to our Earth Mother. It's very encouraging to see allies of Indigenous people stand together with our Land Guardians in the northwest to protect the salmon against dam construction and against mining in places like *Peehee Muh'huh* (Thacker Pass) in Nevada, where Cheyenne, Bannock, Shoshone, Arapaho, and other Indigenous activists are resisting lithium mining by General Motors with $650 million invested.

Summary and Conclusion

This chapter has provided extensive documentation of the history of chemical geoengineering, the deification of western capitalist technologies that flagrantly disregard the sanctity of the Earth, Air, Water, Sun, Moon, and all beings, and how capitalist militarism and violence has sown the seeds of destruction over the past few centuries in particular, evidenced in ecocide, climate catastrophe, and fragile life collapse. The chapter further highlighted how Indigenous peoples' ancient ancestral and spiritual knowledge and wisdom, while dismissed as unviable and unrealistic ways of living by the industrialist capitalist world, signify the key to the preservation of life and to living in balanced, sustainable ways so that all life forms and their inalienable rights to life are preserved and not persistently destroyed.

The only way forward is to return to the ways of the Earth Mother, our real Home, regardless of where and who we are. What language we speak, how much money or material commodities we possess, what gender or class we are, what religion we subscribe to, or whatever philosophy we hold dear do not matter in the final analysis, because our Mother Earth is home to and for us everywhere. Like the rest of the natural world, we need to fulfil and follow the ancient ancestral instructions for living in harmony and decency. There is no other way than the traditional way. As Hataali Jones teaches, "Wherever you stand on the Earth, here or anywhere, you're home." "*Ninguna persona es ilegal!*" many chant in Spanish in U.S. border towns and everywhere in Turtle Island—no person is illegal. Neither legal nor illegal migrants exist because we are all unconditional children of the Earth. No non-citizens, either, since we are all citizens of our Mother Earth. The Earth and the infinitely powerful Spiritual Universe assure us of protection in living out this timeless truth.

The final chapter will discuss the urgent need for decolonization and liberation of our persons, societies, beings, and educational institutions from capitalist ideologies and enslaving debt that have destroyed young learners here and everywhere. Such is essential for proposing ways of living with reverence for our dear Mother Earth and the Infinite Spiritual Universe in a post-capitalist world with capitalism's inevitable collapse in the 21st century, equipped with the educational knowledge from this chapter, preceding ones, and the next, followed by the closing epilogue.

5

Capitalism's Persistent Devaluing of Public Education and the Need for Restructuring and Reorienting Education

> "All of these things were designed ... to take the Aboriginality out of Aboriginal kids."
> —Mick Dodson, Social Justice Commissioner, Report of the National Inquiry into the Separation of Aboriginal and Torres Strait Islander Children from their Families, in *Bringing Them Home: Separation of the Aboriginal and Torres Strait Islander Children from Their Families, 1997,* https://www.youtube.com/watch?v=Sl82VMuuKI0

> Education is a term often used synonymously with schooling, though I don't see them as one and the same. For the most part, schooling is institutional; offering a mandated curriculum as decided and approved by federal and state governments, delivered by people who have the qualifications deemed to be legitimate (Bishop, 2020). Think about it, every day we send our young people to school. In Australia, we have to, it is compulsory. Six hours a day, five days a week, 40 weeks a year, 13 consecutive years. For Indigenous people, this is a long time to be sending our babies away to an institution we know to have caused great harm to many of our communities.
> —Michelle Bishop, "Indigenous Education Sovereignty: Another Way of 'Doing' Education," *Critical Studies in Education*, November 30, 2020.[1]

Introduction

This final chapter on capitalism's colonization of education—in the U.S. in particular and the globalized world in general—is part of a concerted educational effort and struggle to highlight the gravity of this crisis facing us daily. The globalized capitalist culture has entrapped us to the point where we have accepted without scrutiny an educational system that is structured and oriented toward serving capitalism and materialistic accumulation. However, it is well beyond high time to pause for a while, take stock, and critically and clearly analyze what has happened to us all regardless of where we are located in the social spectrum. We have been conditioned by an educational ethos that is Earth-ignorant, anti-impoverishment and anti-working class, patriarchal, elitist, money and profit-oriented, and capitalist engineered, shaped, and controlled. The educational system, in essence, has denied the diversity of knowledge systems and philosophies of Indigenous people, whose lands we all live in, Black people, and other communities of color.

This chapter will discuss the role of billionaire and ruling class capital in defining what we call education so that we become cogs in the enslaving capitalist machine. This system has destroyed every segment of our society, from the little ones to the elderly and struggling-to-retire folks, many now saddled with debt while being forced to embrace an educational orientation that has little to do with the core ingredients for life:

- knowing our respective histories, cultures, and languages so that we can understand our own cultural identities and purposes in life;
- honoring, respecting, and valorizing the Earth Mother, the infinitely powerful Spiritual Universe and sacredness of all life;
- living in harmony and coexistence with all life, including the four-leggeds, birds, insects, plants, trees, mountains, hills, valleys, rivers, rocks, stones, oceans, streams, and the rest of the natural world;
- sharing and caring about one another while being rooted in ancestral language and culture;
- acquiring personal and community knowledge of how we need to live in a world where life is fraught with perennial conflicts and already facing extinction in so many quarters.

The final section of the chapter will demonstrate how a restructuring and reorientation of the current educational system and curricula are needed, so that we are all equipped and empowered to face the ever-increasing problems facing our beautiful but torn asunder world, especially for the children and young ones of our diverse societies.

Capitalist Control of U.S. Higher Education and the College Student Loan Debt Complex

Koch Industries, owned by billionaire brothers Charles and David, is heavily invested in controlling the political process in the U.S. as well as the educational sphere. Majority shares in the corporation are owned by the Koch brothers with investments in various areas such as fossil fuels and energy, refineries, chemical production, paper products, and global commercial trade and finance. The multinational corporation has directly funded organizations with the goal of placing certain U.S. leaders in office, including Donald Trump, who was elected in 2016.[2] The Koch family is the third wealthiest family in the country, with $135 billion in net worth. The Charles Koch Foundation has funded conservative think tanks and climate change denying forums on public and private university and college campuses around the U.S., pumping over $458 million into these institutions. This fulfills Charles Koch's 1974 plan of charting "the educational route" in promoting singular capitalist ideology.[3]

To date, the Charles Koch Foundation has funded Harvard University, Duke University (Center for the History of Political Economy), the University of Arizona in Tucson (Center for the Philosophy of Freedom), George Mason University, the Catholic University of America, Florida State University, Creighton University in Nebraska, Troy University in Alabama, and other institutions, providing various levels of grant funding in six figures or more to shape deregulation economic policies and promote certain political and philosophical ideologies.[4] The Koch family has amassed a fortune from unlawful activities that pillaged oil from Indigenous lands in Turtle Island beyond the quota allowed, fracking in the tar sands of Alberta (which has lethally contaminated the Athabascan River), poisoning Indigenous people's water and food and killing tens of thousands of fish and other forms of life in the river, and engaged in other lethal pollution activities that flout environmental laws.[5] The government has looked the other way when it comes to requiring legal accountability on the part of Koch Industries, and many environmental charges leveled at the conglomerate have been conveniently swept under the rug by the EPA. The corporate financial and energy behemoth controls "at least four oil refineries, six ethanol plants, a natural-gas-fired power plant and 4,000 miles of pipeline," and is one of the largest carbon polluters in the country, more than General Electric and International Paper combined, as investigative *Rolling Stone* reporter Tim Dickinson notes. Its massive extraction and refining of fossil fuels operations (it provided five percent of the U.S. refined oil production in 2014) and other energy sources are not confined to pollution, but also includes corporate negligence that saw the incineration deaths of two teenagers in Texas from an explosion in Koch's pipeline in 1999 as an example. However, the U.S. government is equally culpable for allowing such practices to continue.[6]

Some noted Ivy League schools have divested from fossil fuels as part of combating climate change, like Harvard University, which sits on a $35 billion endowment, the largest in the world. The endowment accrued from acquiring cheap real estate from South Central Los Angeles as the result of depressed housing prices and forced sales of mostly Black homes due to the CIA-influenced intense drug wars there in the 1980s. Yet, many such institutions have continued to receive funding from the Koch Foundation for the sponsorship of lectures that promote fossil fuel deregulation while claiming to be "environmentally responsible." The Koch establishment lobbies against state legislation across the country that requires more stringent fossil fuels standards.[7] Jasmine Banks from the organization *UnKoch My Campus* highlights the irony when she writes, "The money these universities accept from the Koch network is, in some ways, more damaging and insidious than the universities' investment in fossil fuel companies," including ExxonMobil.[8] In May 2016, the state government of Arizona joined this capitalist ideological funding maze when it provided $2 million to the University of Arizona in Tucson, to which the Koch Foundation added another $1 million to fund the right-wing deregulation and capitalist-promoting Center for the Philosophy of Freedom.[9] Arizona State University (ASU) in Tempe received $2 million in state funding to found the Center for Political Thought and Leadership, supplemented with $1 million from the Koch Foundation. This was in the aftermath of the establishment of the Center for the Study of Economic Liberty at ASU, which had already received $3.5 million from the same foundation.

Through his Foundation to Promote Open Society, billionaire George Soros has funded Bard College, Georgetown University, Wesleyan University, the University of Maryland, Fordham University, Ohio State University, and Willamette University, among others.[10] While these may sound "purely" philanthropic, they also conceal the real agenda behind these large tax write-off contributions that aligns with both the national security state apparatus and the promotion of "free-market capitalism" as the ideological bedrock for U.S. higher and K-12 education. Bill Gates has engaged in similar practices so that much of the public sector, particularly disadvantaged communities needing scholarship funding, computer technologies, etc., have become beholden to such billionaires, making it virtually impossible to criticize the root of their disadvantaged status: monopoly capitalism led by the billionaire elite.

Education in the U.S., including public taxpayer education particularly in the last few decades, has essentially become the province of control by millionaires and billionaires through lucrative funding, especially in areas of research and curriculum. School districts, too, often receive substantial technology learning grants from this elitist fringe. It's no surprise, then, that U.S. college presidents have become "CEOs" for all practical purposes, most receiving close to a million to multi-million-dollar salaries befitting of corporate executives, and some hailing from non-academic backgrounds. A breakdown by the *Chronicle of Higher Education* in its 2018–2019 report on college presidential salaries revealed that over 15 university presidents functioning as CEOs received more than $1 million, including institutions promoting social justice like the New School for Social Research in New York, for example.[11]

Chief executives at private institutions like Savannah College of Art and Design, Johns Hopkins University, Thomas Jefferson University, High Point University, Washington University in St. Louis, Columbia University, University of Pennsylvania, Princeton, Duke University, University of Dayton, Harvard University, Brandeis University, Vanderbilt University, Macalester College, Texas Christian University, Northwestern University, Massachusetts Institute of Technology, Boston University, Northeastern University, Worcester Polytechnic Institute, Bentley University, and Tufts University—all considered prestigious private schools—often receive over a million dollars, even while tuition costs are tens of thousands of dollars and rising. In some instances, these compensation packages are more than 50 times the average faculty salary, like at Columbia University.[12] Many of these Ivy League schools are located in areas where economic conditions of the working classes, like in New York (Columbia University near Harlem, for example) and Chicago (the southside or northside), particularly communities of color, are dire, plagued by unemployment, underemployment, and depressed and low incomes. Public universities, too, follow the same corporate capitalist chief executive officer salary trend, even in states where incomes are low in comparison with national median levels, like Kentucky, Alabama, and Arizona, with 17 chief executives receiving close to or more than a million dollars.[13] In November 2023, the University of Arizona was shocked to hear that it had run into a deficit of over $240 million, reflecting mismanagement and fiscal irresponsibility. In response, the administration is considering freezing tuition discounts, lowering financial aid and hiring freezes, even while high-salaried administrators continue to remain ensconced in positions of vast economic privilege compared to the majority working classes of southern Arizona. Financially strapped students (often, but not exclusively, of color) are now forced to pay the price for the folly of the authorities supervising such higher educational institutions.[14]

Big Ten College Presidents' Salaries, 2020[15]

University	President	Salary
Northwestern University	Morton Schapiro	$1.6 million
Rutgers University	Jonathan Holloway	$1.2 million
University of Nebraska	Walter E. Carter, Jr.	$934,600

University	President	Salary
Purdue University	Mitch Daniels	$901,450
Ohio State University	Kristina M. Johnson	$900,000
Penn State University	Eric Barron	$855,228
University of Michigan	Mark Schlissel	$852,346
University of Illinois	Timothy Killeen	$835,000
Michigan State University	Samuel Stanley	$800,000
University of Iowa	Bruce Harreld	$790,000
University of Minnesota	Joan Gabel	$640,000
Indiana University	Michael McRobbie	$639,000
University of Wisconsin	Tommy Thompson	$489,000
University of Maryland	Darryll Pines*	$383,000

*Current salary not available: Pines was hired as president in 2020, but was a dean at the same school and made $383,000

One wonders how such unaccountable state spending can be justified in states and places where college tuition has grown astronomically and hundreds of thousands of middle and working-class students—millions over the years cumulatively—are forced to take out tens of thousands of dollars in loans annually to receive a college education because both state and federal educational funding allocations have been radically reduced, particularly since 2008. There is no accountability to working class and marginalized sectors of society from college chief executives and senior administrators. "Higher" education is now clearly demarcated not because of "higher" academic standards and the quality of educational programs, but higher in terms of executive compensation packages at the presidential, vice presidential, and assistant vice presidential levels. The price of such profligacy is that 45.3 million students are enchained by college debt for life in many instances, as reflected in the over $1.766 trillion cumulative student debt in the U.S. today.[16]

The tightening debt noose around college students' necks signifies a blatant contravention of federal governmental legislation, given that many of the public universities in the U.S. are considered "land-grant" colleges, whose purpose was to "provide students with affordable access to career-oriented higher education in the areas of agriculture, science and engineering, military science, and the liberal arts."[17] Yet, the academic establishment and the "higher education" complex is essentially an exclusive province of life in most capitalist societies, generally elitist, and advances the interests of capitalism, embodying both Eurocentrism and capital accumulation structures. In Turtle Island and other settler-colonial nations like Australia, Aotearoa, Azania/South Africa, Namibia, and Hawaii, universities and colleges were founded and built on dispossessed Indigenous peoples' lands, highlighted by Kiowa nation writer and editor-in-large at *Grist*, Tristan Ahtone, and spatial data analyst Maria Parazo Rose in their work on *Land Grab Universities*.[18] U.S. universities and colleges are fully built on the genocide of Indigenous Indians and continue mega-billion profit extraction based on exploiting dispossessed lands for mining, fracking, deforesting for timber, etc., captured in this poignant excerpt from an article in *High Country News*:

5. Capitalism's Persistent Devaluing of Public Education 135

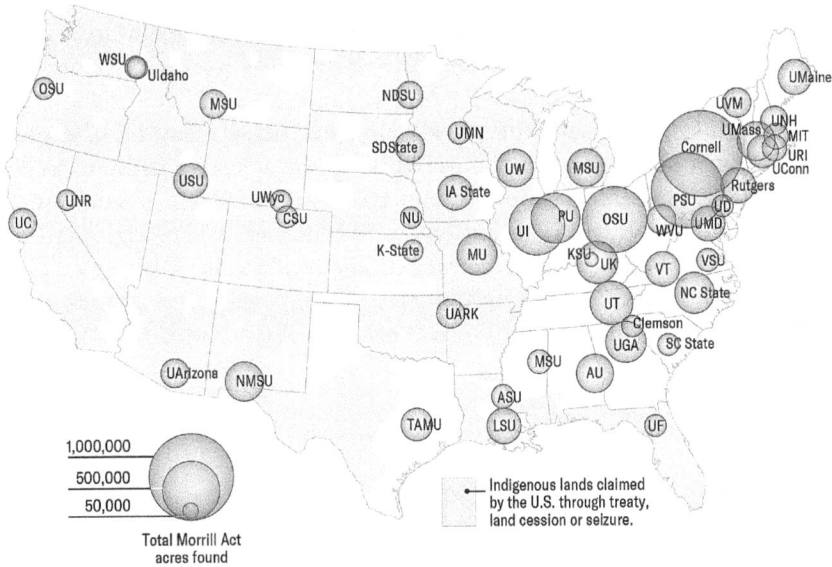

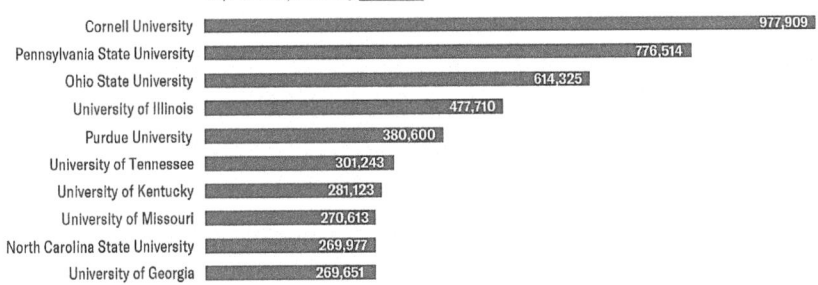

Land-grant universities located in states with large congressional delegations received more acreage (Margaret Pearce for *High Country News*. landgrabu.org).

On August 29, 1911, a Yahi man known as Ishi came out of hiding near Oroville, California. He had spent decades evading settlers after the massacre of his community in the 1860s and had recently lost the last of his family. Whisked off to the University of California's anthropology museum, he was described by the press as the "last wild Indian." Ishi spent his final years living at the museum. When he wasn't explaining his language to researchers or making arrow points for visitors, he swept the floors with a straw broom as a janitor's assistant. In return, he was paid $25 a month by the same university that sold thousands of acres of his people's land out from under him while he hid out in forests and river canyons.[19]

The article explains that, in 1862, President Abraham Lincoln signed the Morrill Act (after Vermont representative Justin Morrill) "that distributed public domain lands to raise funds for fledgling colleges across the nation," yet these now entrenched educational institutions of "higher learning" refuse to acknowledge how they came to be so prosperous after possessing dispossessed Indigenous Indian lands.[20]

Instead, the forced removal of Indians by the U.S. government for cultivation of these colleges is portrayed by these institutions as a generous "donation," celebrated as "abundance" for education for U.S. students to share in the world, conveniently obscuring that these stolen and dispossessed lands functioned as "seed" money for "higher education's" spawning. The horrific and starkly obscene and criminal act of expropriation and confiscation of Indigenous lands covered 11 million acres, larger than the combined land areas of Massachusetts and Connecticut, parceling out 80,000 portions in 24 states. Two hundred and fifty Indigenous nations were thus deprived of legitimate ancestral land rights, and 160 of the land confiscations entailed widescale violence against Indians.

Yet, 12 states still hold Morrill-allocated unsold and unused lands and receive lucrative returns from mineral rights and mining. Fifty-two U.S. universities have

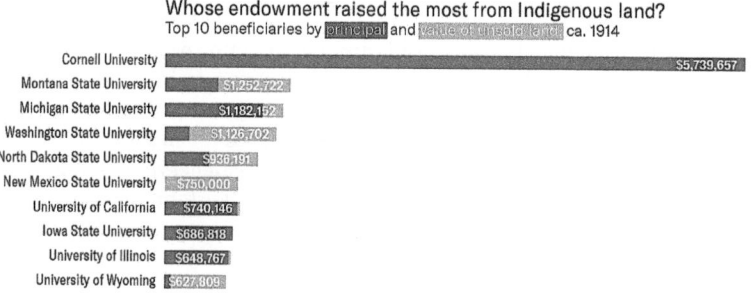

Land-grant university endowments benefited when their acreages were sold and the proceeds invested (Margaret Pearce for *High Country News*. landgrabu.org).

their startup endowments from these stolen land possessions. Over 25 percent of the dispossessed land grabs involved paying no compensation to the respective Indigenous people (a meaningless, valueless $400,000 was paid to other Indigenous nations for millions of acres) who were dispossessed of their lands after living on them from time immemorial and were fully illegally acquired through violence and outright theft void of treaties ratified by the U.S. government.[21] Arizona's "land-grab" university in Tucson certainly follows the exploitative money-making pattern under the cover of being a "land-grant" university, making millions in profits from corporate investments, while raising student tuition each year and still retaining Indigenous cultural and sacred belongings in its museum in violation of the Native American Graves Protection and Repatriation Act (NAGPRA). Worse, in a blatant contradiction of their respective mission statements of promoting knowledge for a diverse world, numerous research universities in the U.S. are involved in nuclear weapons production and promotion, including the leading flagship universities, namely:

1. Aiken Technical College
2. Amarillo College
3. Augusta Technical College
4. Augusta University
5. California Institute of Technology
6. Carnegie Mellon University
7. Cornell University
8. George Washington University
9. Georgetown University
10. Georgia Institute of Technology
11. Johns Hopkins University
12. Kansas State University
13. Massachusetts Institute of Technology
14. Metropolitan Community College
15. Missouri University
16. New Mexico Institute of Mining and Technology
17. New Mexico State University
18. Northern New Mexico College
19. Pittsburg State University
20. Purdue University
21. Roane State Community College
22. Stanford University
23. Texas A&M University
24. Texas Tech University
25. University of Arizona
26. University of Arkansas
27. University of California
28. University of California–Berkeley
29. University of California–Davis
30. University of California–Los Angeles
31. University of California–San Diego
32. University of Colorado–Boulder
33. University of Florida
34. University of Illinois at Urbana-Champaign
35. University of Kansas
36. University of Michigan
37. University of Missouri–Kansas City
38. University of Nebraska
39. University of Nevada–Las Vegas
40. University of Nevada–Reno
41. University of New Mexico
42. University of Notre Dame
43. University of Rochester
44. University of South Carolina–Aiken
45. University of South Carolina–Salkehatchie
46. University of Tennessee
47. University of Texas at Austin
48. University of Utah
49. University of Wisconsin–Madison[22]

These universities are aware of the very lethal dangers of nuclear weapons and nuclear war yet are engaged in supporting such developments in their contracts with military industrial contractors and research laboratories.[23]

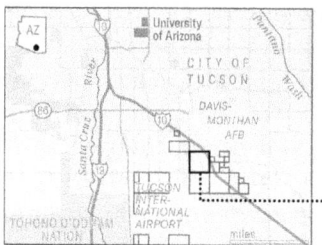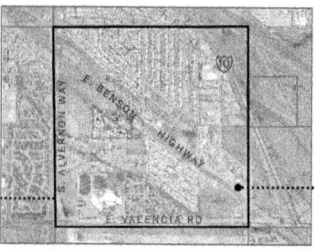

A closer look at a Morrill Act parcel on the ground in Tucson, Arizona. *Left*: Morrill Act parcels selected for the University of Arizona's endowment. Lands: Seized from the Ende (Apache) (Western Bands), Oct 1, 1886. *Center*: Parcel: Township 15S, Range 14E, Section 10. Parcel ID: AZ140150S0140E0SN100AAZ. Size: 640 acres. Ende (Apache) people paid: $0. LGU endowment principal raised: $2,004.39. *Right*: The Apache Tears Motel was once a historic roadside stop for motorists off Benson Highway in Tucson, which featured a kitschy statue of a cross-legged Indian donning a headdress. Today, the motel offers weekly and monthly stays and is a stone's throw away from the sprawling Apache Village RV Park. Sources: Andrews 1918; Royce 1896–1897; GLO, BLM; OpenStreetmap, USGS (map by Margaret Pearce for *High Country News*. landgrabu.org).

When one travels across the massive land mass of Turtle Island (as this author has a few times over the past decades), while viewing millions of acres of land occupied by ranches, farmers, schools, colleges, and universities, corporations, malls, the military, commercial real estate, and the like, one realizes that Indigenous people are forced to eke out an existence on the periphery, a glaring structural injustice of historical colonialism. Shoshone grandmother and decolonization resister Mary Dann, who challenged the U.S. government in a lawsuit demanding sovereignty as declared in the Treaty of Ruby Valley of 1823 that accorded most of the state of Nevada as Shoshone Land, rejected the term "reservations" as the places where Indigenous people live, since reservations, she asserted, refers to travel bookings and set asides for social activity, and Indians were not "set asides."

The concern for the well-being of young people and their success, healthy development, and ability to make effective and constructive contributions to society's well-being as in other societies (both socialist and capitalist ones) is absent in the U.S. because of unbridled and insatiable greed. China, Cuba, North Korea, Germany, Denmark, Finland, Norway, Brazil, Uruguay, Argentina, Panama, Egypt, Malaysia, and Morocco are some of the examples of countries that offer relatively free and accessible college education. Cuba offers free tuition even to international students. It's indeed a capitalist scandal that a country such as Cuba—with very low average income but very decent free education, health care, and virtually free housing compared with large capitalist countries—has educated thousands of doctors, engineers, scientists, technicians, health science experts, and academics from all over the world, including from the U.S. Needy and low-income medical and science students, particularly of color, graduate with no debt, even though Cuba has been struggling under the longest lasting comprehensive U.S. economic sanctions (64 years) and the worst financial, medical, and trade embargo ever.

Stress from debt in "higher" education is the leading cause of ill health among college-aged youth and young people, and even older learners. While college executives command "high-end market salaries and compensation packages," their teaching staff concomitantly receive meager salaries, reduced benefits for part-timers, and experience the most humiliating of teaching conditions. These include large classes, often online, to garner maximum high-tuition revenues according to rates determined by university executives. Blatant capitalist and classist disparities are reinforced in "higher" education, with 70 percent of instruction at many universities and colleges performed by part-time teaching faculty compelled to do moonlighting jobs to supplement meager incomes. Significantly, 293,029 graduates of master's degree programs, three times more than from 2007 to 2010, and 33,655 Ph.D. U.S. graduates, especially in the humanities and social sciences, were receiving food stamps and other forms of welfare, comprising part of almost 50 million such people in 2011. Many of these academic graduates were and are dependent on part-time menial or "on-line self-employment" gigs to survive.[24]

Meanwhile, graduate and undergraduate public funding per student has radically dropped with reduced scholarships and grants over the past 20 years, even as graduate compensation packages have been radically reduced. Graduate students are paid a fraction of professorial salaries while doing even more teaching and functioning as minion faculty with no ability to "bite the hand that feeds them," determined by bureaucratic and often ill-trained, anti-intellectual deans and department heads. Thus, in addition to college and university CEOs, often racist, classist, and high six-figure salaried administrators like provosts, deans, department chairs, and assistants to vice presidents become part of the managerial apparatus of top-down hierarchical educational capitalism. These administrators ultimately serve the capitalist system as managers and middle women/men and coax students into college admission on the basis that "a college education was the surest path to achieving the American Dream."[25] The undisclosed unfolding nature of growing student indebtedness has, on the contrary, generated nightmares to the point where heavily indebted student borrowers have opted to remain single so as not to burden life partners with strangulating lifelong debt from college study.

State funding at major universities has radically declined particularly since the early 2000s and dropped deeper since the Great Recession. Capitalism uses education as the medium and the federal and state governments as the conduit for generating profit. The Center on Budget and Policy Priorities (CBPP) noted the effect of this irreparable, damaging situation on the youth of the nation in a report in October 2019: "Deep state cuts in funding for higher education over the last decade have contributed to rapid, significant tuition increases and pushed more of the costs of college to students, making it harder for them to enroll and graduate."[26] Though educational costs in the U.S. and elsewhere in the capitalist world have risen astronomically, the former has never addressed and redressed the cataclysmic college education crisis. Funding levels from federal and state governments dropped sharply by $6.6 billion in 2018 from 2008 pre–Great Recession levels:

- 41 states spent less per student.
- On average, states spent $1,220, or 13 percent, less per student.
- Per-student funding fell by more than 30 percent in six states: Alabama, Arizona, Louisiana, Mississippi, Oklahoma, and Pennsylvania.

Between school years 2017 to 2018, after adjusting for inflation [using the CPI-U-RS Index—Consumer Price Index for Urban Consumers Research Series-Bureau of Labor Statistics]:

- 27 states spent less per student. In 15 of these states, funding also fell the previous year.
- 23 states spent more per student.
- Overall, per-student funding essentially remained flat.[27]

The deep state cuts in "higher education" are indicated in the table below, with Arizona topping the slate, with public two-year and four-year college funding allocations

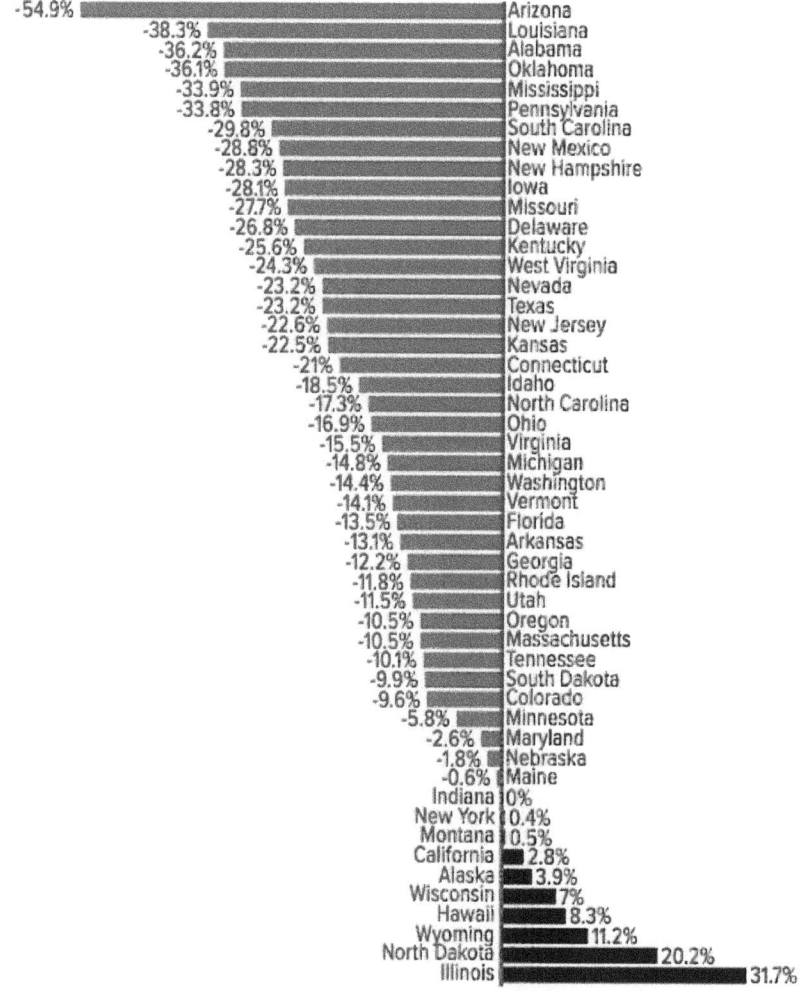

CBPP analysis using SHEEO State Higher Education Finance Report and BLS CPI-U-RS

CENTER ON BUDGET AND POLICY PRIORITIES | CBPP.ORG

Tuition has increased sharply at public colleges and universities. Percent change in average tuition at public, four-year colleges, inflation adjusted, 2008–2018 (Michael Mitchell, Michael Leachman, and Matt Saenz, "State Higher Education Funding Cuts Have Pushed Costs to Students, Worsened Inequality," Center on Budget and Policy Priorities, October 24, 2019).

dropping from 75 percent in 2008 to 34 percent in 2018, making it increasingly difficult for most high school graduates to enroll in college.[28]

The excruciating result of this racist and classist economic oppression and exploitation of young people and coercing hundreds of thousands of students into financial dependency is an issue about which the nation is fully aware. Of the total $1.77 trillion national student debt, $1.6 trillion is in federal loans and the rest private, growing almost $500 billion over the past five years. This public debt entrapping our young college graduates and their families is still less than the $1.9 trillion that was unaccounted for in the audit of the Pentagon in November 2023, continuing a pattern of failed Pentagon audits since 2018 when such audits began.[29] Forty-five million student graduates are mired with federal loans, and 2.4 million are held captive by private loans, with the average student owing $28,950.[30] One of every five people in the U.S. continues to be trapped in college debt, with half of the total owing $20,000 or less, seven percent owing $100,000 (often graduate students, but including undergraduates, too), and one-third owes $10,000 or less.[31] Sixty-seven percent of college student debtors are under age 40, and the largest age group is between 25 and 34—often the time of life when people in this group consider marriage, having children or raising a family. Student debt cripples such dreams and aspirations. U.S. states in the east are the "highest student debt states," and many in the west are the lowest, with debt level differences almost double between both regions:

States with Highest and Lowest Average Student Debt[32]

Highest Debt States	*Lowest Debt States*
New Hampshire, $39,938	Utah, $18,344
Delaware, $39,705	New Mexico, $20,868
Pennsylvania, $39,375	California, $21,125
Rhode Island, $36,791	Nevada, $21,357
Connecticut, $35,853	Wyoming, $23,510

In 2021, the following states had the largest student loan debt:[33]

States with the Most Student Loan Debt in 2021

State	*Balance* (in USD Billions)	*Borrowers* (in Millions)
California	140.1	3.9
Texas	113.7	3.5
Florida	95.9	2.5
New York	90.3	2.4
Georgia	65.7	1.6
Pennsylvania	62.5	1.8
Ohio	60.9	1.8
Illinois	60.0	1.6
Michigan	49.9	1.4
North Carolina	46.8	1.3

Astoundingly, 55 percent of students at the nation's public colleges were forced to take out loans to complete their degrees, almost as much as the 57 percent of students at private colleges who also had to resort to these "predatory" loans to graduate, even with financial aid from both types of institutions. The ploy that many student recruiters, financial aid officers, registrars, and another admissions staff use is putatively, "sacrifice now for the rewards later." Globally, this is a similar deceptive practice of "restructuring debt" by imperialist banking financial institutions like the WB, IMF, and private banking and financial conglomerates through the extension of loans to countries with exorbitant rates of interest, which absurdly then extend further loans to pay for the interest on the original loans. Resultantly, countries of the underdeveloped world, especially those in Africa, Latin America, and Asia, are kept in perpetual debt slavery, recalling that the entire western capitalist world was fully built on slavery and colonialist occupation and extraction. The U.S. is the world's largest debtor nation, at $33 trillion. A total of $92 trillion in debt afflicts nations globally, yet some 3.3 billion

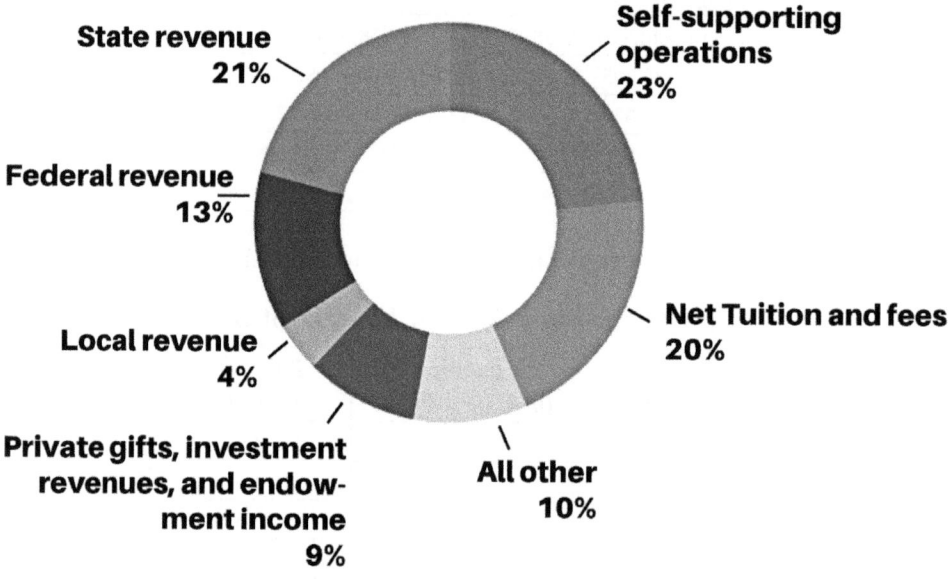

Chart documenting the increasing reduction of state and federal funding for college education and the rising dependence on private sources. "Note: Data include operating and non-operating revenue received by public higher education institutions. Just under 1 percent of all such institutions report their funding using the standards of the Financial Accounting Standards Board and may not include Pell Grants under federal revenue" (Michael Mitchell, Michael Leachman, and Matt Saenz, "State Higher Education Funding Cuts Have Pushed Costs to Students, Worsened Inequality," Center on Budget and Policy Priorities, October 24, 2019).

people in the underdeveloped world are forced to spend precious income on servicing debt first and spend more on debt payments than on health care and education.[34] Africa alone has $1.8 trillion of the $26.9 trillion in debt owed by underdeveloped countries.[35]

Caitlin Zaloom has documented the student debt complex in a *Time* magazine article and book, *Indebted: How Families Make College Work at Any Cost*:

> Middle class parents feel obligated to send their children to college, but the only way to give them that opportunity is to pay for it, and the price is dear. This demand propels them into a bewildering maze of financial policies and programs run by the government, financial firms, and universities. The path is so convoluted that I felt it needed a new name: the "student finance complex."[36]

Yet, the cumulative U.S. educational cost debt crisis is not confined to young people alone, as is often assumed. There are 9 million people aged 50 and over who carry billions in student debt, as well as 2.4 million people over 62 who are either retired or planning to retire who owe an average of $40,750, much more than 25–34-year-old student debtors owing an average of $33,570.47.[37] Such data underscores how all segments of society, including the elderly, are also afflicted by devastating debt, with people over the age of 65 now constituting the fastest growing segment of student loan debtors. Twenty percent of all student loan debtors are over 50. Many seniors are forced to work regardless of the condition of their health in grocery stores and other places to seek supplemental income just to pay off such debt.[38]

The Devaluing and Corporatization of K-12 Education[39]

The result of the accelerating privatization and capitalist direct control of public education, even K-12, has seen hundreds of public schools around the country close. Such closures are part of monopoly capitalism's design of seeing every sector of life, including crises, as profit-making opportunities, a class and race educational colonization even of children and young peoples' learning, growth, and communities.[40] The child has indeed become a full-blown commodity, as Marx described child labor in Europe during his day, highlighted in chapter 2. Many of these latter tendencies and trends are funded by the very same people and billionaire corporations responsible for the crises of social suffering, impoverishment, and injustice in society and the world.

The ruling class control of education at the K-12 level follows the capitalist profit-race-class-based model, with the constant chant and lament propagated by educational and social ideologues: "We have failing schools." Thus, schools are closed, and in many instances, like in Philadelphia in the early 2000s, New Jersey, Washington, D.C., Kansas, Florida, and many other states, states and school districts turned to the for-profit EdisonLearning, Inc. (formerly Edison Schools, Inc.) and other such profit-based organizations like K12 Inc., which runs nine virtual schools in California through its subsidiary, California Virtual Academies (CAVA). CEOs at CAVA receive multi-million-dollar salaries funded by taxpayers, expanding what is a growing phenomenon stemming from the "We have failing schools" ideology. Factually, many of these private for-profit schools underperform overall compared to most public schools, but typical of capitalism's ideological manipulation, desperate and frustrated low-income families and communities support public funding of such flimsy educational outfits to "save" their children. This ploy has constantly been directed toward brainwashing the public and deceiving the economically and socially vulnerable working class, such as Indigenous, Black, and Brown families living in depressed metropolitan districts around the country, including Chicago, New York, New Jersey, Pennsylvania, etc., and obscures the disparities of race and class in these areas and cities where people of color are predominant. The ideology of high income community schools being superior and those in low income areas as inferior is rooted in capitalism's and racism's institutionalization and protection of the very wealthy and upper middle classes, at the cost of the very economically disadvantaged and racially colonized. Obscure capitalist elitist forces control and manipulate the country at large. The experiences of the neediest children and families are hardly raised, even while homelessness in many urban school districts persists. Ruling class propaganda advocates the abolition of failing schools and urges public and state funding for privatized and even for-profit education. The foundational obfuscation centers around institutions that are "public" and the consistent capitalist demand of the abolition of public-owned-or-run enterprises in every sphere of life on the grounds of "inefficiency" and "failure." Thus, the "public" becomes fully steered by capital mediated through state and city officials and legislatures, mayors and city council members, school superintendents, and other key educational managerial administrators. The core issues of why schools "fail" and what kinds of benchmarks are employed to assess the viability of schools and learning are never raised since white ruling class and middle-class state "standards" and "benchmarks" are used to evaluate cultures and learning in contexts where most people are of very low and close to poverty income levels and are marginalized on the grounds of race, class, and gender.

One reads constantly about the granting of shady, lucrative, even multi-million-dollar contracts to private "consultant," "software," and "educational and financial experts," so that executive school administrators in turn promote these "experts," part of an unholy alliance of self-seeking economic interests. Public school superintendent or executive compensations run into the hundreds of thousands, like a CEO of the Chicago schools, Paul Vallas, a "financial expert," who received $225,000 as compensation when moving to become CEO of "troubled" Philadelphia schools in 2002.[41] Though he terminated a lucrative additional $18 to $24 million consulting contract to join Edison, the latter company nevertheless went on to become the leading public for-profit school corporation in Philadelphia even though it was opposed by the local Philadelphia Federation of Teachers Union and the 4,000 member Philadelphia Student Union. The state of Pennsylvania, which was run by former Homeland Security Secretary and former governor Tom Ridge in the early 2000s, disregarded the total community opposition. The state subsequently took over the schools under a 1998 law where "the state could take over the city's education system if it experienced any of a long list of financial problems."[42] From the mid–1990s and for quite a few years, Edison received multi-million-dollar contracts with 100 public school districts, earning millions, but eventually losing millions. Profit-oriented capitalist education operates on the basis that vital public taxpayer funds should not be invested in public schools and districts where the majority of residents are economically depressed and of low income, often communities of color. It stands to reason then that students and school learners from such backgrounds who derive from histories of impoverishment, familial fragmentation, social upheaval, and economic depression generally perform poorly on tests crafted mostly by elitist educational administrators in state governments. Test scores are promoted heavily by for-profit and corporations and in turn championed by most U.S. state officials and legislators, substantiating the interwovenness of the political apparatus and the capitalist economic system.

Further, the capitalist system does not consider the people who spend hours teaching in the classroom each weekday and regularly laboring over weekends to prepare for the following week valuable, since no compensation for the extra hours invested in class preparation is provided in contracts. Though teacher salaries increased by a few thousand in many states ($10,000 in New Mexico in 2021, for instance), these increases were way below the rate of inflation, so that today, teachers are earning less than what they earned in real income in the 1990s and early 2000s considering astronomical inflation growth.

On April 26, 2022, the National Educational Association (NEA) issued a report that though the average U.S. teacher salary in 2021–2022 was $66,397, an increase from $41,770 in 2020–2021, there was a marked real inflation-considered reduction of $2,179 from ten years ago.[43] State, city, and national governmental education officials are all involved in the capitalist racket of boasting about teacher salary increases and raises from previous years, without honestly revealing the fact that K-12 teachers today are really earning *less* than what they received a decade ago. Such data is worrying for all educators who continue to feel undervalued since their tireless and industrious work in teaching our children is worth so little. During the height of the Covid-19 pandemic in 2020–2021, thousands of stressed teachers left the profession because many felt that the deficiency of salaried support against the backdrop of being exposed to the virus and associated illness in classroom settings was just not worth it. Over 600,000 K-12 teachers left their teaching positions from January 2020 through March 25, 2022.[44]

Contrary to what we have all been led to believe by the capitalist ruling establishment in the U.S. with the appeasing salary increases over the past years, actual teacher salaries have dropped to their lowest levels since the Great Recession, as noted below.[45]

The lack of cultural and racial diversity among U.S. public school teachers persists where almost 80 percent of many school districts are staffed by Euro-American teachers, mostly female. While the educational contributions and dedication of many such teachers are extremely valuable considering their commitment to children's learning, it concomitantly reflects the deep chasm between the majority of school learners, who are from communities of color (often in 70–80 percent of the cases), and teachers from the dominant culture. Such learners are deprived of instructional multi-ethnic racial diversity, which is particularly disempowering for students failing to have a single teacher of color during their 12 years of schooling.

In Arizona, for example, over 60 percent of students are of color, but three-quarters of teachers are white.[46] This huge teacher diversity vacuum inevitably results in such students performing poorly because they lack self-esteem for many structural societal and economic reasons, and don't enjoy the right of working with and learning under teachers from their particular cultural grouping. Such is particularly negatively debilitating for Indigenous, Black, and other children of color. The serious absence

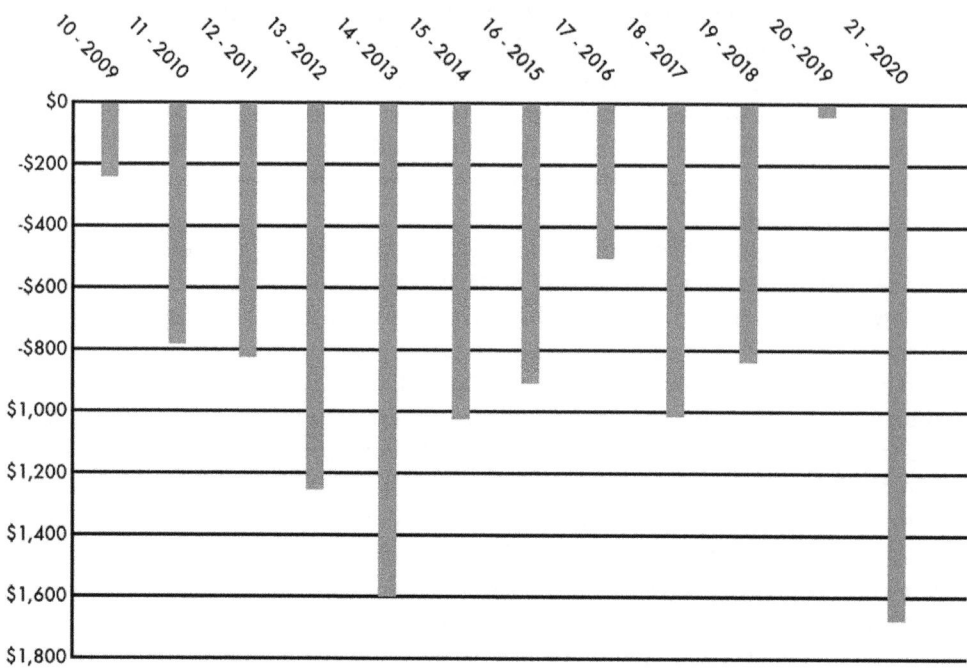

Real Earnings Did Not Increase in Recent Years Considering Inflation Growth & Adjustment

Teachers' real earnings did not increase in recent years considering inflation growth and adjustment (Casey Ontiveros; data from *NEA Today*, April 26, 2022, https://www.nea.org/nea-today/all-news-articles/average-teacher-salary-lower-today-ten-years-ago-nea-report-find).

of male teachers of color exacerbates this crisis. The National Bureau of Economic Research noted that for students of color, having teachers of color determines whether such students aspire and eventually enroll in college. Vanderbilt University researchers reported that Black teachers were more inclined to refer Black students to talented and gifted programs at K-12 schools than non–Black teachers. The inverse is the phenomenon of Black students being disproportionately suspended for "acting out" or "acting up" by mostly white teachers.[47] For Black boys especially, such unfair suspensions and disruption in the learning process stem from such teachers having little understanding of Black families (especially those from the working class) and of Black male learners in particular.

Though the pandemic caused myriad social, cultural, personal, and economic problems in the U.S. and the world, and unemployment peaked to tens of millions in the spring of 2020, the surprising development was that millions of people, in a range of professions, did not return to their previous places of employment. There was a mass exodus of workers across the spectrum, including those in the health care workforce, resulting in dire shortages at the country's hospitals, clinics, and medical treatment centers. The Bureau of Labor Statistics Data on Monthly Job Quits from April 2021 through April 2022[48] reflect such trends:

Quits levels and rates by industry and region, seasonally adjusted(*)

Industry and region	Levels (in thousands)						Rates(**)					
	Dec. 2022	Sept. 2023	Oct. 2023	Nov. 2023	Dec. 2023(P)	Change from: Nov. 2023–Dec. 2023(P)	Dec. 2022	Sept. 2023	Oct. 2023	Nov. 2023	Dec. 2023(P)	Change from: Nov. 2023–Dec. 2023(P)
Total	4,091	3,646	3,628	3,524	3,392	-132	2.6	2.3	2.3	2.2	2.2	0.0
INDUSTRY												
Education and health	629	550	570	592	528	-64	2.5	2.1	2.2	2.3	2.0	-0.3
Educational services	55	58	74	56	64	8	1.4	1.5	1.9	1.4	1.6	0.2
Health care and social assistance	574	492	496	536	465	-71	2.7	2.3	2.3	2.5	2.1	-0.4
Leisure and hospitality	806	814	727	709	730	21	4.9	4.9	4.3	4.2	4.3	0.1
Arts, entertainment, and recreation	77	86	71	71	83	12	3.2	3.5	2.8	2.8	3.3	0.5
Accommodation and food services	729	729	656	639	647	8	5.2	5.1	4.6	4.5	4.5	0.0
Other services	177	120	113	112	108	-4	3.1	2.0	1.9	1.9	1.8	-0.1
GOVERNMENT	233	186	180	193	195	2	1.0	0.8	0.8	0.8	0.8	0.0
Federal	21	18	16	18	18	0	0.7	0.6	0.5	0.6	0.6	0.0
State and local	213	168	164	176	178	2	1.1	0.8	0.8	0.9	0.9	0.0
State and local education	103	87	80	92	95	3	1.0	0.8	0.8	0.9	0.9	0.0
State and local, excluding education	110	81	84	83	83	0	1.2	0.9	0.9	0.9	0.9	0.0
REGION(***)												
Northeast	539	489	516	509	504	-5	2.0	1.7	1.8	1.8	1.8	0.0
South	1,748	1,632	1,592	1,447	1,507	60	3.0	2.8	2.7	2.5	2.6	0.1
Midwest	835	738	771	802	688	-114	2.5	2.2	2.3	2.4	2.0	-0.4
West	969	787	750	766	692	-74	2.7	2.1	2.0	2.1	1.9	-0.2

*The quits level is the number of quits during the entire month.

** The quits rate is the number of quits during the entire month as a percent of employment.

*** The regions are composed of the following: Northeast: Connecticut, Maine, Massachusetts, New Hampshire, New Jersey, New York, Pennsylvania, Rhode Island, and Vermont; South: Alabama, Arkansas, Delaware, District of Columbia, Florida, Georgia, Kentucky, Louisiana, Maryland, Mississippi, North Carolina, Oklahoma, South Carolina, Tennessee, Texas, Virginia, and West Virginia; Midwest: Illinois, Indiana, Iowa, Kansas, Michigan, Minnesota, Missouri, Nebraska, North Dakota, Ohio, South Dakota, and Wisconsin; West: Alaska, Arizona, California, Colorado, Hawaii, Idaho, Montana, Nevada, New Mexico, Oregon, Utah, Washington, and Wyoming.

(P) Preliminary

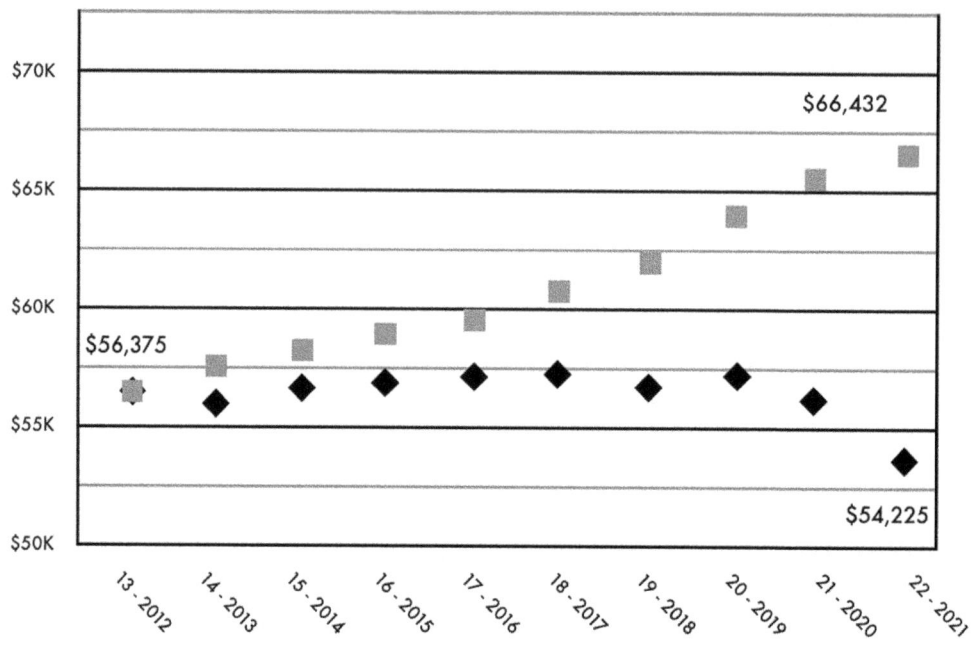

**Average Teacher Salaries Lower Today
Than a Decade Ago Adjusting for Inflation**

Average teacher salaries lower today than a decade ago adjusting for inflation (Casey Ontiveros; Data from *NEA Today,* April 26, 2022, https://www.nea.org/nea-today/all-news-articles/average-teacher-salary-lower-today-ten-years-ago-nea-report-find).

The data on such job departures remind us that most folks within the U.S. are now increasingly opting for entering other humanitarian and life-fulfilling vocations. The aspiration to maximize income positions (notwithstanding the skyrocketing cost-of-living nationally from the pandemic into the present) is not a priority as it used to be. In many cases, such persons are spending more time with spouses, children, grandparents, and community members. While capitalism's overarching enfolding of every segment of life is unmitigated, people are resisting in one form or another—in this case, rejecting monetary preoccupation as the center of their lives despite economic hardship.

The online course instructional norm at all levels of education today, while useful for those in distant places, has becomes a mirage for locally based students and learners ensconced in an impersonal high-tech environment that makes learning a rote process where colorful PowerPoints, slides, and videos become substitutes for substantive learning and personal and societal growth. The deep lack of literacy led by the U.S. and in some countries of the capitalist world, exacerbated by the 2-year-old Covid-19 pandemic, for example, and the associated temporary shuttering of schools, universities, libraries, and various public and private educational spaces have irreparably generated a learning malaise that will be transferred to future generations.

Indeed, the exposé of the farce of higher education in recent years and online "robotic" educational culture reflects the "Great Reset" that founder and executive chairman of the World Economic Forum, Klaus Schwab, has proposed. For Schwab, crisis is "opportunity."[49] Each year, 3,000 billionaires and multi-millionaire elites gather in Davos, Switzerland, to ostensibly address the serious problems and lethal crises facing the world, such as the intensification of impoverishment, failing health care systems, hundreds of millions of job losses, and environmental and climate catastrophe.[50] The "Davos Man" has been Schwab's persistent creation, reinventing images of the wealthiest of the world by providing a place of "refuge" to ventilate their "empathy and sensitivity" to global suffering and pain.[51] Education, too, especially "higher in cost" education, has in turn become entrapped in the tentacles of the "Davos Man."

Revaluing, Restructuring, and Re-Orienting Education

Institutionalized education has always been colonial in foundation, nature, culture, and orientation. Children and learners of color have been and are consistently denied the innate and inalienable right to ancestral language, culture, and community self-determination. As echoed in this book, language is the indispensable foundation of culture, culture is the expression of language, and education signifies the interwovenness of each to practice learning. Edward Koiki Mabo, from northeast Australia, whose landmark successful challenge of Australia's colonial land laws was affirmed in the Mabo court decision, stated in this regard:

> I am being continuously told that the language that the kids bring from home should be the language that they should be taught in. It is a language that they understand. But it is rather unfortunate to hear in our case that we have to employ so-called qualified people to teach us in *their* language. It is a language that we do not speak at home at all.[52]

Like the hegemonic role of Christianity's crusaders signified by the persistent and itinerant missionary, educational institutions that went alongside religious proselytism ensured that the colonized would be violently subdued and enslaved through protracted ideological indoctrination. That legacy of indoctrination, particularly within western educational curricula at colleges and universities in Turtle Island and elsewhere, persists into the present day. The education system is intrinsically designed to break the child and learner of color, rather than build, as noted Black historian Carter G. Woodson eloquently described in his classic, *The Miseducation of the Negro*.[53] Western educational institutions are fully aware of this core learning and growth principle but willfully obscure such truths. The same educational experts have correctly pointed out that learning in one's ancestral language and mother tongue is a *sine qua non* for foundational learning yet cooperate with colonialism's deprivation of such language in educational environments.

For Indigenous peoples globally, the ability to speak fluently in one's ancestral language is the key in the empowerment of every learner on Earth regardless of "level" or "status." Parents, particularly mothers, in conjunction with fathers, are principally responsible for instructing their children in their ancestral language. Community elders expand this instruction through illuminating ancient ancestral stories on cosmologies (illustrated in the first chapter of this book), cosmogonies, and empowering knowledge for living and addressing personal and community problems in a complex world. The role of spiritual healers and seers is fundamental in this educational journey, particularly through ceremonial practices as the child and learner grows. One's fluency in one's mother tongue is *the stepping stone* prior to engaging and becoming versatile with other languages, as major industrialized and formally educated nations, China and Korea, for example, have historically demonstrated. Africa, owing to its far-reaching western colonialist history and neo-colonialist status today, still remains mired in a Eurocentric mold in her educational structure and orientation, with some positive attempts like Tanzania and the practice of *Ujamaa* (cooperative economics) and the institutionalization of Kiswahili from the 1960s through 1985 as the *lingua franca*, even in college. The Arabic-speaking countries of northern Africa have retained their cultural and linguistic core in education. Former Tanzanian president Julius Nyerere and the ruling *Chama Cha Mapinduzi* (CCM) party positively struggled to reject urban Eurocentric capitalism, advocated Pan Africanist socialist culture and self-reliance, and attempted to establish Indigenous collective villages even for urban residents, yet their success was short-lived. The causes were many: western governments and financial "aid" institutions withdrew from Tanzania, women were not at the helm of such programs, policies from central governmental authorities did not augur well with Indigenous communities, food production was stymied, and the relationship between the mainland and the east coast island of Zanzibar was strained. However, the very positive achievement unparalleled on the African continent was the country's highest literacy levels in an African language (70%) and the institutionalization of Kiswahili and development of educational materials in the language at a tertiary level. Though the west and many internal critics in Tanzania scoffed at the *Ujamaa* philosophy of village collectivization and self-reliance and dismissed both the programs and their effects overall, the irony is that since the early part of the 21st century, politicians and communities alike have called for the restoration of the elements of this Indigenist philosophy.[54]

While much has been written about educational transformation, subversive education, resistance education, even "deschooling" education by well-intentioned writers, particularly by radicals like renowned Brazilian educator Paulo Freire, bell hooks, Ivan Illich, John Mink, and others, published in thousands of books and journals over time, the perspective of Indigenous cultures is quite different from "radicalizing" the western educational colonial system.[55] Traditional healer and teacher Hataali Jones Benally instructively notes on "educational transformation," "there is no change … the Creator who created you at birth is always with you and never leaves you … even when you kick the bucket." Every person has a distinctive spirit endowed by the Creator that remains with that person throughout life and never changes. The changes and transformation we experience are external and never alter our essential original spirit. Hataali Jones adds, "regardless of which school you attend and how much of 'education' you receive, it will never tell you who you are."

In Indigenous cultures and ways of living, it is first and foremost the ancestral language, the heritage, culture, and family that is the pillar of knowledge and education, all connecting individually and intrinsically to community. Initiation rites are always axiological for language, cultural identity formation, wisdom, knowledge depth, individual self-esteem, and community integration in this regard. African-centered academies in Turtle Island have been highly successful in reclaiming Indigenous African languages like Kiswahili and in instituting initiation rites for young Black learners.

Examples include Nubian Village Academy in Richmond, Virginia; Kilombo Academic and Cultural Institute in Decatur, Georgia; and Ile Omode in Oakland, California.[56] Literary novelist and African activist Tsitsi Dangarembga reminds us, too, that European languages in Indigenous contexts are incapable of understanding and translating the idioms and nuances of Indigenous languages and that whenever Africans, for example, need to discuss issues in intimate depth and detail, people resort to Indigenous languages.[57] The globally-known writer Ngugi wa Thiong'o rejected publishing his African novels in the English language in 1980, and to mark the break, published *Caitaani utharaba-ini* (*The Devil on the Cross*) in Kikuyu in 1980.[58]

The manner that the impoverished and the working-class families in the U.S. constantly struggle to have an educational success story and aspire and dream of their loved ones (especially for first-generation college students) being admitted to a "prestigious" school that is often owned or managed by millionaires and billionaires is a tragedy of colonization and capitalism. In some sense, the scholarships "awarded" to "disadvantaged students" by such schools like the Ivy League ones mentioned in this chapter are part of an indoctrination ploy that is designed to appease the marginalized and further brainwash members of the oppressed classes with emulation of elitist and ruling class values. Resultantly, the oppressed class of students at these schools eventually becomes so enamored with being enrolled at such schools that they in turn promote the school's philosophy and ruling class ideology. Capitalism and colonialism have always thrived because of the cooptation, cooperation, and collaboration of the fringe oppressed group that is willing to maintain a subordinate status under the tutelage of the dictates of the oppressor group. Capitalism needs to allow such "success" of a minuscule segment of the oppressed so that the overall oppressive system and structure conveys legitimacy in the eyes of the oppressed. Frantz Fanon, Aimé Césaire, Walter Rodney, Steve Biko, and Kwame Nkrumah are examples of this analysis of neo-colonialism in the world.[59] Biko, revolutionary founder of the Black Consciousness Movement in Azania/South Africa in the 1960s and one of Africa's most dynamic and dedicated leaders, was brutally tortured and murdered by apartheid police in 1977. He explained the educational and cultural ramifications of colonization in Azania/South Africa, Africa, and the world:

> The attitude of some rural African folk who are against education is often misunderstood, not least by the African intellectual. Yet, the reasons put forward by these people carry with them the realization of their inherent dignity and worth. They see education as the quickest way of destroying the substance of the African culture.[60]

In the final analysis, decentralized communities and structures within which Indigenous people have lived from time immemorial through clans and village circles are the key to decolonization, where ancestral languages are taught by mothers and parents, and the entire community is fully involved in both individual and community education, rooted in particular family lineages. For learning to be effective and constructive, not rote and routine, but dynamic, alive, and meaningful, all children must necessarily be taught in their respective mother tongues as a scientific principle. Family gatherings, initiation rites and ceremonies from birth through adolescence, marriage, reaching elder status, and returning to the spirit world are essential steps in young learners' growth. All family and community elders, mothers and fathers, siblings, grandparents, and healers are thus collectively involved in grounding the learner in her or his language and culture. Once such deep roots are firmly anchored and established, the learner is then encouraged to engage with processes outside the community and in the broader society. In this manner, learning is a collective, circular, sharing, and caring process. Malidoma Somé's careful illumination of the initiation restoration rites he underwent (with permission to describe some and not all of his experiences of such secret and deep spiritual knowledge) in his Dagara tradition at age 20, after having been beaten, abused, and brainwashed at French Jesuit schools for over 16 years, in Burkina Faso, west Africa, is a classic text in this regard. In *Of Water and the Spirit: Ritual, Magic, and Initiation in the Life of an African Shaman,* he provides a poignant narrative of the power of Indigenous initiation and education. Somé emphasizes that the "prestige" and "wealth" traditionally associated with money and material possessions in the west is essentially determined by the number of people around a person or community for the Dagara, in radical contradistinction to profit-pursuing and materialistic-oriented western education.

The knowledge about the complex processes of the Earth and other Nature Forces of the Infinite Spiritual Universe transmitted to the learner is mediated through the family and community, who are also teachers who have a special interest in their young loved one's growth and the need to understand life and its purpose in multifarious ways. This circle and cycle of learning and teaching continues endlessly and is trans-generationally transferred to children and grandchildren in preparation for sustaining the next seven generations, our grandchildren's grandchildren and beyond. All of our relatives in the rest of the natural world, of which we humans are an integral part, the four-leggeds, birds, insects, and all other creatures of the Earth, Sky, and Sea do the same in their life's journeys, like the quail protecting and teaching their little babies

Quail family with mother looking over chicks (author photo).

from birth about what's needed for protection and preservation, so that they in turn can continue the ancestral line once the elders have passed on. Decolonization education is always fully engaged with the natural world in its endless diversity and kaleidoscope of life, never looking down on this beautiful world but fully respecting and learning from the creation, weaving the learner into this amazing tapestry called Life.

Decolonization education is rooted in Indigenous languages and cultures fluently spoken or recovering/recovered (considering the widespread loss of Indigenous languages especially here in Turtle Island and other European settler-colonies). It is concretely and classically exemplified by the Yuchie Language Project in Glenpool, Oklahoma, where young people are learning the Yuchie language from elders through schooling and a variety of educational and cultural projects, understanding that *real learning* is a life-long process, not temporally 4 years of college and 2–4 years for graduate school.[61] The Yuchie Language Project, like the Salish School of Spokane, Washington, cultivates life-long respect … respect for the Earth, the Universe, Elders, Mothers, Fathers, Healers, Teachers, the Language, and all Life, caring and sharing as the core curriculum, with diversity of areas and experiences as part of the life-long educational journey that's directed toward living holistically and always being in wholeness and balance, *Hoz'ho* in Diné, dipping into the infinite knowledge pool of the Spiritual Universe. It is through such powerful Indigenous language recovery, restoration of mother-tongue instruction at all levels, not hierarchically, but in a circular manner, like the Earth Mother, that the decolonization and revitalization of education will inevitably materialize over time and space. Only then will we be able to expand our traditional communication with and understanding of our relatives, the four-leggeds, birds, insects, plants, trees, mountains, hills, valleys, rivers, oceans, streams, and the rest of the natural world, and vice-versa. The Paga community in Ghana still practices this in harmony with the crocodiles living there, and as Vine Deloria explains in his *The World We Used to Live In*, where butterflies speak with women and rocks reveal truths.[62]

Summary and Conclusion

Education is about life and preparing persons and communities to live their joys and aspirations with affirmation of their histories, heritages, languages and cultures; anything else posing as education is an ideological ploy to deceive and colonize people. For too long, what we call education has been about materialism and status and moving up the hierarchical ladder of success. This is capitalist colonializing education in essence and has in effect destroyed particularly the wonderful children and youth of today of which we who are adults and elders experienced in our respective life journeys. Education is about community empowerment and holism, sharing and caring, and preparing people to do their best in whatever circumstances they find themselves, without stigmatization and marginalization. All children, all youth, all women and men, are gifted and talented; nobody is bereft of such, even the differently abled. There is no single yardstick for what the west uses as a sole criterion to measure ability and chance of success, I.Q. or intelligence quotient. For Indigenous peoples and cultures of all time, there are different intelligences and they are manifest in a myriad of ways, rooted in the Earth, the heart, in sharing and caring for life, not in the singular individual obsession with materialistic and financial success. Most importantly, the Indigenous ancestral language of the learner is critical, and if unknown, needs to learned, re-learned, and reclaimed, so that ancestral knowledge is communicated and absorbed for the continuity of the ancestral line. This chapter has provided several illustrations of these assertions. Ultimately, our Mother Earth remains the best and most informed holistic and rooted teacher, and we all need to cast aside our false inflated egos, become humble, and listen to Her wisdom and learn from her creatures, like our relatives in the quail community who love their children and sacrifice their all for their defense and protection, so that when we return to Her, she will respond, "Well done, our dear Child!" Indigenous and decolonized education is precisely about our Mother, Mother wisdom, and the re-grounding of ourselves to live and share whatever belongs to our Mother anyway with joy and beauty, as our caring becomes the best example of both learning and teaching. The epilogue will summarize these Indigenous principles for the ongoing journey away from capitalist self-pursuit, egotism, and chauvinism so we may return to who we really are and can become ourselves again as Earth intended.

Epilogue
Whither Capitalism?
The Earth Mother Prevails for Life

When all the trees have been cut down,
when all the animals have been hunted,
when all the waters are polluted,
when all the air is unsafe to breathe,
only then will you discover you cannot eat money.
—Ancient Cree Prophecy

We are the natural nurturers of the Earth Mother. The Earth Mother needs our help, she needs our prayers. We need to educate the women of the world that prayer works.
—Agnes Baker-Pilgrim, *Takelma*[1]

If the world is upside down the way it is now, wouldn't we have to turn it over to stand up straight?
—Eduardo Galeano (referring to the map by Moroccan cartographer al–Idrisi in the 12th century that depicted South America "upside down" compared to the Eurocentric cartography that placed Europe on top).[2]

The Earth is our original, ever-powerful, and life-giving Mother, nourished by the Grandfather Sun that provides light, the Father Sky that releases rain and precipitation (what Dagara elder Bakhye described as the "erotic ritual between heaven and Earth"), and the Grandmother Moon that balances the tides of the vast teeming Sea, to maintain the balance of the web of life.[3] All life is made from the dust of the Earth, to whom all life returns following the end of one's physical life. But there's no real death, as Indigenous peoples reiterate everywhere: all of the Earth's children simply assume another reborn life, and the original physical form enters the Earth as body and bones when buried.

All life is sacred and integral to the Creation as chapter 1 explains in its illumination of diverse creation stories from various Indigenous peoples, the Dagara as exemplary, who hold that "every person is in an incarnation, that is a spirit who has taken on a body ... our true nature is spiritual."[4] This view is in radical contradistinction to capitalism, which believes that our nature is to be selfish and make money. Thus, for Indigenous people globally and for all time, all of the Earth is sacred, not just ceremonial or sacred sites, compelling all life to live in harmony and co-existence so that we can all live collectively because all have a right to life as the Creator has innately accorded. Food for all life requires sacrifice of plants, trees, four-leggeds, birds, fish, and insects. For Indigenous people, being vegan or vegetarian provides no moral higher ground because plants—fruit and vegetables—are all sacrificed so that others may live. Unsurprisingly, Indigenous people are not ideologically vegetarian.

It is our Earth Mother who will save us, since we humans cannot "save" our Mother, who birthed us and all life. The now widespread spraying of lethal chemicals into the atmosphere to dim the light of the Sun and reduce the heat rays of the Sun (especially in Turtle Island, as delineated in chapter 4) is a case in point. Even some conscious western scientists have issued warnings about such myopic practices because they question the aftermath and ask, "What happens after the exhaustion of spraying such chemicals?" and in response, caution that effects from engineering the weather are unpredictable and that it could become so much hotter, defeating the objective of the "Sun-dimming" project in the first place.[5] The Sun is not a toy to be manipulated, but a divine force of life, honored and venerated by all Indigenous peoples and cultures, like the Lakota of the Dakotas, for example, who perform complex sun dance ceremonies each summer, just as the Moon, Earth, Stars, and Celestial Beings are all held as ultimate and manifestations of indomitable and indefatigable sacred power. These are *Nítch'i Naalkidí* (Spirit in Diné), *Umoya Ngcwele* (Holy Spirit) in Nguni languages, *Naǧí* in Lakota, *goh@nTonA* in Yuchie, the Spirit of Life who permeates our existence whether we realize it or not. All are life-giving. "Science" in response to the intensifying crisis of climate collapse and water shortages encompassing at least 2.3 billion people according to official UN data in May 2022, of whom a quarter are children, is totally impuissant in the face of such infinite power possessed by these ever-present divine forces in the Sky.[6]

Beautiful Mother Earth from space (NASA).

Capitalism is by nature an anti–Earth and anti-life hegemonic anthropomorphic system in its totality because it sacralizes profit agglomeration and maximization, making the utilitarian medium of commercial and financial exchange deified, as highlighted by Karl Marx and delineated in chapter 3. It makes an apotheosis of money and profit, and materialistic accumulation the focus of life, while derogating the Earth and the entire natural world to expendable proportions for profit. It is fundamentally contradictory because it makes something utilitarian, money, a deity and idol, and accords the real divine power from which and within and on which we live, the Earth, capitalist commodity value.

In the past twenty-three years of the 21st century, the capitalist system has been particularly unstable and unsustainable, providing cheaply accessible electronic wealth on Wall Street. The top 10 billionaires shamefully more than doubled their cumulative monetary wealth from $700 billion to $1.5 trillion, and billionaires in the U.S. collectively made $1.8 trillion from profits during the Covid-19 pandemic in 2020–2021, while millions died here and around the world and many more were severely affected by the lethal virus. Such avarice reaped profits "at a rate of $15,000 per second or $1.3 billion a day—during the first two years of a pandemic that [saw] the incomes of 99 percent of humanity fall and over 160 million more people forced into poverty."[7] These billionaire capitalist oligarch elites ruling the world are monetarily worth six times more than 3.1 billion people, the overwhelming majority of whom are both abysmally impoverished and monetarily depressed.[8] The monetary wealth of the U.S. billionaire class saw an increase of wealth by $6.5 trillion, and the top 1% held an astounding $45.9 trillion in such wealth, mostly through rapidly rising stock prices and manipulated financial markets in 2021, almost twice the total of the U.S. gross domestic product.[9]

Four billion people lost incomes and any form of economic and social security due to the shut downs of major economies and employment avenues in 2020–2021 from pandemic restrictions virtually in every country.[10] Such insouciance to suffering and death on the part of the ruling economic and financial elites is evidence of the roots of the capitalist system: predications on the basest instincts of lascivious greed and patriarchal, egotistical lust for domination and control of the peoples and lives on Earth, using the accumulation of debt on the part of the working and lower classes to maintain their servitude to the ruling elites. The capitalist system epitomizes a parasitic relationship where the parasite lives on the host perpetually, and in the scenario where the host dies, the parasite follows suit. Karl Marx dissected the capitalist system in his massive three volume work *On Capital,* along with numerous other writings of the 19th century. He explained that the capitalist exists only because of the worker and the latter's subjugated and forced exploitation. Adam Smith's philosophical ideology of the "invisible hand" redistributing cumulative wealth was based on individual self-seeking interest without really intending such redistribution, detailed in his 1759 work, *The Theory of Moral Sentiments,* and in his 1776 publication, *Wealth of Nations.*[11] Smith concocted a chimerical assurance to the impoverished and vast exploited working classes in Europe that the goodness of the rich feudal and landed aristocracy would filter redistribution to the needy. He even invoked "providence" in countenancing the division of the world into rich and poor:

> They [the rich] consume *little more than the poor* [emphasis added] and in spite of their natural selfishness and rapacity, though they mean only their own conveniency, though the sole end which they propose from the labours of all the thousands whom they employ, be the gratification of their own and insatiable desires, they divide with the poor the produce of all their improvements.... They are led by *an invisible hand* [emphasis added] to make the same distribution of the necessaries of life.... When providence divided the earth among a few lordly masters, it neither forgot nor abandoned those who seemed to have been left out in the partition.[12]

Such a quote underscores the duplicitous role that capitalism has always played from its inception, led by the U.S. today. The examples abound.

Cuba, an island nation of under 12 million, is still suffering and reeling from totalitarian U.S. empire economic blockades since its socialist revolution of 1959; North Korea, divided by the U.S. empire, is struggling for adequate food and medical supplies after being relentlessly bombed to the ground by U.S. bombers in the 1950s, leaving about 5 million killed; Venezuela is still grappling with financial, trade, food, medical, and political blockades that resulted in 40,000 people dying (but fortunately now forging constructive trade and commercial partnerships with major Asian and African countries); Nicaragua lost thousands of lives from the U.S.-funded Contra war against Nicaragua in the 1980s[13]; Afghanistan still mourns almost 200,000 dead and suffers widespread impoverishment from unjust occupation and colonization by the U.S. military for two decades, and is still illegally denied its own $9 billion international bank reserves, consequently depriving 95 percent of Afghan people of food and medical provisions, one of the worst food and social crises in the world[14]; Iraq suffers from 2 million killed, maimed, and made homeless from the U.S. wars of occupation of 1991 and 2003; Syria continues to suffer from U.S. occupying forces and Israeli military bombing that has killed and maimed hundreds of thousands; Iran lost thousands of lives, especially during the pandemic from lack of medicine and food supplies due to U.S.-orchestrated sanctions and trade embargos; Yemen, subject to consistent military bombing by Saudi Arabia using U.S.-made arms that killed thousands of people in the 2000s; Africa, where the U.S. has imposed punitive sanctions and economic embargos on 15 countries and engages in violence from its 39 AFRICOM military bases in 22 countries, including 11 in devastated Somalia; Russia and Belarus, sanctioned by the U.S. and western Europe over the Ukrainian-Russian conflict, which has aggravated social hardships in both regions; and unjust trade and military measures against China, the leading product supplier to the U.S., all serve as classic cases of imperialist capitalism dehumanizing so much of the world.

The core corruption of capitalism was evident from the outset of the ideology from the mid–18th century when the rich exploited the marginalized working classes and portrayed such action as philanthropic. The western capitalist countries and people consume and possess 80 percent of the world's resources while being a tiny minority of the world, with the top 1 percent of U.S. society controlling 27 percent of the U.S. monetary wealth in 2021, more than the middle class combined, an uptick from 17.2 percent in 1989, when the middle 60 percent controlled 36.4 percent. Middle-class ownership of real estate dropped to 38 percent from 44.3 percent over the past 34 years, indicating another fallacy of Smith's ideology. U.S. billionaires grew 62 percent richer with increased wealth of $1.8 trillion during the Covid-19 pandemic, more than the middle class combined. These facts regarding wealth disparity are all part of the litany of falsified Smithian ideologies.[15] Four billion people globally, mostly of color and historically colonized and contemporarily marginalized and repressed, are suffering their worst economic and financial hardship today, largely driven by the economic and social restrictions and shutdowns during the Covid-19 pandemic engineered by the world billionaire elite class. Pfizer, Moderna, and BioNTech made $1,000 in profits per second in Covid-19 vaccine sales in western capitalist countries, while the most impoverished nations were unable to afford these vaccines due to strangulating debt to capitalism and paltry health care budgets, with Africa in particular only administering two percent of vaccines to the continent.[16] The total pre-tax profits of Big Pharma, represented by the three companies above, were a stunning $34 billion, all the while receiving $8 billion in taxpayer funds. The pandemic restrictions were never about protecting the health of the world, particularly the impoverished majority, but about profits for the billionaire corporations and their associates like Bezos, Gates, and Musk.[17]

The devastating AIDS pandemic in the 1980s and 1990s is another glaring perfidious capitalist effect that resulted in tens of millions of deaths in Africa, Asia, and the rest of the underdeveloped world, with women in particular suffering and dying at the highest rates. Most were generally impoverished, exposed to the most stressful political, economic, and social environments, particularly war and military conflict, and further denied fully funded treatment in virtually all cases, even while hundreds of billions of dollars were spent on AIDS research and debt service to western banks. Medical doctor and researcher Nancy Turner Banks has documented such in her work on AIDS, and notes that "the principle [sic] methods it [Big Pharma] has chosen to buttress this instability is through war and war substitutes. Creating an 'AIDS' crisis was a war substitute."[18] She adds:

> Pharmaceutical companies, under the control of the eugenics dynamic and the petrochemical industry, have been on the *winning* side of history the last fifty years ... this industry, dominated early on by Rockefeller interests who were tied directly to the chemical and banking cartels, knew how to take control of the market by controlling politicians and the flow of information.[19]

Similarly, the same deceptive pharmaceutical conglomerate and western capitalist strategies were employed in the horrific Ebola pandemic in west Africa in 2014 that killed thousands and destroyed families across the region, substantiated by Chernoh Alpha M. Bah in *The Ebola Outbreak in West Africa: Corporate Gangsters, Multinationals & Rogue Politicians*.

The continued violent police attacks on Black people as outlined in an earlier chapter—like Eric Garner's brutal suffocation in New York City in 2014; George Floyd's 8-minute merciless strangulation by police in Minneapolis in 2020; Breonna Taylor's execution by police in her own apartment in Louisville, Kentucky, in the same year; and racist attacks on Blacks, Turkish people, Muslims, and other people of color in western Europe—corroborate persistent colonialist racism that's endemic to capitalism. Racism remains unmitigated in the U.S. empire in 2023:

- the obscene disproportion of incarcerated Black people, who constitute 40 percent of the prison population, with many serving torturous sentences like 250 years and 75 years for sex offenses and on other fabricated charges, often receiving life-without-parole sentences. Egregious examples of both Indigenous and Black people mercilessly incarcerated abound. Mumia Abu-Jamal, former Black Panther journalist and world-renowned author of 14 books, is the most heinous example, wrongfully imprisoned, after being on death row from 1982 to 2011 and now serving life without parole; Jamil Abdullah Al-Amin, formerly H. Rap Brown, framed for the killing of a Georgia police officer killed by Otis Jackson who confessed to the killing in 1995, remains confined in prison on life without parole since 2002; Leonard Peltier, framed for the murder of two FBI agents at Pine Ridge in South Dakota in 1975, was finally released conditionally for health reasons after 48 years in prison; Douglas Stankewitz, Indigenous

Mono Indian serving life after having been framed for murder in 1978 at 18 years of age, remains confined at San Quentin prison in California; and detainees are still held at Guantanamo Bay without being charged;
- the lack of decent housing, clean and accessible drinking water deprivation (Flint, Michigan, the Diné/Navajo nation in the southwest, Jackson, Mississippi, New Orleans, following Hurricane Katrina and the forced displacement of over 100,000 Black people, mostly impoverished, who were unable to return);
- U.S. governmental perfidy in honoring the 370 peace treaties made with Indigenous nations, *of which every single one has been broken* and the U.S. government refuses to accord the inalienable right of Indigenous nations to self-determination and independence in accordance with the UN Declaration on the Rights of Indigenous People and protection of particularly sacred Indigenous ceremonial places like *Dook'oos'liid* (San Francisco Peaks in Flagstaff, Arizona) and *Chi Chil Bildagoteel*, sacred to Indigenous Chi Endé (Apache) people in the southwest;
- the so-called border (*frontera* in Spanish) between the U.S. and Mexico still remains militarized and is violently policed with killing and detention of desperate migrants and others fleeing political persecution and climate-induced impoverishment in Latin America;
- exposure of 80 percent of communities of color to toxic waste sites and lethal contamination of water and air as a result of oil refineries and lead-based paint in housing complexes;
- woefully inadequate health care, educational funding, and employment opportunities for people of color and the impoverished are normative; and
- the flawed foundation of the U.S. legal system that's based on British feudal law (with many democratic rights inspired by the ancient Iroquois system of cultural rights) and the total failure of the U.S. Supreme Court (only the Creator is supreme in Indigenous cosmology) to uphold any consistency and protection of the rights of the Indigenous people, Black people, and other people of color, people in prisons, women, and those in the U.S. working class.

Importantly, traditional Indigenous people, elders, healers, and cultural bearers, especially women, do not make any claims of rightness or morality of position. Our elders simply state what can be stated and share what can be shared with those outside our particular Indigenous cultures through permission of the ancestors and leading senior spiritual elders. Collectively, thus, all of us, four-leggeds, birds, insects, sea creatures, plants, trees, mountains, hills, valleys, waterways, and human beings can humbly engage with each other in mutual respect to solve the very lethal problems we face. As the Indians of the Amazon rainforest explain, "We cultivate the forest with the forest." All traditional Indigenous peoples are rooted in Indigenous languages, cultures, and ceremonies for the Earth, Sun, Moon, and the rest of the divine Spirit Forces and use the same to address the issues within and against life. Indigenous peoples are the Guardians of the Earth Mother, as Bosmun, Papua New Guinea elder and community leader Melchior Ware noted pointedly:

> Life comes from the land ... we are only guardians. The river, the environment and biodiversity sustains our life. We in return, regard the land, the environment and the river as sacred.[20]

The Earth Mother does not belong to us, neither do any of the spirit forces, or even members of our family, principally because we belong to them and are here on the Earth borrowed from the endless Spiritual Universe, living our lives with spiritual purpose, after which we "return home," not die, as the Indigenous people of Australia, Papua New Guinea, Africa, Asia, and the western hemisphere consistently teach. Hataali Jones reminds us again, "we are in heaven … right here on the Earth," regardless of what we feel, where we are, how we live, what we do, and whatever happens with us.

There is no "punishment" as is the case with eastern High God religions. The universe is essentially *Spirit* and all life is endowed with the divine spirit; "evil" or negative spirits are part of the universe and will never be extinguished in some ultimate incendiary apocalypse. All Indigenous cultures from time memorial have understood this truth and the universe as a fundamentally spiritual place where the fragile balance between diverse spirits requires constant ceremonies to maintain the balance of "good" and "evil" forces.

The Minianka culture of Mali, for instance, has a very complex cosmological account on how such spirit divisions came to be. Creator God *Kle* was even visible. However, this harmonious existence and co-existence between the two realms was radically and horrifically disrupted when one of the Minianka youth in the village became isolated and felt deeply dissatisfied with the community pettiness and envy and left the village. He moved to a nearby bush to cultivate his fields away from the community, and while there, a spirit being encountered him and was amazed at the shortness of his hair because he had never seen anyone with such hair ever before. The spirit being was frustrated because every time he walked through the bush, his hair would become tangled in the woods and trees. He thus asked the villager whether the latter would cut off his hair and the youth agreed and cut off the spirit being's hair. Things were fine until the being went to the stream to drink water and saw his own reflection. He was shocked that he appeared so ugly and so unlike his spirit being community. He returned to the spirit community, but the community rejected him because they said he looked like the humans and demanded that he have his hair replaced so that he could become the same spirit being again. Thus, he approached the youth again. The youth became totally confused because he was unable to replace the hair, so he consulted the village council, which strategically advised the youth to promise to restore the hair of the spirit being so long as all footprints and handprints on the youth's fields were removed, which the spirit being community rejected. They both realized that a major conflict was brewing between the visible and invisible beings. To avoid a serious wrestling contest that would be fatal for one of the two, Creator *Kle* stepped in and asked the youth what he would like to become, to which the youth replied that he would like to become like his father. Thus it was that *Kle* sided with the youth because he had not willfully violated the spirit being, and resolved that each realm would be separate: the visible and invisible. Yet some of the spirit beings grew angry that *Kle* had sided with the youth, and in turn, they decided to attack any human being with weaknesses and faults, introducing evil deeds on the parts of humans into the world.[21]

The didactic lesson from this intriguing Minianka narrative is that the presence of negativity of what we describe as "evil" will always be present, and even the Creator(s) is/are unable to eradicate it. Yaya Diallo, a drummer healer from the Minianka community, instructively summarizes this inevitable and persistent phenomenon and the antidote:

Human madness—going wild—refers us back to these primal problems.... The breakdown of harmony in the human community is the precursor of any conflict between humans and spirits. It can lead to possession. Harmony and solidarity among humans is the best protection from that form of mental disturbance recognized in the culture as spirit possession.[22]

We are called to do our best and to live lives in the manner that the Creator and Creation originally had designed; Indigenous cultural ceremonies like birth ceremonies, initiation, seasonal ceremonies, like at winter or summer solstice, Water, Sun, Lightning, and healing ceremonies are all part of the complex process of discovering and realizing one's purpose in the world.

Yet, capitalism, like neo-colonialism and militarism, is the most challenging to uproot in the 21st century, and the question is how do we engage in such decolonization to restore transformative Earth-centered and life-sustaining cultures? The billionaire elite and their military associates possessed by the "madness" Yaya Diallo just described in the previous chapter and in this epilogue are extremely difficult to uproot. Liberalist "fair and progressive taxes" will not undo the real cancerous cause: capitalism.

Capitalism is fundamentally incapable of being reformed because it's hegemonic and ruled by exploitative, ruling, elitist billionaires, millionaires, conglomerates, and military industrial corporations. For instance, BlackRock, the $10.4 trillion asset management conglomerate, the largest of its kind in the world, together with some of the largest asset management and retirement investment companies, "Vanguard, State Street, Fidelity, Capital Group, Wellington, JPMorgan Chase, Morgan Stanley, Newport Trust Company, Longview Asset Management, Massachusetts Financial Services Company, Geode Capital, and Bank of America," are all the top shareholders in the five largest military industrial contractors that received a whopping $196.5 billion in revenues in 2022. Topping this list was Lockheed Martin with $63.3 billion in military-related revenues, followed by RTX (formerly Raytheon) with $39.6 billion, Northrup Grumman with $32.4 billion, Boeing with $30.8 billion and General Dynamics with $30.4 billion. Each of the CEOs of these companies received between $63 million and $66 million in compensation from 2020 to 2022. Retirees with investment portfolios in the large asset management companies listed above are often forced to see their portfolios inevitably invested in weapons production whether they like it or not.[23] This fact substantiates that violence, war, and conflict are "good for business," as a CEO of one of the military industrial companies stated in so many words.

Historically, the wealthiest in every capitalist society in the world have always paid the minimum in taxes, and billionaire "philanthropists" like Bill Gates and Warren Buffet have always disguised their ill-gotten wealth through funding health, educational, and social uplift projects, which are simply written off their hoarded and banked billions. Of the wealthiest U.S. world billionaires, Elon Musk (worth $230 billion in May 2022) paid a dismal 2.1 percent of income in taxes from 2013 to 2018; Warren Buffett, fifth wealthiest, paid just 0.1 percent of income in taxes; and Mark Zuckerberg, owner of Meta and 15th wealthiest, paid 1.1 percent of income in taxes. Others like Charles and David Koch generally paid 1.3 percent and 1.4 percent in income taxes respectively. In some instances, they have received tax rebates, and in others, federal stimulus payments designated for U.S. families hard hit by the pandemic (like 18 billionaires did in 2020–2021). Jeff Bezos, worth $18 billion in 2007, claimed he lost income, paid zero in taxes, and received $4,000 as a tax credit for his children.[24]

The U.S. Federal Reserve that ostensibly "supervises" the five largest banks and their assets respectively, J.P. Morgan Chase ($2.87 trillion), Bank of America ($2.16 trillion), Wells Fargo ($1.75 trillion), Citigroup ($1.65 trillion), and U.S. Bancorp ($488.02 billion) and other major banking corporations is a classic case of the "fox guarding the chickens."[25] Collectively, these entities, individuals, and corporations function as government for all practical purposes because they determine how much people are compensated for their labor, the number of jobs created, interest rates, levels of consumer debt permissible, whether people can afford to purchase homes and automobiles, receive loans, rent apartments, etc., and the direction of not just the U.S. economy, but the global economies that are dependent on the U.S. Federal Reserve for banking, stocks and bonds investments, loans, and interest rates. These gigantic banking entities that are diffused into the banking establishment in most nations around the world were principally responsible for the 2008 subprime mortgage crisis that saw billions of people lose livelihoods, jobs, incomes, homes, retirement funds, and other assets of middle-class and working-class people.

In the 2008 Great Recession, the U.S. Federal Reserve extended an unfathomable $29 trillion to these mammoth, bankrupt financial institutions because the government claimed "they were too big to fail" (try getting a zero interest loan of $1,000 if you're deeply in debt and receive low income).[26] The Federal Reserve was also responsible for protecting these and other illegitimate banking conglomerates in the September 2019 repo blowup and once more provided trillions to them to conceal their financial crimes, of which J.P Morgan, Citigroup, Union Bank of Switzerland (UBS), Royal Bank of Scotland (RBS), and Barclays have all been charged for criminal activities involving manipulation of foreign exchange rates for individual banking financial benefit and collectively paid $5 billion in fines. Yet, these banking mammoths continue deceptive practices in one shape or another.[27] In "both 2008 and 2019, the US Fed made *trillions* of dollars in cumulative loans at below-market interest rates to the trading units of these megabanks in order to resuscitate them and cover up its own failure to properly supervise the banks."[28] These banks are now sitting on an explosive financial time bomb, *$234 trillion in derivatives*, of which the largest five hold 80 percent, similar to the ones that triggered the 2008 subprime mortgage, financial, and credit crisis, causing a potential global financial and economic debacle yet again.[29] The Notional Amounts of Derivative Contracts (Holding Companies) in the Fourth Quarter of 2021 from the Office of the Comptroller of the Currency Report on Bank Trading and Derivative Activities reveals how vulnerable the U.S. economy is to derivative debt with trillions of dollars in the balance.[30]

The U.S. Federal Reserve is essentially a private bank collective (supposedly formed to serve the broader public economic and financial interest) consisting of 12 major reserve banks in Boston, New York, Philadelphia, Cleveland, Richmond, Atlanta, Chicago, St. Louis, Minneapolis, Kansas City, Dallas, and San Francisco. It is responsible for setting interest rates by which banks lend each other money and was formed as a result of the passage of the U.S. Federal Reserve Act of 1913. The "Fed" is also responsible for influencing the reserve bank of every country in the world, for determining the viability of banking mergers, and, to date, it has given the green light in virtually all cases, so that it has "never seen a merger it didn't like!"[31]

The ruling political establishment is interwoven into the banking industry, all part of the tiny minority governing elite as explained earlier, and legislative decisions made either at the White House or Congress inevitably foster benefit for the large banks and capitalist conglomerates that control the U.S. empire and the rest of the world. Thus, we constantly witness the revolving door between large financial conglomerates and the U.S. Treasury, with the classic example of Henry (Hank) Paulson, who served as Treasury Secretary from 2006 to 2009 following a stint as CEO at Goldman Sachs from 1999 to 2006, and Robert Rubin, who served as Treasury Secretary under Bill Clinton after working as a Citigroup board member and advisor to "senior management" at Citigroup, receiving $126 million in cash and stocks over a decade. As a critic asked:

> Was Mr. Rubin really hired for his expertise in the industry and the advice he could provide to senior management or to use his knowledge of government to pave the way for Citi to grow in size and influence to the point where it clearly became "too big to fail" in the recent crisis?[32]

Similar pointed questions were raised in Simon Johnson and James Kwak's 2010 book, *Thirteen Bankers: The Wall Street Takeover and the Next Financial Meltdown*.[33] This is endemically how capitalism has always been controlled and operated: like a Mafia syndicate. The examples are innumerable, as Michael Lewis has documented in *Flash Boys* and on CBS's *60 Minutes*, demonstrating that capitalist greedmongers have done it all.[34] To top everything, these corporations and banks are still protected by the Federal Reserve, which is charged with protecting U.S. consumers and supposedly serving the nation. The laughable artificial "creation" of money through Quantitative Easing (QE), where the Federal Reserve purchased bonds and other financial assets (generating trillions that were, in turn, handed over to the banking conglomerates to "stimulate the economy" since 2008), simply reproduced the same system of banking hegemony and profiteering. Anecdotes in Christopher Leonard's *The Lords of Easy Money: How the Federal Reserve Broke the American Economy* serve as sample illustrations.[35]

Banking conglomerates, under the protective auspices of the U.S. Federal Reserve in the U.S. empire, and central banks following suit in other countries, where majority shares are owned by the billionaire and millionaire class, shape the activities of massive corporations even where other smaller shareholders have interests. The Bank of England is a classic case of such colonial capitalism that provided a haven for British securities and the London financial system that hoarded colonial financial reserves from gold, diamond, and other precious metals industries from its colonies in Africa, Asia, and the Caribbean, and was key in protecting the British imperialist economy from collapse following the first intra–European war. These established the foundation for the capitalist imperialist British system today.[36] Banking is the cornerstone of capitalism because banks are in business for receiving deposits and approving loans, the latter of which accrues the debt that is at the heart of the capitalist system and consequently drives the capitalist world so that the majority of the world is held in debt bondage and peonage.

Former Wall Street executive and former Lehman Brothers banker Nomi Prins explains lucidly in her exposé of the banking scandal operations that the U.S. Federal Reserve, the European Central Bank, the Bank of Japan and other national central banks "have fabricated or 'conjured' money to fund banking activities at the people's expense," and following the 2008–2009 subprime mortgage crisis, "these illusionists have created money, altered the nature of the financial system, and orchestrated a de facto heist that enables the most powerful and central bankers to run the world."[37] Following the repeal of the Glass-Steagall Act that required that banks either engage solely in deposits and loans or create "securities and merge companies and speculate" by none other than Bill Clinton (the "people's" president), large banking conglomerates were unconstrained to have "free reign" in manipulating and rigging the banking system in the U.S. empire and around the globe.[38] These translated into "free money loan offers," which led to the biggest U.S. and global financial crisis since the Great Depression of 1929. Prins elaborates in instructive detail:

> Eight years after the crisis began [i.e., 2016], the Big Six US banks—JP Morgan Chase, Citigroup, Wells Fargo, Bank of America, Goldman Sachs, and Morgan Stanley—collectively held 43 percent more deposits, 84 percent more assets, and triple the amount of cash they held before. The Fed has allowed the biggest banks on Wall Street to essentially double the risk that devastated the system in 2008....
>
> But in the banks' moment of peril, the Fed unleashed a global policy of injecting fabricated money into the world-wide financial system. This flood of cheap money resulted in the issuance of trillions of dollars of debt, pushing the global debt to $325 trillion, more than three times global GDP. By mid-2017, the total assets held by the G3 central banks—the US Fed, the European Central Bank (ECB), and the Bank of Japan (BOJ)—through conjured-money QE programs had hit more than $13.5 trillion.[39]

These very same banks that are now sitting on $234 trillion in derivatives are extremely vulnerable entities in the endless pathways of debt instruments, so that one spark that breaks the link in this essentially fragile and unstable system brings down the entire global banking and financial system and capitalism. The impoverished majority is strangled by growing debt like a noose around its neck, often feels trapped, and declares bankruptcy in resignation, especially small farmers and business owners here and everywhere, since they can never afford to repay high interest loans to their capitalist creditors. This is real capitalism at work.

It is neither transhistorical nor universal that all people have been and still are motivated by self-interest and greed, since virtually all of the world's Indigenous people and our ancestors have always been and still are defined by our collective belonging to the land—ancestral land in particular—rather than the inverse, and individuality was always and is collectively defined, "*Umuntu Ngabantu Abanye* and *Motho Ke Botho Ka Batho*"—Nguni and Sesotho languages of Africa, for example, and the trans–Africa proverb, "I am because we are, We are, therefore I am."

Adam Smith's writings highlighted earlier became the pillar of capitalist philosophy and were used by the early European capitalists to justify the propagation of such fraudulent and mythic ideologies so that the ideology would eventually become embraced by the nations of the world. However, it was the Russian Bolshevik Revolution of 1917 led by Vladimir Lenin, and the communist revolution in China in 1949 led by Mao Tse-Tung and the Chinese Communist Party, still ruling China over 70 years later, that marked a radical challenge to capitalist economics and ideology, revolutions that triggered the Cold War between the west and capitalism on the one hand, and the east with communism and socialism on the other.

China today advocates a "Socialism with Chinese Characteristics," incorporating traditional Chinese cultural notions like the principle of *Guanxci*, which is concerned with "moral economic understandings of interpersonal relations," upholding compassionate family and mutually respectful and reciprocal values as historically practiced in traditional Indigenous Chinese society.[40] China's historic uplift of 800 million people from deep impoverishment over the past few decades using the Marxian paradigm is commendable because such numbers are unprecedented in recent history. Aiwa Ong makes the important point that while China appears to have "ceded" to global transnational corporations, especially from the U.S., such were very strategic moves to establish a strong Chinese nation with economic wherewithal. Such moves were prompted by China's increasing engagement and involvement with the world in areas of economic development and assistance, trade and investment, environmental preservation and food sustainability, and medical research cooperation, etc., particularly with nations of the underdeveloped "three-quarters" world, where most people live. Though up to 60 percent of the Chinese economy may be in private corporate ownership (about which the Chinese government is fully aware and is currently addressing so that this ownership is radically reduced), the government has total discretion and power to dismantle any of these large Chinese corporations with immediate effect, abolishing such ownership overnight, unlike U.S. and western capitalist society, which is led by hegemonic entrenched capitalist monopolies and oligarchs.

Further, in terms of ecological restoration, China is planting 6.6 million hectares of new trees throughout the country, an area equivalent to the size of Ireland, investing ¥538 billion thus far.[41] China has planted 19 million trees over the past decade to address and redress the widespread effects of climate change, floods, soil erosion, and deforestation, constituting 25 percent of all trees replanted and the most by far in the world, exceeding average global tree planting of 5 to 7 million hectares from 2007 to 2017. This ongoing reforestation project covers 80 million hectares (198 million acres of land).[42] In 2019, China planted 7 million hectares (17 million acres) of trees with 23 percent of the country reforested.[43] Yet, the government is also hungry for metals for its booming industries and is engaged in mining operations domestically and globally, which have also caused consternation among Indigenous peoples in various places, including *Chi Chil Bildagoteel* (Oak Flat) in Arizona, as well as in Papua New Guinea, Latin America, and elsewhere. China has experienced significant and ongoing Earth heating and climate change disaster and catastrophe, with no signs of abatement, even while water sources there especially for drinking and consumer usage grow increasingly scarce. Such facts inevitably undermine expanded production. The 10,505-ft. Yùlóng Xuěshān Mountain glacier near the ancient town of Lijiang in Yunnan province is melting as we observed in 2013, and rivers are under constant drought threat, including the large Yellow and Yangtze Rivers, with 28,000 rivers that have dried up since 1990, like much of the world.[44] Sooner rather than later, China and all of us will need to come to grips with the limitations of industrialism that depends on resource extraction and bleeding of the Earth because the Earth Mother (*Diqiu Muxin* in Chinese) can only take so much. The continued use of oil and plastics endemic in Chinese industrialized production and in other parts of the world has produced lethal ecological and environmental consequences in the long run, as chapter 4 detailed. China, like all nations of the world, will need to tap more fundamentally into her own 54 Indigenous peoples' cultures and ancestral wisdom in addressing these critical issues facing her and the world.

The relative success of versions of socialism, Marxism, and a pseudo "synthesis with Indigenous cultural philosophies" in some places like Cuba, where Africanisms are still present in the eastern part of the island (evident and witnessed during African-Atlantic academic conferences there by the author in 2000), Venezuela, Nicaragua, and Chiapas in Mexico have been noted. Many eastern European countries where illiteracy was eradicated, education and health care were free, employment was provided, and other public services were fully subsidized by the socialist state were models, too, of limited success even in the face of state-controlled contradictions. Others like Ethiopia under Mengistu and some in eastern Europe like Rumania's Ceausescu regime in the 1970s and 1980s deteriorated into repressive oligarchical structures. To the credit of socialism's orientation and function during its tenure following the intra–European war of the mid–1940s, eastern Europe was relatively successful in meeting rural and urban peoples' needs in income levels, even surpassing that of its western European neighbors, and ensured that there were no extremes in wealth accumulation and impoverishment. Much of these accomplishments were undermined by pressure from the North Atlantic Treaty Organization (NATO) under the auspices of the U.S. empire and the expansion of NATO through the present day that saw the Soviet Union under Mikhail Gorbachev launch *perestroika* (restructuring of the economy towards market liberalization) and *glasnost* (transparency) in 1985. This shift and eventual erosion of the eastern European socialist bloc was celebrated by the west but unpopular in the Soviet Union and many other neighboring nations since many of the economic and social protections ensured under the previous system were abandoned, along with centralized economic planning.[45] It's no surprise today that there's a nostalgia in many quarters in eastern Europe among workers, for instance, for the previous socialist system that guaranteed employment, a decent income, subsidized housing and food, free education, and free health care, much of which are not as available today; instead, widespread unemployment, the rising cost of food and living expenses for health, education, and other social services, and breakdown of rural communities have resulted in the past few decades leading to millions migrating, over two million from the former German Democratic Republic (GDR) alone.[46]

Indigenous cultures and ways of life persist in Venezuela, Bolivia, Nicaragua, and the Chiapas region of Mexico and in varying degrees in the hills of Ethiopia and other parts of Africa, Asia, and the Americas, where socialist rhetoric may be heard. Venezuela, especially under the revolutionary leadership of Hugo Chávez from 1999 to 2013, and continued today with Nicholás Maduro's leadership, pursues some socialist-leaning policies in areas of land allocation, housing, health, oil production, and education. The Wayuu, Warao, Waica, Makiritare, Guahibo, Pemon, Yanomami, and other Indigenous peoples of Venezuela were formally recognized as part of the 44 major communities by the Chavista revolution of 2000. Chávez was radically critical of capitalism's economic injustice and violence, and as a well-read leader drew upon the works of anti-capitalists like Eduardo Galeano (Chávez gave Barack Obama a copy of *Open Veins of Latin America* at a Summit of the Americas meeting in Trinidad and Tobago in 2009), Hungarian Marxist philosopher István Mészáros (*Beyond Capital*), and others as he sought to understand and lead the implementation of the complex strategies necessitated to uproot colonialist capitalism's exploitative history in Venezuela and Latin America. Eco-Marxists John Bellamy Foster, Brett Clark, and Richard York highlight this point in *The Ecological Rift*.[47] Chávez was nevertheless rooted in Indigenous and African ancestral knowledge, forming the Indigenous parliament and the Magna Carta with constitutional articles 119 and 120 when he became president in 1999. These articles ensured the constitutional right of Indigenous peoples to their ancestral lands as "inalienable rights" and protections that "the use of natural resources in the Indigenous habitats by the state shall be made without harming the cultural social and economic integrity of the same (lands) … and … subject to prior information and consultation with the respective Indigenous communities."[48] He was most concerned, too, about preserving and protecting the Earth and "saving the conditions of life on this planet."[49] The push toward moving from away oil production (particularly as Indigenous people in Venezuela oppose oil extraction in the ancestral lands of the Orinoco Basin and other areas, for example) moved the government to institute the Indigenous Parliament constituted by the over 30 Indigenous nations in the country, visited by this author in 2007.

Bolivia under the first Aymara president, Evo Morales (overthrown in a U.S.-supported coup in 2019 for control of lithium resources), now led by Luis Arce under the same *Movimiento al Socialism* (MAS) party, has socialist and Indigenous Aymara leanings. Nicaragua under the Sandinistas from the late 1990s through the present has some form of constitutional protection for the "Rights of *Pacha Mama* (Earth Mother)" and the "Rights of Nature," which was communicated to us during our 2012 stay there.[50] The pressing need to meet society's material and economic needs while protecting *Pacha Mama* against mining and oil extraction is persistent. Venezuela, Bolivia, and Ecuador, with ecologically conserving governments from Chávez to Maduro in Venezuela, Luis Arce in Bolivia, and then socialist leader Rafael Correa in Ecuador from 2007 to 2017, have generated significant tensions in the collective Indigenous struggles to liberate Latin America from its colonialist-capitalist past. Evo Morales, even in facing opposition from Indigenous people during his tenure as president for allowing lithium mining for economic revenues, maintained such practices, as does the current government for economic reasons, given that Bolivia has suffered significant mass impoverishment and underdevelopment. The lingering question, of course, is how long such extractive mining policies in the name of revenue generation for economic need can maintain a land and people through a constantly drying and turbulent climatic world?

In this context, while Marxist philosophy is well-intentioned toward economic and social justice, it falls short because materialism is still its basis and orientation, with the redistribution of the fruits of human labor on the Earth as the principal objective in overcoming the capitalist root cause of human self-alienation, alienation of labor production from the laborer herself or himself, and alienation from others. Vine Deloria asserted in his critique of Marxism that it "accepts uncritically and ahistorically the worldview generated by some ancient Western trauma that our species is alienated from nature and then offers but another version of Messianism as a solution to this artificial problem."[51] An incisive criticism, but not entirely without merit. Indigenous people, on the other hand, view the Earth and the spirituality of life and all relationships with the rest of the natural world as *ultimate,* not anthropocentric human labor as Marx did, and historically have never experienced such self-alienation or cultural alienation. Dina Gilio-Whitaker also observes that:

> The very thing that distinguishes Indigenous peoples from settler societies is their unbroken connection to ancestral homelands. Their cultures and identities are linked to their original places in ways that define them as reflected through language, place names, and cosmology or religion. In Indigenous worldview, there is no separation from people and land, between people and other life forms, or between people and their ancient ancestors whose bones are infused in the land they inhabit. All things in nature contain spirit (specific types of consciousness), thus the world is seen and experienced in spiritual terms.[52]

Gilio-Whitaker echoes Tink Tinker's view of the "personhood of all 'things' in creation/creating and the space-land-basis of life, rather than linear temporal categories as western cultures have constructed":

for Indigenous people land and all its elements have agency by virtue of their very life energy in a way that they do not in Western cultures. Humans are only part of the natural world, neither central to nor separate from it.[53]

Greg Cajete explains that Indigenous science "is evolutionary ... unfolding ... through the general scheme of the creative process of first insight, immersion, creation, and reflection," where it "is a reflection of the metaphoric mind and is embedded in creative participation with nature."[54] Creativity in this paradigmatic structure emerges from chaos and constantly unfolds, underscoring that the reality of the universe is in fact fluid, endless and open-ended, never fixed and static, so that human consciousness is akin to what western science describes as quantum physics. Lame Deer similarly notes that Indigenous knowledge is always multi-sided and Indigenous people view reality from a variety of angles.[55]

Materialism has never been absolutized by Indigenous cultures principally because of the finite course of physical life and materialist reality constantly in flux. The world has not always been defined in fundamental economic, financial, and materialist terms and pursuits as Marx historicized. In this vein, "linear progress," as Marx theorized, is antithetical to Indigenous philosophical, cyclical concepts of time. The seasonal nature of life with endless rotations and revolutions of the Earth, Sun, Moon, and seasons demonstrates the circularity of life, which is intrinsic to us all.[56] Anthropomorphism has no place in Indigenous thought because it signifies egotism, chauvinism, and arrogance, and manifests itself violently in the expendability of other forms of life. To his credit, Marx spent time in his later years in Algiers, read William Howitt's *Colonization and Christianity,* William Prescott's *The History of the Conquest of Mexico* (1843) and *The History of the Conquest of Peru* (1847), Edward Gibbon Wakefield's editing of *England and America* (1834), Lewis Henry Morgan's *League of the Ho-dé-no-saunee or Iroquois* (1851), Thomas Raffles' *History of Java* (in 2 volumes) (1830), Thomas Fowell Buxton's *The African Slave Trade and the Remedy* (1840), and Herman Merivale's *Lectures on Colonization and Colonies* (1841). He became fully aware of the role that these processes played in capitalism's development and evolution.[57] Paul Burkett, in *Marx and Nature: A Red and Green Perspective* (1999) ("red" referring to communism, not Indigenous people), surmises that Marx was growing in the direction of an ecological preservation advocate, which he may have expressed and felt towards the latter part of his life. Yet, as Irving Howe writes, Marx still "drew upon German philosophy, French political thought, and English economics—...the standard formula employed by his disciples" and was thus steeped in this monocultural literary sphere.[58]

Capitalism is destined to inevitably implode and collapse, as we see its internal corrosion daily, exemplified in the most precious of life's ingredients, Water, becoming increasingly scarce and drying up here and everywhere. *Mni Wakan, Mini Wiconi—Water is Sacred, Water is Life,* the Standing Rock Sioux Nation taught the world from 2016 especially. The Earth, in turn, along with the sacred power, Water, will never allow the creatures on Earth and in the Water to be destroyed by capitalist myopia and pathology, Indigenous wisdom reminds us. Capitalism will thus necessarily and ineluctably collapse since it will have neither Water nor Earth to sustain its gluttony. Its daily decline is becoming more apparent as we live particularly now through ecological collapse, climate change, drought, unemployment, and financial crises, exacerbated by military conflicts and war, as with the Russian-Ukraine confrontation in 2022. The

2019–2021 and ongoing Covid-19 pandemic shifted capitalism into "ventilator dependency," and its breathing capacity is being rapidly and progressively depleted.[59] Capitalism is an essentially crisis-ridden and crisis-generating system, as *Black Agenda Report* commentator Danny Haiphong notes:

> The capitalist, imperialist, and white supremacist roots of the American Empire are rotting from within. Capitalism is at a dead-end and is faced with an impossible choice: either accept reforms that cut into finance capital's ill-gotten profits or terrorize workers and oppressed people for everything they have.[60]

Capitalism is living on a ventilator, as the measure was deployed during the pandemic. It's always the land, the Earth, that's real, never money or capital, though capitalism's unconstrained violence forces people to earn and have money to eat.

The constant racist furor and animosity by the repressive regimes of the world, especially in Europe and the U.S., regarding enforcing imprisonment, detention, or deporting "refugees" to protect "the homeland" is yet another moral indictment on colonial capitalism. The landless, dispossessed, destitute, and impoverished people of the world are forced to come to the centers of western imperialism, the U.S. and the EU, because in of all of these western-supported repressive underdeveloped countries from which these refugees come, capitalism has failed miserably, and has never been inherently congruent with the ancient Indigenous histories and cultures of these people in virtually most cases. No culture in the world has ever made materialist accumulation and profit the sacred, quintessential essence of life and core foundational principle and basis of life … except western colonial capitalism. The aberration is capitalism, not "refugees." Capitalism generates refugees because of militarism and industrialist-caused climate change. Japan has opted to follow the capitalist trajectory, dispossessing its own Indigenous Ainu and Hokkaido Island people to adhere to this destructive industrialist path, rejecting the ancestral cultures of the land. Yet Indigenous people, the primary Guardians of the Earth Mother, continue to hold Land and the Earth and life in totality as sacred, just as the Xhosa nation among the Nguni people of Southern Africa continue to revere *nkunzi* (the bull) as sacred, as irreplaceable value, what western societies refer to as "capital."

Indigenous Language, Land, and Cultural Reclamation and Utility Toward Global Integration and Cultural Self-Determination

Ngũgĩ wa Thiong'o raised the question of internalized colonization in his text, *Decolonizing the Mind: The Politics of Language in African Literature,* expressing his dismay at Indigenous African writers at an African Literature conference at Makerere University decades ago, including noted ones like Chinua Achebe, Amos Tutuola, Gabriel Okara, Biragu Diop, and others. He asked incisively, "how did we arrive at this acceptance of 'fatalistic logic of the unassailable position of English in our literature'"?[61] The same can be said of French, Portuguese, and Spanish, from their colonial entrenchment in parts of Africa and other parts of the colonized world. How did African intellectuals and scholars assume the TINA syndrome in publication language, "There is no alternative," in their often-published works on decolonizing Africa? Similarly, with cultures and economies, why do many scholars of color also continue to uphold the TINA syndrome in resignation?[62]

Indigenous scholars in Turtle Island are resisting such internalized linguistic and cultural colonization. Sushpa (Richard Grounds), who learned his Yuchie ancestral language with great effort, even resigning from a professorship to collectively establish the Yuchie Language Project and school in Glenpool, Oklahoma, and Inés Hernández-Ávila, who reclaimed the Nez Perce/Niimiipuu language and developed a project on revitalizing Indigenous culture through language reclamation, are leading examples.[63] Margaret Pearce, Potawatomi geographer and cartographer, withdrew from academia to demonstrate how massive dispossession of lands in Turtle Island has been intentionally obscured by academia, reconstructing this mapped history, as discussed in the previous chapter.

It's essential that Indigenous people marshal our own resources in concert with other non–Indigenous allies to develop instructional materials for children at kindergarten level and beyond, underscoring that decolonization is first and foremost about children coming to understand who they are at the incipient stages of childhood and anchored in their ancestral language. Such actions are preconditions for cultural teachers becoming specialists in practicing decolonization in their respective communities and ensuring that the linguistic-cultural ancestral line is perpetuated for the next seven generations and beyond: our grandchildren's grandchildren.

The Yuchie Language Project and School, an Indigenous immersion language program in Oklahoma highlighted in the preceding paragraphs (in which this author has participated and continues to support), is certainly a relevant decolonization point of departure in this regard. Bringing elders fluent in the language and employing them in teaching the language to children a few years old and beyond is utilizing Indigenous cultures and traditions to preserve the lifeblood of Indigenous people. Without the ancestral language, the identity of a people is forever dislocated and rootless. The Salish School of Spokane is another example of the successful struggle to reclaim, expand fluency, and train teachers in the Salish language. The determination of the executive director, LaRae Wiley, to speak her Salish language robbed from her and so many Salish people through colonialism is like Sushpa's (Richard Grounds') perseverance in the resolve to become fluent in Yuchie and launch the Yuchie Language Project and School.[64] The medium for communication, hearing, and speaking with the land is through the Indigenous ancestral language of the area, as Diné teacher, healer, and world champion hoop dancer Hataali Jones Benally, Salish cultural and language educator LaRae Wiley, Yuchie nation leader Sushpa, and Indigenous activists and educators repeatedly stress. The production of the łčxʷl'stim'i? n'səl'xčin'/ Salish Revitalization video, *The Language This Land Remembers*, is a classic example.[65] Indigenous Fisher River Cree Nation social work scholar Michael Hart makes the same point in his interview with Margaret Kovach, scholar from the Pasqua Salteaux and Néhiyaw First Nation in southern Saskatchewan:

> There are concepts that we have in Cree that don't have English translations ... what I am doing now is looking at different things that I have been through with Elders, with traditional teachers, and trying to understand the underlying teaching, what values are being demonstrated. What am I supposed to do, and how am I supposed to act. I will try to transfer these pieces into the new context ... what we have to do now in terms of decolonizing ... my intent is to focus on our own ways ... back to language and place ... it can be hurtful not to speak the

language in terms of the peoples ... if there is not enough [Cree] speaking, then we [have] lost that aspect of our future, not just within us, but as peoples ... it needs to be retained.[66]

Ancestral language reclamation for many Indigenous nations of Turtle Island is a prerequisite for the restoration of dispossessed Indigenous lands and the key to realizing the global inalienable right of all Indigenous nations to self-determination and independent nationhood. The restoration of Indigenous languages as mediums of struggle is equally intrinsic. "Land Back" is what Indigenous peoples of Turtle Island demand, and the clarion call grows.

Capitalism enforces specious depictions and descriptions of reality to ensure its optimal benefit, influence, domination, and the subjugated's inferiorization, marginalization, and self-victimization. Indigenous Maori pioneer decolonization scholar Linda Tuhiwai Smith notes:

> The whole process of colonization can be viewed as a stripping away of *mana* (our standing in our own eyes), and as an undermining of *rangatiratanga* (our ability and right to determine our destinies).[67]

Such views echo W.E.B. Du Bois' *The Souls of Black Folk* (1903), where he asserts that the Black person is constantly conditioned into viewing herself or himself always in the mirror of others, through the eyes of the oppressor.[68] Academic colonization in "higher education" particularly and education generally follows the same pattern, adhering firmly to a structure colonizing Indigenous peoples and languages with "Research" being "an important part of the colonization process because it is concerned with defining legitimate knowledge."[69] While deceptively and strategically acknowledging all colleges and universities sit on dispossessed Indigenous lands (Turtle Island, Australia, Aotearoa-New Zealand, Azania/South Africa, for example), all without exception still refuse to repatriate Indigenous mediums and historical and/or spiritual material items in violation of international law and even U.S. federal law, NAGPRA.[70]

The *Kanyen'kehà:ka* (Mohawk) warriors in *Kanesatake* Resistance of 1990 in Oka, in the northern region of Turtle Island, who challenged the Canadian military and police in a 78-day standoff from July 11 through September 26, are just one example of such ongoing "Land Back" Indigenous resistance movements. The Indigenous Hawaiian movement and struggle for national sovereignty and for the recovery of the *Aina*, all dispossessed and colonized lands, has intensified in recent years, heightened with the resistance to billionaire Meta owner Mark Zuckerberg, who shamelessly sued Indigenous Hawaiians for living on their own lands so that he could enjoy his 700-acre estate on the dispossessed Big Island.[71] The movement of the Indigenous Kalinga people against the Chico River Hydro electrification complex of four dams in the Philippines is yet another example of defense of the traditional and scared Kayakayya River. Elder Macli-ing Dulag, a leading land rights defender and warrior, stated it succinctly, instructive for all Indigenous and justice-seeking people:

> If we fight, we die honorably. I exhort you all then: *KAYAW!* Struggle!
> And because we are (willing to) fight now, our children may win and keep this Kalinga land. And the land shall become even more sacred then, nourished by our sweat and blood. Then we who sacrificed that they may live and be secure and happy shall abide with them and nurture the generations, guarding the fields, the *pappayaw, the ili*, blessing their lives till endless time.[72]

Such courageous and defiant Indigenous resistance in defense of ancestral lands serves as inspiration to us: when those with the fewest resources win, they ensure blessings in ongoing generations.

The peoples and cultures closest to the Earth Mother are the ones who can lead us all out of these crises, while working respectfully and mutually with others from *all* backgrounds who care about the Earth and preservation of all life so that we are all stronger together. The Mole Lake Ojibwe Nation in Wisconsin who planted *manoomin*, wild rice, for millennia and beyond, for example, have demonstrated how "Traditional ecological knowledge prioritizes gratitude and forward-thinking solutions, helping bring back the human connection with nature and the land."[73] Such ancient practices can resolve the seemingly unsolvable conundrum of climate change and function as possible solutions to "Western thinking" that works to "separate people from nature, removing the responsibility to protect it, and making solutions seem out of reach."[74]

Much life is being lost, but the faithful Earth Guardians are always here, in whatever form the Infinite Spiritual Universe deems best. We all need to collectively come together since we all have the same root: the Earth Mother and the Spiritual Universe. It's Indigenous prayers and ceremonies that will reconnect us deeply to the Earth Mother and the Spiritual Universe, as Indigenous elders and healers have embodied constantly. We need to live simply and humbly, so that all others and life can simply and humbly live.

Four important principles are critical in this journey of self and community realization and liberation:

i. We need to decenter our individual selves and eradicate our illusory egos and decenter Eurocentric culture and the associated patriarchal subjugation of women;

ii. For those of us living in Turtle Island especially, we need to decenter the violence and materialistic obsession of capitalism;

iii. We need to *recenter* the Earth as Mother and the Eternally Powerful Beginningless and Endless Spiritual Universe as the sole Power of life; and

iv. We need to accept the leadership of Indigenous peoples in steering us all back to the path of Earth Mother and restoration of relationships with all life, requiring and supporting the return of all dispossessed and colonized Indigenous lands for such restoration for the healing of all life.

Such a reorientation and shift are essential prerequisites in permanently uprooting the chauvinistic and patriarchal anthropomorphic dysfunctionality that plagued the totalitarian Roman empire and people subsequently. This movement will refocus our entire global cultural, economic, educational, social, and political systems toward the rest of the natural world, which has begged us to stop the incessant insanity of anthropomorphism. All of the plants, trees, birds, insects, four-leggeds, sea life, micro-organisms, rocks, hills, mountains, valleys, and waterways have always desired conjunctive relationships with us so that we can share the fruits offered by our Mother Earth in mutuality and reciprocity. As the youth of Soweto, Azania/South Africa sang so mournfully in June 1976 following the merciless police killing of almost a thousand students from the uprising against the colonial Afrikaans language and the apartheid system, "*Senzeni na? Senzeni na? Senzeni na? Senzeni na?*" "What have we done? What have we done? What have we done?" What have we done to our Mother Earth?

In this mode of liberation, too, it's critical to work industriously to eviscerate the false consciousness from capitalist violence towards the Earth and other life forms. It's important to shy away from condemnatory language that demeans buzzards, vultures, bats, and owls and the like as "predators." Similarly, calling all crocodiles, elephants, wolves, coyotes, tigers, and other large creatures who are relatives "beasts" when they pursue other creatures for food in consonance with their natural way of living and part of the cyclical balance of the creation is unhealthy and demeaning and only functions to widen the chasm between the creatures of the "wild" and ourselves, something we desperately need to break within ourselves. These creatures of Creation have not destroyed the ecology and environment, but capitalist greed and extraction have. Formal science education, particularly in medicine, needs to halt the torture and dissection of our creature relatives and instead observe the creatures about whom we need to have a scientific understanding in their natural settings, from which Indigenous people have traditionally learned science, which we have done with ants, and the late renowned eco-biologist E.O. Wilson has done the same.

Concretely, the wisdom teachings of our healers and elders are the most appropriate for the final pages of this epilogue, and you, dear reader, are invited to join this endless Circle of Life and Revitalization. Theda New Breast, elder and teacher from the Blackfeet Nation and daughter of well-known elder and social worker Betty Cooper, with whom we worked in the 1980s, proposed this eleven-point way of living, which she explained could be modified and used according to the language and culture of various indigenous communities and people everywhere in whatever context they lived:

i. Be rooted in indigenous spirituality connecting us to all forms of life on Earth, with Mother Earth and the Spirit Powers in particular places;

ii. Show kindness to others that celebrates all life as sacred and assist those who need guidance or request such assistance in their life journeys;

iii. Show respect for others and who they are as they are and meant to be by the Creator, Creation, and Ancestors;

iv. Live with purpose in whatever you do, understanding that you have been placed on the Earth for some particular role in a particular place and time and need to do your utmost to understand and fulfill your spiritual purpose as the Creator, Creation, and Ancestors intended. For Indigenous communities worldwide, initiation and puberty ceremonies are the mediums by which such purposes are determined and understood (Some, 1995);

v. Always be who you are. Never attempt or struggle to erase your identity to be someone else because others insist on or coerce you into rejecting your self-identity and community;

vi. Never sell your Ancestors for anything. Always adhere to the ancestral path and honor the ancestors in you even when seemingly insurmountable obstacles compel you to stray from the path;

vii. Take on tasks independently and never wait to be instructed to do so, honoring your spiritual creativity and originality while always being considerate of the broader community;

viii. Be helpful in whatever way you can, an active member of society as opposed to standing or sitting around waiting for others to do what you could be doing;

ix. Use your community and personal medicine to always heal others and situations provided by the Mother Earth and the Spirit Powers. For instance, the

Blackfoot nation has devil's club, which can be used to cure all kinds of illness like unstable blood sugar, diabetes, arthritis, high or low blood pressure, coughs, pneumonia, tuberculosis, fevers, and the like. The Earth provides everything for our wellness. Always share your medicine, but never dictate the terms by which you share such healing medicine. Allow every person to use the medicine in the manner in which she or he desires;

    x. Transfer knowledge to others with patience. When demonstrating use of technology, repeat processes several times if necessary and practically demonstrate how the particular technology should be used. Demonstrate your patience if necessary by fasting for four days if clarity on situations is needed;

    xi. Be aware of your surroundings and be alert always. Just as the newborn baby is aware of being in the cradle and feeding on the mother's breast milk, be aware and conscious of your environment as you live your life journey.[75]

The Indigenous Kiowa scholar N. Scott Momaday echoes the power of the Earth Mother, our ultimate teacher, healer, refuge, and protector:

> We humans must revere the Earth, for it is our well-being. Always the Earth grants us what we need. If we treat the Earth with kindness, it will treat us kindly. If we give our belief to the Earth, it will believe in us. There is no better blessing than to be believed in.... The Earth is alive, and it is possessed of spirit. Consider the holy tree. It can be allowed to thirst. It can be cut down. Worst of all, it can be denied our faith in it, our belief. But if we speak to it, if we pray, it will thrive.[76]

Similarly, Tadodaho Elder and Faith Keeper Leon Shenandoah has wisely taught us:

> The Creator is not a he or she.... Being a "Human Being" is being closer to the Creator. When a "Human Being" is close to the Creator, then they just know things. It's called the "knowing." "Human Beings" don't know all things. Each one is given a different ability. Some know some things. Others know other things. But no one knows it all. That's why we need each other. When we come together we can know more things. That's what participating in the ceremonies is all about. When we are together in the circle, we are one. As one we can be more in "the knowing." In the circle we get closer to the Creator. We know that nothing is impossible with the Creator.... I want to be living in a way that Creator will be happy with me.[77]

Indigenous artist Nakoma also echoes this love for our Earth Mother in his classic painting and poster, *The Circle Is My Path—From where I came I will return*, that states:

> The Earth is my floor, and it is sacred. The Sky is my cathedral, my altar all about. I must cry when Mother Earth is hurt. If Nature is damaged or destroyed forever I cannot live, life cannot be sustained. The elements of the Sacred Circle are meant to be joined, beating to a common pulse. If one dies, it is the same as family dying.

From our Earth Mother we all came, and to her we inevitably return. Honor our Mother!

Ajo! Hózhó Naasha ... Balance and Beauty ... (Diné) ... *wAnZeOshelA ... Bless You* (Yuchie) ... *Iintsikelelo ... Blessings ...* (isiXhosa of the Nguni tradition).

Chapter Notes

Preface

1. Diné language for the Earth.
2. Michael E. Marchand and Kristiina A. Vogt (eds.), *The River of Life: Sustainable Practices of Native Americans and Indigenous Peoples*.
3. The piece, "Mother Earth" by *Wuauquikuna*, is accessible on the web (in the bibliography). The Earth Mother names indicate Diné/Navajo, Indigenous Quechuan, Aymara, and other Indian nations in the southern part of the western hemisphere, Nguni languages of southern Africa, Maori, Chinese, and Akan of west Africa respectively. These are just a fraction of the thousands of Indigenous languages and cultures that revere Earth as Mother.
4. See, for instance, Chinweizu, *White Predators, Black Slavers, and the African Elite*; Eric C. Williams, *Capitalism and Slavery*; Gerald Horne, *The Dawning of the Apocalypse: The Roots of Slavery, White Supremacy, Settler Colonialism, and Capitalism in the Long Sixteenth Century*; and Walter Rodney, *How Europe Underdeveloped Africa*.

Chapter 1

1. Richard Erdoes and Alfonso Ortiz (selectors and editors), "Rabbit Boy" in *American Indian Myths and Legends*, 5. The term "man" in the quote has been replaced by "humanity" to indicate gender inclusiveness.
2. Elm, Demus, and Harvey Antone. Translated and edited by Floyd Lounsbury and Bryan Glick. *The Oneida Creation Story*, chapters 1 and 2. See also the Oneida Nation site.
3. Canadian Museum of History.
4. Victoria R. Williams, *Indigenous Peoples: An Encyclopedia of Culture, History, and Threats to Survival*, Volume 4, 1046.
5. Virginia Hamilton, *In the Beginning: Creation Stories from Around the World*, 35–41.
6. Kola Abimbola, *Yoruba Culture*, 51–52.
7. Ibid., 49.
8. Marimba Ani, *Yurugu*, Author's Note.
9. Virginia Hamilton, *In the Beginning*, 104–109.
10. Zohl dé Ishtar, "A-Bombs to Star Wars—The Sixty-Six Years War on Marshall Islanders," in *Paradigm Wars: Indigenous Peoples' Resistance to Globalization*.
11. Percy Bullchild, *The Sun Came Down*, 1–38.
12. Greg Cajete, *Native Science*, 13 ff, and adaptations of the creation accounts in Hallie N. Love's *Watakame's Journey*.
13. "Animals" are in quotes because Indigenous languages traditionally did not view what we consider "animals" today as being removed from the human sphere, but integral to humanity.
14. "Creation of the Animal People" (Okanagan) in *American Indian Myths and Legends*, 14–15.
15. Lily Bennett, Margaret Katherine, Sybil Ranch, and Liz Thompson, *Living with the Land*, 6.
16. Lily Bennett, Margaret Katherine, Sybil Ranch, and Liz Thompson, *Living with the Land*, 5.
17. Willie Gordon (with Jody Bennett), *Gurrbi: My Special Place*.
18. Ibid., 15.
19. Rangimarie Turuki Pere, "A Celebration of Maori Sacred and Spiritual Wisdom," in *Indigenous Peoples' Wisdom and Power*, 148–150.
20. Mickias Musiyiwa, "The Significance of Myths and Legends in Children's Literature in Contemporary Zimbabwe."
21. John S. Mbiti, *Introduction to African Religion, Second Edition*, 34–35.
22. Paula Gunn Allen, *Grandmothers of the Light*, 33. The quote is taken from John M. Gunn (1917), 1977, *Schat-Chen: History, Traditions, and Narratives of the Queres Indians of Laguna and Acoma*, 89.
23. Paula Gunn Allen, *Grandmothers of the Light*, 89.
24. Ibid.
25. Ibid., 66.
26. Ibid.
27. Ifi Amadiume, *Male Daughters, Female Husbands*, 15.
28. Ibid., 10.
29. Oyeronke Oyewumi, *The Invention of Women*.
30. See, for example, Vine Deloria's classic and last book, *The World We Used to Live In*, and N. Scott Momaday's instructive work, *Earth Keeper*, on this subject.

31. Jennifer Schuessler, "Lessons from Ants to Grasp Humanity," *New York Times*, April 9, 2012.
32. Deloria, *The World We Used to Live In*, 213.
33. "Are People Getting Full Facts on COVID Vaccine Risks," *Total Health: Written by Leading Doctors*, August 1, 2021.
34. Mae-Wan Ho, "The New Genetics and Dangers of GMO" in *Seed Sovereignty, Food Security*, 105. The scientists confirming this information are A. Marnef and G. Legube, "m6RNA Modification as a New Player in R-loop Regulation," *Nature Genetics*, 52, 2020, highlighted by Eyrun Thune, "Modified RNA has a direct effect on DNA stability," at the Institute of Basic Medical Sciences, University of Oslo, 2020.
35. Vandana Shiva, *Making Peace with the Earth*, 17.
36. Vine Deloria, *Red Earth, White Lies*, 51–52.
37. *Ibid.*, 52–53.
38. Taiaiake Alfred, *Wasáse*, 32.
39. Vine Deloria, *Red Earth, White Lies*, 53.
40. *Ibid.*, 55–56.
41. Vine Deloria, *The World We Used to Live In*, 197–198.
42. *Ibid.*, 198–199.
43. John (Fire) Lame Deer and Richard Erdoes, *Lame Deer*, 161–163.
44. Malidoma Somé, *Of Water and the Spirit*, 178.
45. John (Fire) Lame Deer and Richard Erdoes, *Lame Deer*, 39.

Chapter 2

1. Robert Bartlett, *The Making of Europe: Conquest, Colonization, and Cultural Change 950–1350*, 60. Available online via Google Books.
2. Christopher Columbus, "Columbus' Journal Entries from August to November 1492" in *Digital History*, ID 1248.
3. Bartolomé de las Casas, *A Short Account of the Destruction of the Indies*, 15, 26.
4. See, for instance, Edward Tannenbaum, *European Civilization Since the Middle Ages* (Second Edition); Hugh Trevor-Roper, *The Rise of Christian Europe*; and R.I. Moore, *The War on Heresy*.
5. See, for instance, John Jackson, *Introduction to African Civilizations*, specifically the chapter on "Africa and the Civilizing of Europe."
6. John Jackson, *Introduction to African Civilizations*, 175.
7. Stanley Lane-Poole, *The Story of the Moors in Spain: A History of the Moorish Empire in Europe*, 1990, v. Though there are constant disagreements between scholars on the Blackness of the Moors, two facts are patently clear: first, "Race" and the concept of race as we have been conditioned and indoctrinated into accepting globally, where Blackness denotes the lowest rungs of humanity and Whiteness the pinnacle, did not exist in the times that the Moors occupied Spain and even in the ancient world, and second, there was unquestionably a Black Moorish presence in Spain amid the diversity of peoples originating from North Africa, as is the case today generally in the African-Arab-Asiatic world. See, for instance, Cozmo El, *Moor Vol I. and II: What They Didn't Teach You in Black History Books*, which depicts various examples of the Black Moorish presence, as well as Stanley Lane-Poole's work, *The Story of the Moors After Spain*, 2017, previously titled *The Barbary Corsairs*, 1890.
8. Kirkpatrick Sale, *The Conquest of Paradise: Christopher Columbus and the Columbian Legacy*, 32, 29–36.
9. *Ibid.*, 29.
10. *Ibid.*, 38. Sale quotes from Lauro Martines' *Power and Imagination: City-States in Renaissance Italy*, 216.
11. Lauro Martines, *Power and Imagination*, 196.
12. *Ibid.*, 197.
13. *Ibid.*, 197.
14. *Ibid.*, 191.
15. Kirkpatrick Sale, *The Conquest of Paradise*, 39.
16. *Ibid.*, 40.
17. *Ibid.*, 42. Sale refers to British science historian Peter Matthias, who attacked the ideology of European science and its obsessiveness with "increasingly seeking to improve, wanting to experiment," and considered such ideology as demonic and unbefitting of "real science." Mathias's book, *Science and Society*, 78, provides an illuminative background on the deification of "science, technology, and ecological annihilation."
18. Kirkpatrick Sale, *The Conquest of Paradise*, 42.
19. Joukowsky Institute for Archaeology and the Ancient World, Brown University, Providence, Rhode Island, "Thirteen Things: Gunpowder: Origins in the East," 2009, accessed on November 5, 2021.
20. *Ibid.*, Gerald Horne, *The Apocalypse of Settler Colonialism: The Roots of Slavery, White Supremacy, and Capitalism in Seventeenth Century North America and the Caribbean*, 25–26, 42, 45, 73, 126, 155.
21. Chinweizu, *The West and the Rest of Us: White Predators, Black Slavers and the African Elite*, 3.
22. Kirkpatrick Sale, *The Conquest of Paradise*, 83. The Domesday Book was a survey of the feudal and social structures and policies and practices in English and Welsh society ordered by William I in 1086.
23. Keith Thomas, *Man and the Natural World: A History of the Modern Sensibility*, 15–16.
24. *Ibid.* Thomas notes the British colonial dispossession of the Indigenous people of Massachusetts and other areas of Turtle Island that were invaded is based on the works of Samuel Purchas, *Hakluytus Posthumus or Purchas his Pilgrims*, xix, 218–224; Chester E. Eisinger, "The Puritans'

Justification for Taking the Land," lxxxiv, 1948; Roy Harvey Pearce, *Savagism and Civilisation*, 1965, 21; and Alden T. Vaughan, *New England Frontier*, 110–12, conferring from John Locke, *Two Treatises of Government*, edited by Peter Laslett, 308–10, 315–16 (I paragraphs 32–35, 42). See also Peter Laslett's edited book, *Locke: Two Treatises of Government*.

25. Kirkpatrick Sale, *The Conquest of Paradise*, 86.

26. *Ibid.*, 87.

27. Klement Tockner, Urs Uehlinger, and Christopher T. Robinson, *Rivers of Europe*, 2009, chapter 7.

28. Hayley Munguia, "The 2,128 Native American Mascots People Aren't Talking About," *FiveThirtyEight*, September 5, 2014.

29. Thin Lei Win, "Fighting Global Warming, One Belch at a Time," Thomson Reuters Foundation, July 19, 2018.

30. See Winona LaDuke's excellent book, *All our Relations: Native Struggles for Land and Life*, 1999, specifically chapter 7, "Buffalo Nations, Buffalo Peoples."

31. Keith Thomas, *Man and the Natural World*, 26–27. The quote in the citation is taken from William Rabisha's *The Whole Body of Cookery Dissected*, 1661, 241. For the carving reference, see Hannah Woolley's *The Queen-Like Closet* (5th edition), 1684, 258; *The Letters of the Earl of Chesterfield to his Son* (1901), i, ii, 89, 94, 118, 269; John Trusler, *The Honours of the Table* (2nd edition), 1791, 24ff; and John Hodgkin, "Proper Terms II," *Trans. Philological Soc.*, 1907–10, on page 312 of *Man and the Natural World*. See also the chapter on the Lakota in Winona LaDuke's book, *All our Relations*.

32. Kirkpatrick Sale, *The Conquest of Paradise*, 89.

33. Lewis Mumford, *Technics and Civilization*, 5.

34. *Ibid.*, 13–14.

35. Karl Marx, *Capital: A Critique of Political Economy, Volume 1: The Process of Capitalist Production*, 371 ff.

36. See, for instance, chapter 8 of Vine Deloria's *The World We Used to Live In*, and MariJo Moore and Tracey A. Demeyer's *Unraveling the Spreading Cloth of Time*, especially Part 1.

37. See Scott Timke's instructive article, "West Africa is the Latest Testing Ground for U.S. Military Artificial Intelligence," *Mint Press News*, April 22, 2021.

38. See, for instance, Helen Rawlings' work, *The Spanish Inquisition*, 2. Other important and relevant works are Charles Lea, *The Inquisition of Spain, Volumes I, II, III, and IV*; Emmanuel La Roy LaDurie, *Montaillou: The Promised Land of Error*, which delves into the Cathar controversy and Papal persecution of this community; Henry Kamen, *The Spanish Inquisition: A Historical Revision*; Toby Green, Philip Havik, and F. Ribeiro da Silva's edited work, *African Voices from the Inquisition, Volume 1: The Trial of Crispina Peres of Cacheu, Guinea-Bissau 1646–1668*, on the manner that enslaved Africans were subject to the Inquisition in Portugal during the Portuguese enslavement of Africans and colonial occupation of Brazil, including the life and experience of Chrispina Peres, who had lived in Portugal as a productive trader for a few years, but was then tried for sorcery and deported to Guinea-Bissau, where she was born in the 17th century; and James H. Sweet, *Domingos Alvarez: African Healing and the Intellectual History of the Atlantic World*, which documents that painful trial of Domingo Alvarez, a person from Benin who was kidnapped and enslaved in Brazil, but fought and realized his freedom, only to be short-lived by the Portuguese Inquisition that deported him to Portugal.

39. Eduardo Galeano, *Open Veins of Latin America*, 12.

40. *Ibid.*

41. Pope Alexander VI. *Inter caetera* (May 3, 1493). In *European Treaties Bearing on the History of the United States and its Dependencies to 1648*. Ed. Frances Gardiner Davenport. Washington, D.C.: Carnegie Institution of Washington, 1917. Encyclopedia Virginia: Virginia Humanities.

42. Robert J. Miller, Jacinta Ruru, Larissa Behrendt, and Tracey Lindberg, *Discovering Indigenous Lands: The Doctrine of Discovery in the English Colonies*, 21. Miller cites the works of Henry Reynolds, *The Law of the Land*, 173; Lynn Berat, *Walvis Bay: Decolonization and International Law*, 118; Colin G. Calloway, *Crown and Calumet: British-Indian Relations 1783–1815*, 9; and Alex C. Castles, "An Australian Legal History" in *Aboriginal Legal Issues: Commentary and Materials*, 10, 63, for his explanation.

43. Luis Rivera Pagán, *A Violent Evangelism*, 7. The source is Consuela Varela, "Prólogo," *Textos y Documentos Completes: Relaciones de Viajes, Cartas y Memorials*, 140.

44. Here the assertion basis is from some of the important books on the genocide of Indigenous peoples in the Western hemisphere following the European—particularly Iberian—invasion from the 15th century. These are principal and reliable historical and analytical texts and include William Deneven, *The Native Population of the Americas in 1492*; Henry F. Dobyn's *Their Numbers Became Thinned: Native American Population Dynamics in the Eastern North America*; and Russell Thornton's *American Indian Holocaust and Survival: A Population History Since 1492*. Other important books and articles on the subject of genocide by European colonial invasion in the western hemisphere are: David Michael Smith's paper, "Counting the Dead: Estimating the Loss of the Indigenous Holocaust, 1492-Present"; Andrew Woolford, Jeff Benvenuto, and Alexander Laban Hinton, *Colonial Genocide in Indigenous North America*; James Wilson, *The Earth Shall Weep: A History of Native America*; and David E. Stannard, *American Holocaust: The Conquest of the New World*.

45. James Loewen, *Lies My Teacher Told Me*, 64–65.
46. Cynthia Chambers, "'We Are All Treaty People': The Contemporary Countenance of Canadian Curriculum Studies" in *Reconsidering Canadian Curriculum Studies*, edited by Nicholas Ng-A-Fook and Jennifer Rottman, 23–38.
47. See, for instance, Jack Beeching, *The Chinese Opium Wars*, and James Bradley, *The China Mirage*.
48. Gerald Horne, *The Apocalypse of Settler Colonialism*.
49. Pietro Martire d'Anghiera, *De Orbe Novo, the Eight Decades of Peter Martyr d'Anghera*.
50. Bartolomé de las Casas, *A Short Account of the Destruction of the Indies*. 24–25.
51. Eric Williams, *Documents of West Indian History, Vol.1, 1492–1655*, 67.
52. Andrés Reséndez, *The Other Slavery*, 23–24. Though Reséndez has penned a relevant educational book on the enslavement of Indigenous people in the western hemisphere, his uncritical use of the terms "New World" and "slaves" dilute the transformative educational thrust of the book. "Indigenous world of the western hemisphere" and "enslaved people" would be more accurate to reflect the process of Europeans kidnapping and shipping off Indigenous and African peoples into slavery for the western European accumulation of capital. His source endnotes (18) for the Portuguese and Columbus' presence in west Africa are found on page 336 of his book.
53. Amerigo Vespucci, *Mundus Novus*, translated by George Tyler Northrup, 1, from *Encyclopedia Virginia: Virginia Humanities*.
54. Eduardo Galeano, *Open Veins of Latin America*, 32.
55. *Ibid.*, 30.
56. Celso Furtado, *The Economic Development of Latin America: A Survey from Colonial Times to the Cuban Revolution*, 13.
57. Sidney W. Mintz, *Sweetness and Power: The Place of Sugar in Modern History*, xx.
58. Trevor Burnard and John Garrigus, *The Plantation Machine: Atlantic Capitalism in French Saint Domingue and British Jamaica*, 3.
59. *Ibid.*, 39.
60. *Ibid.*, 38.
61. *Ibid.*, 37–38.
62. Gerald Horne, *The Apocalypse of Settler Colonialism*, 24.
63. *Ibid.*, 78.
64. See, for instance, Jean Eddy Saint Paul, "Assassinations and Invasions: How the U.S. and France Shaped the Long History of Haitian Turmoil," *The Conversation*, August 27, 2021; and Nikolas Barry-Shaw and Dru Oja Jay, "New Documents Detail How Canada Helped Plan 2004 Coup d'État in Haiti," *The Breach*, July 15, 2021.
65. Rod Prince, *Family Business*, 18.
66. B. C. Daurelle, "Haitian Textile Workers Strike Against U.S. and International Sweatshops," *Left Voice*, February 18, 2022.
67. See, for example, the foundational work of C.L.R. James, *Black Jacobins*, and Charles Arthur and Michael Dash (eds.), *A Haiti Anthology: Libète*.
68. Abihijit Mohanty, "Extracting a Radioactive Disaster in Niger," *DownToEarth*.
69. Eric Williams, *Capitalism and Slavery*, 51–52.
70. Philip Foner, "The International Slave Trade" in *African Americans in the U.S. Economy*, edited by Cecilia A. Conrad, John Whitehead, Patrick Mason, and James Stewart, 12–13.
71. Allan McPhee, *The Economic Revolution in British West Africa, Second Edition*. The citation in the paper is taken from Joseph Inikori's classical study on slavery and the industrial revolution, *Africans and the Industrial Revolution in England: A Study in International Trade and Economic Development*, 403–404.
72. Joseph Inikori, *Africans and the Industrial Revolution in England*, 404.
73. Chinweizu, *The West and the Rest of Us*, 273.
74. Julian E. Kunnie, *The Cost of Globalization: Dangers to the Earth and Its People*, 265.
75. Gerald Horne, *The Apocalypse of Settler Colonialism*, 47, 54.
76. See, for instance, Eugene Genovese, *Roll, Jordan, Roll: The World the Slaves Made*; Albert Raboteau, *Slave Religion*; John Hope Franklin and Loren Schweninger, *Runaway Slaves*; Herbert Aptheker, *American Negro Slave Revolts*; and Sterling Stuckey, *Slave Culture*.
77. See, for instance, Edward Baptist, *The Half Has Never Been Told: Slavery and the Making of American Capitalism*; Cecilia Conrad, John Whitehead, Patrick Mason, and James Stewart (eds.), *African Americans in the U.S. Economy*; and Walter Johnson, *River of Dark Dreams*. The author was part of a team of Indigenous Ndeé Neé (Apache) peoples of the Southwest that filed a shadow report for reparations with the Committee for the Elimination of Racial Discrimination (CERD) of the United Nations for historical acts of genocide by the Holy See, the Papal authorities that sanctioned conquest, colonization, and confiscation of Indigenous lands in the Western hemisphere, and the subsequent forced conversion of Indians to Christianity, and for those who did not, extermination. The systematic extermination of Indigenous Indians and the exceptionally high mortality rate of enslaved Africans in the Americas and the Caribbean is by definition "genocide," referring to a "mass slaughter of people." The reference to the 100 million Indigenous people in the Americas at the time of initial European invasion is from Roxanne Dunbar-Ortiz, *An Indigenous Peoples' History of the United States: ReVisioning American History*, on the first page of chapter 1 in the eBook version entitled "Follow the Corn."
78. David Eltis and David Richardson, *Atlas of the Transatlantic Slave Trade*, 19.
79. Gerald Horne, *The Apocalypse of Settler Colonialism: The Roots of Slavery, White Supremacy, and Capitalism in Seventeenth Century North America and the Caribbean*, 47, 7–8. In his endnotes in this

regard, Horne cites the work of Wendy Warren, *New England Bound: Slavery and Colonization in Early America*, 4, 5, and Graham Ellison, *Destined for War: Can America and China Escape Thucydides' Trap*, 239. He points out that Ellison referred to historian Niall Ferguson and cited six "killer apps" that were key in elevating the hegemonic dominance of European nations: "competition; scientific revolution; property rights; modern medicine; consumer society; and work ethics," yet strangely and consciously elided any serious reference to slavery, colonialism, and white supremacist ideology that empowered both oppressive systems.

80. Gerald Horne, *The Apocalypse of Settler Colonialism*, 47, 7-8. In his endnotes in this regard, Horne cites the work of Wendy Warren, *New England Bound*, 4, 5 and Graham Ellison, *Destined for War*, 239. He points out that Ellison referred to historian Niall Ferguson and cited six "killer apps" that were key in elevating the hegemonic dominance of European nations: "competition; scientific revolution; property rights; modern medicine; consumer society; and work ethics," yet strangely and consciously elided any serious reference to slavery, colonialism, and white supremacist ideology that empowered both oppressive systems.

81. Gerald Horne, *The Apocalypse of Settler Colonialism*, 47, 8. Horne cites the work of Andrés Reséndez with regard to Indigenous resistance to slavery in what has come to be known as "New Mexico," *The Other Slavery: The Uncovered Story of Indian Enslavement in America*, 5-6, 149.

82. Robert Fogel, *Without Consent or Contract: The Rise and Fall of American Slavery*, and Joe William Trotter, *The African American Experience*.

83. John Lienhard, "Engines of Our Ingenuity, No. 127: Black Inventors." Black inventors have never received legal and proper credit for their inventive genius and their foundational creative contributions to the development and expansion of the U.S. economy and industry, particularly during slavery, but even into the present where racism and white supremacy continue to be the normative lens through which history is recounted and the world is perceived. See Portia James, *The Real McCoy: African American Invention and Innovation, 1619–1930*; James Michael Brodie, *Created Equal: The Lives and Ideas of Black American Innovators*; and Dorothy Yancy: "Four Black Inventors with Patents," *Negro History Bulletin* 39 [1976]: 574.

84. Joe William Trotter, *The African American Experience*, 151.

85. Andrés Reséndez, *The Other Slavery: The Uncovered Story of Indian Enslavement in America*, 65.

86. Ibid., 66.

87. Ibid., 196.

88. M. Kat Anderson, *Tending the Wild: Native American Knowledge and the Management of California's Natural Resources*, 89.

89. Ibid.

90. Horace Bell, *Reminiscences of a Ranger: Or, Early Times in Southern California*, 48-49.

91. M. Kat Anderson, *Tending the Wild*, 90.

92. J. H. Carson, *Recollections of the California Mines: An Account of the Early Discoveries of Gold, with Anecdotes and Sketches of California and Miners' Life, and a Description of the Great Tulare Valley*, 69.

93. See Mark Dowie's informative book, *Conservation Refugees,* and Métis writer Mark David Spence's book *Dispossessing the Wilderness: Indian Removal and the Making of the National Parks*, 1999. Though a couple of decades old, it is still worth reading.

94. David Treuer, "Return the National Parks to the Tribes," *The Atlantic,* May 2021.

95. Roxanne Dunbar-Ortiz, *An Indigenous Peoples' History of the United States*. See also her 2021 book, *Not a Nation of Immigrants*.

96. Karl Marx, *Capital: A Critique of Political Economy, Volume 1: The Process of Capitalist Production*, 751.

97. See, for instance, Walden Bello's excellent article, "Post-9/11 'Nation-Building' in Afghanistan and Iraq was Nothing But Destruction," *Common Dreams,* September 12, 2021. Bello explains that the U.S. occupation of the Philippines was facilitated and systematically imposed on the Filipino people by "U.S. colonial authorities and Protestant missionaries" who provided the blueprint for "liberal democracy," U.S. empire style. He also notes that this "experiment in liberal democracy" was based on the U.S. experience in the last years of its colonial occupation of the country toward the final years of the 19th century when the U.S. assumed colonial reins from the Spanish.

98. International Indigenous Peoples' Movement for Self-Determination and Liberation (IPMSDL), "Stop the Attacks against Lumad! No to Campus Militarization," IPMSDL website, February 15, 2021.

99. Chinweizu, *The West and the Rest of Us: White Predators, Black Slavers and the African Elite*, 6.

100. Gerald Horne, *The Apocalypse of Settler Colonialism: The Roots of Slavery, White Supremacy, and Capitalism in Seventeenth Century North America and the Caribbean*, 53-54.

101. Ibid.

102. Walter Rodney, *How Europe Underdeveloped Africa*, 82.

103. Jack Beeching, *The Chinese Opium Wars*, 66-67.

104. Ibid., 75-76.

105. Kristyn Harman, "Explainer: The Evidence for the Tasmanian Genocide," *The Conversation*, January 17, 2018. Harman mentions two important recent books that substantiate the article's contentions and following James Bonwick's 1870 text, *The Last of the Tasmanians*, Tom Lawson, *The Last Man,* and Nick Brodie's *The Vandemonian War*. Sven Lindqvist's *Exterminate All the Brutes,* 1997, 2021, is another instructive source for understanding European colonial genocide.

106. See, for instance, Georges Nzongola-Ntalaja,

The Congo from Leopold to Kabila, 2002, 2003; Chinweizu, *The West and the Rest of Us*; and Adam Hochschild, *King Leopold's Ghost*.

107. This point was acknowledged and addressed by a meeting between the African Atlantic group members in Havana, Matanzas, and Santiago de Cuba and the Cuban Minister of Culture as part of the schedule for the African Diaspora Conference in July 2000, of which the author was part.

108. Oliver Stone's 2016 documentary *Ukraine on Fire* provides an excellent illumination of the history of U.S. and NATO's imperialist interventions in and designs for Ukraine, including the U.S.-orchestrated coup against Viktor Yanukovych in 2014.

109. Amnesty International, "Israel/OPT: Civilian Deaths and Extensive Destruction in Latest Gaza Offensive Highlight Human Toll of Apartheid," June 13, 2023.

110. See Ian Zabarte, "A Message from the Most Bombed Nation on Earth," *Al Jazeera*, August 29, 2020.

111. See Ramsey Clark's poignant book *The Fire This Time*, and the Geneva International Centre for Justice, "Razing the Truth about Sanctions Against Iraq."

112. Kwame Nkrumah, *Neo-Colonialism*.

113. Karl Marx, *Capital: A Critique of Political Economy, Volume 1*, 742–743.

114. *Ibid.*, 94.

115. *Ibid.*, 143–144.

116. Ann C. Tweedy, "From Beads to Bounty: How Wampum Became America's First Currency—and Lost Its Power," *Indian Country Today*, October 5, 2017.

117. Karl Marx, *Capital: A Critique of Political Economy, Volume 1*: 71.

118. *Ibid.*

119. *Ibid.*

120. *Ibid.*, 79.

121. *Ibid.*, 264–265.

122. *Ibid.*, 265. Marx based this assertion on accounts of manufacturing businessmen in England, noted on this page as l.c., 64, p. xiii.

123. King White, "How Big is the U.S. Call Center Market Compared to India, Latin America, and the Philippines?," *Site Selection Group Blog*, Site Selection Group. The author was informed by colleagues upon encountering a call worker at a restaurant in Baguio City in the Philippines whom he attempted to talk with on his work in August 2013 that such workers were under very tight security restrictions by their respective companies and were prohibited from divulging any information about work conditions, the violations of which would result in worker termination.

124. Karl Marx, *Capital: A Critique of Political Economy, Volume 1*, 461.

125. *Ibid.*, 443.

126. See, for instance, Sam Yellowhorse Kesler, "Indian Boarding Schools' Traumatic Legacy, and the Fight to Get Native Ancestors Back," *Code Switch: Race in Your Face: National Public Radio*; "Canada: 600 Graves Found at Indigenous Boarding School," *Al Jazeera*, June 24, 2021; and "Who Are the Stolen Generations?," *Healing Foundation*.

127. Karl Marx, *Capital: A Critique of Political Economy, Volume 1*, 462.

128. *Ibid.*, 430–431.

129. *Ibid.*, 462.

130. *Ibid.*, 635.

131. Eugene Kamenka (ed.), *The Portable Marx*, 224.

132. Aryn Baker/Georgia, "Extreme Heat is Endangering America's Workers—And Its Economy," *Time*, August 2023, 46.

133. *Ibid.*, 50.

134. The author had the privilege and blessing to meet with Cesar Chavez during one of his farm worker rights fasts at the UFW Convention in Fresno in the early 1980s. The outrageous colonization of Mexican farm workers, especially migrant laborers, continues, with a blind eye from the U.S. government. See Daniel Costa and Philip Martin, "Record-Low Number of Federal Wage and Hour Investigations of Farms in 2022," *Economic Policy Institute*, Aug 22, 2023.

135. Lu Xian and Ruo Yan, "After Workers Flee China's Largest iPhone Factory, Activists Demand Accountability from Apple," *Labornotes*, November 10, 2022.

Chapter 3

1. *Climate Home News*, "Kiribati President: My People Are the Polar Bears of the Pacific."

2. See, for instance, David Goodstein, *Out of Gas*, 16.

3. *Ibid.*

4. Rachel Koning Beals, "Every Whale is Worth $2 Million? Why It's Time to Add the Value of Nature to GDP," *Market Watch*, October 28, 2021. Yet, the legacy of confused European cultures entraps the western capitalist nations today since they are unable to wrench free from the yoke of confused self-identity that perceives us human beings as separate from our Earth Mother and "Nature," and further, insist on attaching a price tag to "Nature" unlike all other global cultures, especially in the Indigenous world.

5. Heidi Scott, "Whale Oil Culture, Consumerism, and Modern Conservation" in *Oil Culture*, edited by Ross Barrett, Daniel Worden and Allan Stoeki, 4.

6. New Bedford Whaling Museum.

7. For arms sales, see Dave Lawler, *Axios*, March 14, 2023.

8. See Lance E. Davis, *In Pursuit of Leviathan*, 51.

9. Derek Thompson, "The Spectacular Rise and Fall of U.S. Whaling: An Innovation Story," *The Atlantic*, February 22, 2012.

10. Emily DeLetter, "Endangered North Atlantic Right Whales Are in 'Crisis': Species Approaching

Extinction," *USA Today*, July 19, 2023; Greenpeace, "Commercial Whaling."

11. Emily DeLetter, "Endangered North Atlantic Right Whales Are in 'Crisis'"; Greenpeace, "Commercial Whaling."

12. Emily DeLetter, "Endangered North Atlantic Right Whales are in 'Crisis'"; Associated Press, "Over 50 Whales Found Dead in Scottish Beach in Worst Mass-Stranding in Area, Experts Say," *USA Today*, July 17, 2023.

13. Michelle Carrere (translated by Sarah Engel), "To Fight Climate Change, Some Scientists Say, Save the Whales," *Mongabay*, March 1, 2021.

14. Julian E. Kunnie, *The Cost of Globalization*, 2015, 225.

15. U.S. Energy Information Administration, Washington DC, "Today in Energy," April 27, 2011.

16. U.S. Energy Information Administration, "How Much Oil Does the U.S. Import and Export?"

17. U.S. Energy Information Administration, Washington DC, "Frequently Asked Questions: How Much of Oil is Consumed In the United States?"

18. Neta C. Crawford, "Pentagon Fuel Use, Climate Change, and the Costs of War" in *Costs of War*.

19. Tim Weiner, *Legacy of Ashes*, chapter 9.

20. Anthony Sampson, *The Seven Sisters: The Great Oil Companies and the World They Shaped*, 7.

21. Ibid., 6.

22. Ibid., 9–13. Alan Sherter, "After 146 Years, Rockefeller Family is Exiting the Oil Business," *CBS News*, March 24, 2016.

23. Anthony Sampson, *The Seven Sisters*, 20–22.

24. Edward Miguel and Gérard Roland, "The Long-Run Impact of Bombing Vietnam," *Journal of Development Economics*, Vol. 96, Issue 1, 2011.

25. Tsira Shvangiradze's outstanding article, "The US' Secret War in Laos: The Most Heavily Bombed Country in History," provides a very detailed and poignant illumination of this genocidal U.S. military campaign against southeast Asia. See *The Collector*, Jan 22, 2022.

26. Douglas Valentine, *The Phoenix Program*, 2000.

27. For comparison today, the U.S.-supplied Israeli military regime dropped almost 30,000 tons of bombs on Gaza over a 42-day period in October-November 2023, about twice the tonnage equivalent dropped on Hiroshima in 1945.

28. Ramsey Clark, *The Fire This Time: U.S. War Crimes in the Gulf*, 1992, 60.

29. Ibid., 59–75.

30. Ahmed al-Rubaie, "How the U.S., UAE and Israel plundered Iraq's Antiquities," *The Cradle*, September 8, 2023.

31. Madeleine Albright, *CBS 60 Minutes*, "Punishing Saddam," May 12, 1996.

32. See especially chapter 3, "War Crimes Against Iraq's Civilian Population," in Ramsey Clark's very well documented *The Fire This Time: U.S. War Crimes in the Gulf*, particularly pages 60–75.

33. Julia Payne, "Israel Accepts First Delivery of Disputed Kurdish Oil," *Reuters*, June 20, 2014. Note the propagandistic character of Reuters here in its obfuscation of the truth of the looting of Iraqi oil!

34. Christine Baumeister and Lutz Kilian, *Journal of Economic Perspectives*, 142.

35. Marc Stocker, John Baffes, and Dana Vorisek, "What Triggered the Oil Price Plunge of 2014–2016 and Why It Failed to Deliver an Economic Impetus in Eight Charts," World Bank, Blogs, January 18, 2018. See also Tom DiChristopher, "The Financial Crisis Crushed Record Oil Prices but the Market Is Still Gripped," *CNBC, Energy*, September 12, 2018.

36. Catherine Anderson and Rebecca Engebretsen, "The Impact of Coronavirus (COVID-19) and the Global Oil Price Shock on the Fiscal Position of Oil Exporting Developing (sic-mine) Countries," Organisation of Economic Co-operation and Development (OECD), September 30, 2020.

37. U.S. Energy Information Administration, "Today in Energy: EIA Forecasts Oil Prices Will Fall in 2022 and 2023," January 12, 2022.

38. Though the truth of Ukrainian suppression of the large Russian-speaking community in the Donbass republics of Donetsk and Lugansk following the U.S.-orchestrated coup in Ukraine in 2014 and the associated nullification of the Minsk II Accords in the western capitalist media establishment has been obscured, M.K. Badrakumar, a longtime Indian diplomat, has provided a sobering illumination and analysis of the Ukraine-Russian conflict. See his article on his site, *Punchline*, "India and the Donbass Republics," February 22, 2022. See also Oliver Stone's accurate and graphic coverage of the roots of the Ukrainian crisis, *Ukraine on Fire*, which documents important Ukrainian history from the 17th century through the overthrow of elected leader Viktor Yanukovych in 2014 through 2016, which was temporarily removed from YouTube's parent company, Google, in March 2022.

39. Anne Schmidt, "Here's How Much Money These Environmental Organizations Make," *FOX-Business*, April 22, 2020.

40. Carly Nairn, "Where Does the Money Go in Environmental Grantmaking?" *Nonprofit Quarterly*, August 23, 2023.

41. Steve Pryor, "A Partial List of the 6,000 Products Made from One Barrel of Oil (After Creating 19 Gallons of Gasoline)," July 26, 2016, *Linkedin*.

42. Brett Wilkins, "Microplastics in Clouds Could Be 'Contaminating Nearly Everything We Eat and Drink': Study," *Common Dreams*, September 29, 2023.

43. See Carlo Petrini's excellent book on this subject, *Slow Food Nation*.

44. L. Human, "Farm Workers March Against Pesticides," *GroundUp*, September 9, 2023.

45. Robert Hunziker, "Insect Decimation Upstages Global Warming," *Counterpunch*, March 27, 2018.

46. National Oceanic Atmospheric Administration (NOAA), National Integrated Drought System

(NIDS), "Special Edition Drought Status Update for the Western United States," May 11, 2023.

47. Ella Nilsen, "Exclusive: Experts Say the Term 'Drought' May be Insufficient to Describe What's Happening in the West," *CNN*, February 16, 2022.

48. Matthew Renda, "Feds to Take Water from Wyoming to Bolster Lake Powell," *Courthouse News Service*, May 3, 2022.

49. Eduardo Sotero, "Worst Drought in Decades Devastates Ethiopia's Nomads," *Agence France-Presse*, May 4, 2022.

50. Peter Schwartzstein and Rojita Adhikari (in Nepal), "Merchants of Thirst," *New York Times*, January 12, 2020.

51. Kate Walton, "In Java, Water is Running Out," *The Interpreter*, September 4, 2019.

52. Hayley Smith and Ian James, "To Survive Drought, Parts of SoCal Must Cut Water Use by 35%. The New Limit Is 80 Gallons a Day," *Los Angeles Times*, April 30, 2022.

53. Jamie Ding, "California Urban Water Use Rose 19% in March Despite Worsening Drought," *Los Angeles Times*, May 10, 2022.

54. Kathleen B. Williams, "The Military's Role in Stimulating Science and Technology: The Turning Point," Foreign Policy Research Institute, May 28, 2010.

55. Adam Morton, "Football Pitch-Sized Area of Tropical Rainforest Lost Every Six Seconds," *The Guardian*, June 2, 2020.

56. Graham Readfearn, "Antarctica Warming Much Faster than Models Predicted In 'Deeply Concerning' Sign for Sea Levels," *Guardian*, September 7, 2023.

57. People for the Ethical Treatment of Animals, "Experiments on Animals: Overview."

58. See Dilyana Gaytandhzieva's site, "Experiments Expose U.S. Biological Experiments on Allied Soldiers in Ukraine and Georgia," *Arms Watch*, January 25, 2022.

59. Matt Zampa, "How Many Animals are Killed for Food Every Day?," *Sentient Media*, September 16, 2018.

60. World Wildlife Fund, *Living Planet Report 2020: Bending the Curve of Biodiversity Loss*, 6.

61. United Nations, UN News: Global Perspective Human Stories, "UN Chief Calls for Bold Action to End 'Suicidal War with Nature,'" October 11, 2021. Of course, it needs to be pointed out as all Indigenous nations and peoples have maintained from time immemorial, we two-legged creatures are nature, too!

62. John R. Platt, "Why Don't We Hear about More Species Going Extinct?," *Scientific American*, June 16, 2019.

63. Marlowe Hood, "World Seeing 'Catastrophic Collapse' of Insects: Study," *Science X Daily, Phys Org*, February 11, 2019. See also Francisco Sánchez-Bayo et al., "Worldwide Decline of the Entomofauna: A Review of Its Drivers," *Biological Conservation*, 2019. DOI: 10.1016/j.biocon.2019.01.020.

64. Kelsey Kopec and Lori Ann Burd, "A Systematic Status Review of North American and Hawaiian Native Bees," "Pollinators in Peril," Center for Biological Diversity, February 2017.

65. National Collaborating Center for Environmental Health, *Neonicotinoid Pesticides*, Vancouver, Canada, January 21, 2020.

66. Andrea Germanos, "Biden Urged to 'Be the Hero' to Save American Bumblebee from Extinction," *Common Dreams*, February 1, 2021. Such appeals are ironic given the U.S. capitalist system's ecocidal history in particular.

67. Marlowe Hood, "World Seeing 'Catastrophic Collapse' of Insects: Study," *Science X Daily, Phys Org*, February 11, 2019. See also Francisco Sánchez-Bayo et al., "Worldwide Decline of the Entomofauna: A Review of Its Drivers," *Biological Conservation*, 2019.

68. Deidre Fulton, "Climate Change Threatens 'Rampant Bird Extinctions,' Study Finds," *Common Dreams*, September 9, 2014; and Felicity Barringer, "Climate Change Will Disrupt Half of North America's Bird Species, Study Says," *New York Times*, September 8, 2014.

69. Vandana Shiva, "Introduction: Seed Sovereignty, Food Security," in *Seed Sovereignty, Food Security: Women in the Vanguard of the Fight Against GMOs and Corporate Agriculture*, x.

70. Ove Hoegh-Guldberg, Elvira S. Poloczanska, William Skirving, and Sophie Dovie, "Coral Reef Ecosystems Under Climate Change and Ocean Acidification," *Frontiers in Marine Science*, May 29, 2017.

71. Nathan Rott, "The World Lost Two-Thirds of Its Wildlife In 50 Years: We Are to Blame," *National Public Radio (NPR)*, September 10, 2020. See Alex Michulich's article to highlight some Indigenous/Black critiques (even though the Heather Davis alluded to is not Indigenous) of the "anthropocene" era from the 1960s, "Indigenous Scholars Invite Decolonization of the Anthropocene," *National Catholic Reporter*, October 8, 2019.

72. Jennifer S. Walker, Robert E. Kopp, Christopher M. Little & Benjamin P. Horton, "Timing of Emergence of Modern Sea Level Rise by 1863," *Nature Communications*, 13, article number 966, February 18, 2022.

73. *Ibid.*

74. "Shale Story: Barnett," Texas Oil & Gas Association (TOGA).

75. American Petroleum Institute, "How Much Water Does Hydraulic Fracturing Use?"

76. Elizabeth Ridlington and Kim Norman (Frontier Group) and Rachel Richardson (Environment America Research & Policy Center), "Fracking by the Numbers: The Damage to Our Water, Land and Climate from a Decade of Dirty Drilling," Environment America Research & Policy Center, April 14, 2016, 5.

77. *Ibid.*

78. Railroad Commission of Texas, "Texas Oil and Gas Production Statistics for December 2021"; *Earth Justice*, "Texas and Fracking," September

29, 2015; and Elizabeth Ridlington (Frontier Group) and John Rumpler (Environment America Research and Policy Center), "Fracking by the Numbers: Key Impacts of Dirty Drilling at the State and National Level," Environment America Research and Policy Center, 2013.

79. Earth Justice, "Texas and Fracking," September 29, 2015. See also *Statista*'s article, "Monthly Number of Started Fracking Operations in the United States from January 2019 to March 2020."

80. Elizabeth Ridlington and Kim Norman (Frontier Group) and Rachel Richardson (Environment America Research & Policy Center), "Fracking by the Numbers: The Damage to Our Water, Land and Climate from a Decade of Dirty Drilling," Environment America Research & Policy Center, April 14, 2016, 7, at https://environmentamerica.org/reports/ame/fracking-numbers-0. Compiled by Casey Ontiveros.

81. U.S. Govt. Pollution Exemptions and Exceptions for Oil and Gas. Compiled by Veronica Rodriguez and Casey Ontiveros.

82. "Uranium Spill Still Worries Navajos," *Special to the New York Times,* July 21, 1983, digitized version from *The Times* print archive.

83. Ibid.

84. "Federal Probe Blames EPA for Gold King Mine Spill," *Associated Press,* October 22, 2015.

85. Jonathan Romeo, "Five Years After Gold King Mine Spill, Water Quality Remains a Concern," *The Durango Herald,* August 5, 2020.

86. Susan Montoya Bryan, "New Mexico Reaches $32 Million Settlement Over 2015 Gold King Mine Spill that Polluted Colorado Water," *CPR News,* June 17, 2022.

87. Sam Metz, "Multibillion Dollar Uinta Basin Railway Chugs Along Amid Environment and Derailment Concerns," *CPR News,* August 10, 2023.

88. P. Erickson and P. Achakulwisut, "How Subsidies Aided the U.S. Shale Oil and Gas Boom" report.

89. Ibid.
90. Ibid.
91. Ibid.
92. Ibid.
93. Ibid.

94. Elizabeth Ridlington and Kim Norman (Frontier Group) and Rachel Richardson (Environment America Research & Policy Center), "Fracking by the Numbers: The Damage to Our Water, Land and Climate from a Decade of Dirty Drilling," April 14, 2016.

95. Ibid., 5.

96. Earthworks, "Fracking-Related Earthquakes."

97. Nala Rogers, "2019: The Year Fracking Earthquakes Turned Deadly," *Inside Science,* February 21, 2020.

98. Michael Behar, "Fracking's Latest Scandal: Earthquake Swarms," *Mother Jones,* March/April 2013 Issue.

99. Matt Smith, "Fracking Seems to be Causing Hundreds of Earthquakes in the Country Each Year," *Vice News,* January 8, 2015. See also Mike Hendricks, "Kansas Has a Lot More Earthquakes. Is that Because of Fracking?," *Kansas City Star,* January 28, 2014.

100. Ibid.

101. Seismological Society of America, "Studies Link Earthquakes to Fracking In the Central and Eastern U.S.," *ScienceDaily,* April 26, 2019.

102. Abrahm Lustgarten, "Injection Wells: The Poison Beneath Us," *ProPublica,* June 21, 2012.

103. U.S. Environmental Protection Agency (EPA), "Underground Injection Control: Class II Oil and Gas Related Injection Wells," September 12, 2021. The 40,000 wells figure from 2013 is from Michael Behar's article, "Fracking's Latest Scandal: Earthquake Swarms," *Mother Jones,* March/April 2013 Issue.

104. U.S. Environmental Protection Agency (EPA), "Underground Injection Control: Class II Oil and Gas Related Injection Wells," September 12, 2021.

105. Abrahm Lustgarten, "Injection Wells: The Poison Beneath Us," *ProPublica,* June 21, 2012.

106. Earthworks, "Fracking-Related Earthquakes."

107. Michael Greenwood, "Chemicals in Fracking Fluid and Wastewater Are Toxic, Study Shows," *YaleNews,* January 6, 2016.

108. See, for instance, *RabbitAir,* Southern California Environmental Health Sciences Center, "Pollution in Los Angeles County," n.d, and Black Pinto, "Real Problems, Real Solutions: Examining Pollution in Los Angeles County," *Voices,* Pacific Oak College, Pasadena, CA.

109. University Of California, Los Angeles, "Air Pollution Linked to Premature Birth In Pregnant Women," *ScienceDaily,* August 27, 2007.

110. Comment from Diane Bailey of the Natural Resources Defense Council in *Scientific American,* "Breathe Wheezy: Traffic Pollution Not Only Worsens Asthma, but May Cause It," January 9, 2013.

111. Julie Sze, *Noxious New York: The Racial Politics of Urban Health and Environmental Justice.*

112. Meredith Hoffman, "Live Near a Fracking Site in Pennsylvania? You Might Be Going to the Hospital More than Others," *Vice News,* July 21, 2015.

113. Renee Cho, "The Fracking Facts," Columbia Climate School, *Climate, Earth, and Society: State of the Planet,* June 6, 2014.

114. Ibid.

115. Irin Ivanova, "Who Are the Biggest U.S Methane Emitters?," *CBS News,* August 30, 2019.

116. Sarah Min, "U.S. Oil and Gas Boom Powers Nation's Fastest-Growing Industry," *CBS News,* February 7, 2019.

117. Megan Ewald, "The Oil Pollution Act of 1990: 30 Years of Oil Spills," National Oceanic and Atmospheric Administration, U.S. Department of Commerce, August 18, 2020, March 21, 2022. The Environmental Protection Energy (EPA) source is

at "Summary of the Oil Pollution Act, 33 U.S.C. § 2701 et seq. (1990)."

118. Megan Ewald, "The Oil Pollution Act of 1990: 30 Years of Oil Spills."

119. Andrew Nikiforuk, "The Oil Spill Cleanup Illusion: Why Do We Pretend to Clean Up Oil Spills In the Ocean?" *Hakai Magazine*, July 12, 2016.

120. Tom Saunders, "2023 on Track to Be World's Hottest Year on Record, Temperatures Exceed 1.5C Above Pre-Industrial Levels For First Time," *ABC Australia*, September 10, 2023.

121. Ben Heubi, "How a Company Got Away After Committing the Worst Oil Spill in History," *E&T, Engineering and Technology*, February 4, 2021.

122. Eyitope Kuteyi, "Many Burnt Beyond Recognition in Imo Illegal Refinery Explosion," *Channels TV*, Nigeria, April 23, 2022.

123. Gabriel Ewepu, Abuja, Nigeria, "Imo Explosion: 11 CSOs Accuse Government of Failure to Address Factors Behind Proliferation of Illegal Refineries," *Vanguard News Nigeria*, April 25, 2022.

124. See, for instance, Ken Saro-Wiwa, *Genocide in Nigeria: The Ogoni Tragedy* and Cyril I. Obi, "Globalization and Local Resistance: The Case of Shell Versus the Ogoni" in *Globalization and the Politics of Resistance*.

125. See the People Advancement Center.

126. See the international Indigenous Peoples' Movement for Self-Determination and Liberation (IPMSDL) based in the Philippines.

127. See the Asia-Pacific Forum on Sustainable Development 2022, March 31, 2022, side event organized by the Asia Pacific Network of Environment Defenders and IPMSDL.

128. *Oxfam International*, "Carbon Emissions of Richest 1 Percent More than Double the Emissions of the Poorest Half of Humanity," September 21, 2020.

Chapter 4

1. "1962: 'HE WHO CONTROLS THE WEATHER, WILL CONTROL THE WORLD,' [LBJ]," *NoGeoengineering: Portal Against Climatic and Environmental Manipulation*, August 2, 2018.

2. "Definition, Importance, and History of Organometallics," *LibreTexts Chemistry*.

3. Carlo Petrini, *Slow Food Nation: Why Our Food Should be Good, Clean, and Fair*, 24–25. Petrini cites Piero Bevilacqua's *La Mucca è Savia*, 2002, 22, for the initial use of chemical fertilizers in Europe in the 1840s. See also east Indian environmentalist Vandana Shiva's introduction to *Seed Sovereignty, Food Security: Women in the Vanguard of the Fight Against GMOs and Corporate Agriculture*.

4. Daniel A. Gross, "Chemical Warfare: From the European Battlefield to the American Laboratory," in *Distillation: Using Stories from Science's Past to Understand Our World*.

5. Wang Kexin and Xie Wenting, "U.S. Cluster Bombs Continue to Plague Laos, Vietnam, and Cambodia," *Global Times*, July 18, 2023.

6. Daniel A. Gross, "Chemical Warfare: From the European Battlefield to the American Laboratory."

7. Michael Uhl and Tod Ensign, *GI Guinea Pigs: How the Pentagon Exposed Our Troops to Dangers More Deadly than War: Agent Orange and Atomic Radiation*, 7.

8. *Ibid.*, 25.

9. *Ibid.*, 56.

10. Atomic Heritage Foundation, "Nevada Test Site."

11. Michael Uhl and Tod Ensign, *GI Guinea Pigs: How the Pentagon Exposed Our Troops to Dangers More Deadly than War: Agent Orange and Atomic Radiation*, xii.

12. "Air Force Weather: Our Heritage 1937–2012: Meeting the Challenge for 75 Years," *Air Force Weather: Our Heritage 1937-2012: An Illustrated Chronology* prepared by Air Weather Association, July 1, 2012.

13. See Indigenous Chamorro lawyer and activist from Guam Julian Aguon's critical book, *What We Bury at Night: Disposable Humanity*. The book is briefly summarized in Khury Peters-Smith's article, "Twenty-First Imperialism in the Pacific."

14. James Bradley, *The China Mirage*, 508. See also Walter Isaacson and Evan Thomas' *The Wise Men*, 1986, and Dean Acheson's *Present at the Creation*, 1969, for background information on this subject.

15. James Bradley, *The China Mirage*, 512.

16. American Friends Service Committee, "Federal Budget Priorities," November 19, 2021.

17. Mandy Smithberger and William Hartung, "Making Sense of the $1.25 Trillion National Security State Budget: A Dollar-by-Dollar Tour of the National Security State," Center for Defense Information, The Project on Government Oversight, May 7, 2019.

18. *Ibid.*

19. *Ibid.*

20. American Friends Service Committee, "As Global Military Expenditures Top $1.9 Trillion, 39 Organizations Call for Cuts to the Pentagon."

21. Mandy Smithberger and William Hartung, "Making Sense of the $1.25 Trillion National Security State Budget."

22. Mandy Smithberger and William Hartung, "The Pentagon's Yearly Blank Check," Center for Defense Information.

23. Tracy Raczek, "Geoengineering: Reining in the Weather," Chatham House, February 15, 2022.

24. Zohl dé Ishtar, "A Bombs to Star Wars—The Sixty Years War on Marshall Islanders" in *Paradigm Wars: Indigenous Peoples Resistance to Globalization*, edited by Jerry Mander and Victoria Tauli-Corpuz, 2006, 101–102.

25. "Air Force Weather: Our Heritage 1937–2012: Meeting the Challenge for 75 Years," *Air Force Weather: Our Heritage 1937-2012: An Illustrated Chronology* prepared by Air Weather Association, July 1, 2012, 4–8, 62.

26. *Ibid.*, 7–32, 144.
27. *Ibid.*, 7–35, 7–36, 147–148.
28. *Ibid.*
29. H. Patricia Hynes, "Pentagon Pollution, 7: The Military Assault on Global Climate," *Climate & Capitalism,* February 8, 2015. The endnote number 1 in the quote is from Barry Sanders' book, *The Green Zone: The Environmental Costs of Militarism*, 39.
30. *Ibid.*, 50, 61.
31. *Ibid.*, 51.
32. International Energy Agency, December 2, 2019.
33. International Atomic Agency, "Shaping a Secure and Sustainable Future for All," video, December 2, 2019.
34. Vijay Prashad, "Our Time is Now," # 3.
35. H. Patricia Hynes, "Pentagon Pollution," 7.
36. *Ibid.*
37. *Ibid.*
38. See Paul Atkin's educational article, "Let's Get Out from the Carbon Military Boot Print," *urbanramblings,* June 21, 2023.
39. Neta C. Crawford, "Pentagon Fuel Use, Climate Change, and the Costs of War," November 13, 2019.
40. *Ibid.*
41. Whitney Webb, "On Earth Day, Remembering the U.S. Military's Toxic Legacy," *Mint Press News,* April 22, 2019.
42. "Military Greenhouse Gas Emissions: EPA Should Recognize Environmental Impact of Protecting Foreign Oil, Researchers Urge," University of Nebraska-Lincoln, *ScienceDaily,* July 22, 2010. Also published in *Environment* magazine, July-August 2010.
43. *Ibid.*
44. William Blum's informative book *Rogue State* has substantive documentation about the self-assigned imperialist role of the U.S. as the "world's policeman." There are extensive chapters on the consistent overthrow of socialist and anti-capitalist governments all over the world, training war and torture criminals, engaging in cocaine trafficking through CIA sponsorship in Latin America and the Caribbean, developing and exporting weapons of mass destruction while claiming to oppose them, including extremely dangerous chemical and biological weapons that have killed or maimed millions in the underdeveloped world, particularly east and west Asia, and on the U.S. being the leading destroyer of peace in the world. See also Han Dongping's article, "No Country Can Be the World's Policeman: Debt-Ridden U.S. Should Focus on Itself," *Think China,* September 17, 2020.
45. Reto Pieth, "Toxic Military," *The Nation,* June 8, 1992, 773.
46. Ramsey Clark, *The Fire This Time: U.S War Crimes in the Gulf,* 108.
47. *Ibid.*
48. Lancaster University, "U.S. Military Consumes More Hydrocarbons than Most Countries—Massive Hidden Impact on Climate," *ScienceDaily,* June 20, 2019.
49. Dane Wigington, "Wildfires Serve Geoengineering Agenda," www.geoengineeringwatch.com.
50. Janet Wilson, "Solar Surges in the California Desert. So Why Are Environmentalists Upset?," *Desert Sun,* January 3, 2020. For more on the solar farm destruction by storms and toxic chemical components, see Michael Schellenberger, "If Solar Panels Are So Clean, Why Do They Produce So Much Toxic Waste?," *Forbes,* May 23, 2018.
51. Eva Uguen-Csenge and Bethany Lindsay, "For 3rd Straight Day, B.C. Village Smashes Record for Highest Canadian Temperature at 49.6C," *CBC News,* June 29, 2021.
52. Jeff Masters, "The Top 10 Global Weather and Climate Change Events of 2021," *Yale Climate Connections.*
53. NASA, "NASA Announces Summer 2023 Hottest on Record," September 14, 2023.
54. Jeff Masters, "The Top 10 Global Weather and Climate Change Events of 2021."
55. World Wildlife Fund Australia, "New WWF Report: 3 Billion Animals Impacted by Australia's Bushfire Crisis," July 28, 2020, and Brian Resnick, "An Australian Ecologist Explains Just How Bad the Fires Are for Wildlife," *Vox,* January 10, 2020.
56. Brett Taylor, "Firewatch: California Experiencing Second Worst Fire Season in State's History," *KDRV-TV, ABC News, Newswatch,* September 5, 2021 (updated January 12, 2022).
57. Dane Wigington, "Wildfires Serve Geoengineering Agenda," *Geoengineering Watch,* August 24, 2018, updated in July 2019 during the most serious fires in U.S. recorded history.
58. Cascade Tuholske, Kelly Caylor, Chris Funk, and Tom Evans, "Global Urban Population Exposure to Extreme Heat," *Proceedings of the National Academy of Sciences (PNAS),* October 4, 2021.
59. Peter Sousounis (edited by Meagan Phelan), "The 2011 Thai Floods: Changing the Perception of Risk in Thailand," *Verisk,* April 19, 2012.
60. UN Office for the Coordination of Humanitarian Affairs (OCHA), *Relief Web: Bangladesh: Floods and Landslides,* and Edvin Aldrian, "Indonesia's Capital Jakarta Is Sinking: Here's How to Stop This," *The Conversation,* November 11, 2021.
61. Catastrophic Weather Events Requiring over $20 billion and deaths Data from NOAA and EMDAT, international disaster database, EMDAT, at https://yaleclimateconnections.org/2022D/01/-the-top-10-global-weather-and-climate-change-events-of-2021/, compiled by Casey Ontiveros.
62. Edvin Aldrian, "Indonesia's Capital Jakarta Is Sinking: Here's How to Stop This," *The Conversation,* November 11, 2021.
63. Malavika Vyawahare, "South Africa Declares National Emergency as Flood Toll Crosses 440," *Mongabay,* April 19, 2022.
64. South African Weather Service, Department of Forestry, Fisheries & the Environment, Media Release, "Extreme Rainfall and Widespread Flooding Overnight: KwaZulu-Natal and Parts of Eastern Cape," April 12, 2022.

65. Dane Wigington, *Geoengineering Watch Global Alert News,* April 30, 2022, #351, and see Anja Chalmin, "Geo-engineering Activities on the African Continent," *Geoengineering Monitor,* January 12, 2021.
66. Anja Chalmin, "Geo-engineering Activities on the African Continent," January 12, 2021.
67. European Centre for Medium-Range Weather Forecasts, "Flood List," May 22, 2022; Warren Cornwall, "Europe's Deadly Floods Leave Scientists Stunned," *Science,* July 20, 2021.
68. *Reuters,* "Record-Breaking May Rain in Valencia Triggers Floods," May 4, 2022, yahoo!.
69. Brianna Zamora-Nipper, Community Producer, "Timeline: Inside the 2021 Winter Storm, Power Crisis," February 15, 2022.
70. World Meteorological Organization, "Weather-Related Disasters Increase Over Past 50 Years, Causing More Damage, but Fewer Deaths," August 31, 2021.
71. International Federation of Red Cross and Red Crescent, "India: Monsoon Rains and Floods Final Report DREF n° MDRIN024," July 23, 2020, provided by the UN Office for the Coordination of Humanitarian Affairs.
72. *Moscow Times,* "Nearly 100 Fires Blazing Across Russia's Republic of Sakha," August 7, 2023.
73. *Russian Times,* "Siberian Wildfires Cause Devastation and Deaths (VIDEO)," *Agence-France Presse,* May 7, 2022.
74. *Press TV, Iran,* "Millions Stranded, Dozens Dead as Deadly Floods Hit Bangladesh, India," May 21, 2022.
75. Jake Johnson, "Dozens Arrested as Scientists Worldwide Mobilize to Demand 'Climate Revolution,'" *Common Dreams,* April 7, 2022.
76. "Geoengineering: Parts I, II, and III: Hearing Before the Committee on Science and Technology, House of Representatives, One Hundred and Eleventh Congress First Session and Second Session," November 5, 2009, February 4, 2010, and March 18, 2010, Serial No. 111–62, Serial No. 111–75, and Serial No. 111–88.
77. U.S. Food and Drug Administration website.
78. "Imminent Ozone Layer Collapse: A Dire Warning from a Former NASA Contract Engineer," www.geoengineeringwatch.com, April 13, 2022.
79. Malcolm P.R. Light, "Lucy-Alamo Projects-Hydroxyl Generation and Atmospheric Methane Destruction," American Meteorological Society, January 13, 2016.
80. *Science Daily,* reporting on researchers from Case Western University in Cleveland, Ohio, "Nanodiamonds Made Under Ambient Conditions," October 21, 2013.
81. Julian E. Kunnie, *The Cost of Globalization,* 225–226. See Soumya Dutta, "The Basic Science of Global Warming and Climate Justice" in *Critical Issues of Justice, Equity and the Climate Crisis,* for further detail.
82. Malcolm P.R. Light, "Lucy-Alamo Projects-Hydroxyl Generation and Atmospheric Methane Destruction."

83. U.S. EPA, "Overview of Greenhouse Gases," n. d.
84. This video summary on the HAARP at the *Geoengineering Watch* website is in strong contrast to the University of Alaska's HAARP research projects site that has potential questions raised and answered. See the websites for both in the bibliography.
85. *NBC News,* "Conspiracy Theories Abound as U.S. Military Closes HAARP," May 22, 2014.
86. David Vetter, "Solar Geoengineering: Why Bill Gates Wants It, But These Experts Want to Stop It," *Forbes,* January 20, 2022.
87. Environment Health Trust, "Liability, 5G and Cell Tower Radiation," February 26, 2022. There are very important research articles at this site substantiating the lethal consequences of cellphone towers and wireless technologies for everyone, particularly school-age children. Joel Moskowitz, Director of the Center for Community and Public Health at the School of Public Health, provides a very instructive and illuminating presentation on the same site.
88. *Ibid.*
89. Marguerite Reardon, "How the FAA Went to War Against 5G," *CNET,* January 28, 2022.
90. *Generation Zapped,* written and produced by Sabine El Gemayel, 2017.
91. Environment Health Trust, "Liability, 5G and Cell Tower Radiation," February 26, 2022.
92. *Ibid.*
93. "5G: Environmental Effects, Birds, Bees, Trees," YouTube educational research video on the Environmental Health Trust site.
94. 5GFree.org, Arizona action against 5G, "Josh Del Sol's 5G Solutions Summit with Lena Pu. Watch This," June 5, 2020.
95. Jon Bardin, "Fish Getting Skin Cancer from UV Radiation, Scientists Say," *Los Angeles Times,* August 2, 2012.

Chapter 5

1. Michelle Bishop, "Indigenous Education Sovereignty: Another Way of 'Doing' Education," *Critical Studies in Education,* November 30, 2020.
2. Pam Martens and Russ Martens, "The Money Trail to the January 6 Attack on the Capitol Is Ignored in Last Night's Public Hearing," *Wall Street on Parade,* June 10, 2022.
3. Dave Levinthal, "How the Koch Brothers Are Influencing U.S. Colleges," *Time,* December 15, 2015.
4. *Ibid.*
5. Tim Dickinson, "Inside the Koch Brothers' Toxic Empire," *Rolling Stone,* September 24, 2014. Though Koch Industries disregarded repeated attempts to respond to Dickinson's questions prior to the publication of the article, it did respond to many of the criminal activities documented in the article, which Dickinson responded to one by one. See Tim Dickinson, "Koch Industries Responds

to Rolling Stone—And We Answer Back," *Rolling Stone*, September 29, 2014.

6. Tim Dickinson, "Inside the Koch Brothers' Toxic Empire," *Rolling Stone,* September 24, 2014. Though Koch Industries disregarded repeated attempts to respond to Dickinson's questions prior to the publication of the article, it did respond to many of the criminal activities documented in the article, to which Dickinson in turn, responded to one by one. See Tim Dickinson, "Koch Industries Responds to Rolling Stone—And We Answer Back," *Rolling Stone*, September 29, 2014.

7. Jasmine Banks, "Elite Schools Must Quit Koch Money," *The Progressive,* December 1, 2021. Harvard University's ill-baked profits that funded its endowment is noted in Julian Kunnie's 2015 book, *The Cost of Globalization,* 264. The book documents the intriguing account of shady business dealings involving the CIA and illegal drug-laundering money, as well as Capricorn Holdings, which invested $600 million in DynCorp, a military contractor, that in turn worked with the CIA and its ostensible drug war operations in Latin America, Enron, and the Harvard Endowment. Herbert "Pug" Winokur, who was part of Wall Street's "insider" group, was the CEO of Dyncorp and served as one of the seven members of Harvard's governing board. He was also involved in underhanded criminal dealings at Enron that became embroiled in financial fraud scandals culminating in 2001 and involving over $74 billion lost by investors, including employees' retirement funds, while also laundering drug money. The Harvard Endowment Fund and Enron invested in tandem with each other and made billions from insider stock trading that caused the bankruptcies of major "natural" gas and energy companies.

8. Jasmine Banks, "Elite Schools Must Quit Koch Money," *The Progressive,* December 1, 2021.

9. Alex Kotch, "The Kochs' Mad Science: The Billionaire Barons Have a New Libertarian Laboratory," *Salon,* May 25, 2016.

10. Dave Levinthal, "How the Koch Brothers Are Influencing U.S. Colleges," *Time,* December 15, 2015.

11. Julia Piper and Brian O'Leary, "Executive Compensation at Public and Private Colleges," *The Chronicle of Higher Education,* updated February 15, 2022.

12. *Ibid*.

13. *Ibid*.

14. Sam Parker, "UA Students, Faculty Outraged Over Financial Crisis at Board Meeting," *The Daily Wildcat,* November 17, 2023.

15. Samantha Boyle, "Big Ten College Presidents' Salaries Ranked" in *Daily Illini,* November 5, 2020. Table compiled by Casey Ontiveros and Veronica Rodriguez from original graph by Cassidy Brandt.

16. Matt Schulz, "2023 Student Debt Loan Statistics," *LendingTree,* August 10, 2023.

17. NEA Research Land Grant University Brief No. 1, "Land Grant Institutions: An Overview."

18. Aydelotte Foundation podcast, "Land Grab Universities," with Tristan Ahtone and Maria Parazo Rose, May 18, 2023.

19. Robert Lee and Tristan Ahtone, "Land Grab Universities," *High Country News,* March 30, 2020.

20. *Ibid.*

21. *Ibid.*

22. International Campaign to Abolish Nuclear Weapons (ICAN), "Schools of Mass Destruction: Universities List."

23. Beatrice Fihn, "Universities Across America Profit from Developing Weapons. It's Unconscionable," *Newsweek,* November 13, 2019.

24. Bonnie Kavoussi, "Number of Ph.D. Recipients Using Food Stamps Surged During Recession: Report," *Huffington Post,* May 8, 2012. See also Stacey Patton's article, "The Ph.D. Now Comes with Food Stamps," *Chronicle of Higher Education,* May 6, 2012.

25. Alana Semuels, "Easing Student Debt Makes Economic Sense: So Why Is It So Hard To Do?," *Time* magazine, April 1, 2021.

26. Michael Mitchell, Michael Leachman, and Matt Saenz, "State Higher Education Funding Cuts Have Pushed Costs to Students, Worsened Inequality," Center on Budget and Policy Priorities, October 24, 2019.

27. *Ibid.*

28. Kathleen Kunz, "A Closer Look at Arizona's Funding for Higher Education," *Tucson Local Media,* July 31, 2018.

29. Lindsay Koshgarian, "The Pentagon Fails Its Audit Again—and Again and Again and Again and Again and Again," *Common Dreams*, December 1, 2023.

30. Alicia Hahn and Jordan Tarver, "2022 Student Loan Statistics: Average Student Loan Debt," *Forbes,* June 9, 2022. See also Zack Friedman's article, "Student Loan Statistics in 2021: A Record $1.7 Trillion," *Forbes*, February 20, 2021.

31. Alyssa Fowers and Danielle Douglas-Gabriel, "Who Has Student Debt in America?," *Washington Post,* May 22, 2022.

32. Alicia Hahn and Jordan Tarver, "2022 Student Loan Statistics: Average Student Loan Debt." The Institute for College Access and Success.

33. Zack Friedman, "Student Loan Statistics in 2021: A Record $1.7 Trillion." Compiled by Veronica Rodriguez and Casey Ontiveros; data from Enterprise Data Warehouse.

34. *Al Jazeera,* "Nations Where 3.3bn People Live Spend More on Debt than Health Care, Schools," July 13, 2023.

35. United Nations Conference on Trade and Development (UNCTAD), Africa, "A sequence of shocks beyond its borders diminished Africa's ability to develop and led to fast increasing debt levels," 2023.

36. Caitlin Zaloom, "How the Student Debt Complex is Crushing the Next Generation of Americans," *Time,* October 29, 2019. Zaloom's book is *Indebted: How Families Make College Work at Any Cost*, 2019.

37. Ashley Smith, "Millions of Americans Over 60 Are Lugging Student Loans into Retirement," *13 News Now*, April 4, 2022.

38. U.S. Department of Education, Federal Student Aid, "Federal Student Loan Portfolio: Portfolio by Age and Debt Size." These statistics are discussed in Ashley Smith's article, "Millions of Americans Over 60 Are Lugging Student Loans into Retirement." See also Samantha Fields, "Nearly 9 Million Older Americans Still Have Student Loan Debt," *Marketplace, Morning Report*, June 21, 2021, and Alexander Tanzi and Madison Paglia, "Older Americans Are on the Front Line of the Student Debt Crisis," *Bloomberg*, June 21, 2021.

39. Michael Mitchell, Michael Leachman, and Matt Saenz, "State Higher Education Funding Cuts Have Pushed Costs to Students, Worsened Inequality," Center on Budget and Policy Priorities, October 24, 2019.

40. Aubri Juhasz, "NOLA-PS Know They Need to Close, Consolidate Schools. But which Ones?" *WWNO-New Orleans Public Radio*, 89.9 FM, May 30, 2022.

41. Jay Matthews, "The Philadelphia Experiment: The Story Behind Philadelphia's Edison Contract," *Education Next*, Vol. 3, No. 1, July 14, 2006.

42. Jay Matthews, "The Philadelphia Experiment."

43. Tim Walker, Senior Writer, National Education Association, "Average Teacher Salary Lower Today than Ten Years Ago, NEA Report Finds," *NEA: News Today*, April 26, 2022.

44. Emma Mayer, "More Teachers Are Facing Penalties for Quitting During Pandemic," *Newsweek*, March 25, 2022.

45. Tim Walker, Senior Writer, National Education Association, "Average Teacher Salary Lower Today than Ten Years Ago, NEA Report Finds," *NEA: News Today*, April 26, 2022.

46. *Arizona Daily Star*, "Report: Teachers in Arizona Are Less Diverse than Students," *Associated Press*, March 1, 2020.

47. Ibid.

48. U.S. Bureau of Labor Statistics, "Economic News Release: Quits Levels and Rates by Industry and Region, Seasonally Adjusted," June 2022.

49. Klaus Schwab, "Now Is the Time for a 'Great Reset,'" World Economic Forum, January 3, 2020.

50. Ibid.

51. Peter S. Goodman, "'He Has an Incredible Knack to Smell the Latest Fad': How Klaus Schwab Built a Billionaire Circus at Davos," *Vanity Fair*, January 18, 2022.

52. Koiko Mabo and Noel Loos, *Edward Koiki Mabo: His Life and Struggle for Land Rights*, 1996.

53. Carter G. Woodson, *The Miseducation of the Negro*, 1933, 1990, 2005.

54. Eric Burton, "Socialisms Between Cooperation and Competition: Ideology, AID and Cold War Politics in Tanzania's Relation with East Germany," in *Socialismes en Afrique: Socialisms in Africa*, edited by Françoise Blum, Héloïse Kiriakou, Martin Mourre, et al., in *Éditions de la Maison des sciences de l'homme*/OpenEdition Books, 2021. Dr. Alwiya Omar, Director of African Languages at Indiana University in Bloomington, is an expert in this area and an instructor of this author in Kiswahili.

55. See Paulo Freire, *Pedagogy of the Oppressed*, 2000 (30th Anniversary Edition, originally published 1968); bell hooks, *Teaching to Transgress: Education as the Practice of Freedom*; Ivan Illich, *Deschooling Education*; and John Mink (ed.), *Teaching Resistance*, as examples of radical teaching pedagogy proposals and discussions.

56. Jazmin Towe, "These 4 Afrocentric Schools Teach Kids Their Culture as a Matter of Approach," *Parents*, August 29, 2022.

57. Tsitsi Dangarembga, *Nervous Conditions*, 1988.

58. Ngugi wa Thiong'o, *Caitaani utharaba-ini*, 1980.

59. Frantz Fanon, *The Wretched of the Earth*, 1965, and *Dying Colonialism*, 1967; Aimé Césaire, *Discourse on Colonialism*, 2001 (originally published by *Présence Africaine* as *Discourse sur la Colonialism*, 1955); Walter Rodney, *How Europe Underdeveloped Africa*, 1980; Steve Biko, *I Write What I Like*, 2004; and Kwame Nkrumah, *Neo-Colonialism: The Last Stage of Imperialism*, 1966.

60. Steve Biko, *I Write What I Like*, 75–76. Biko uses the term "Native" since this was the official apartheid regime's term for "full-blooded" Indigenous people, as part of its 10-term racial classification system.

61. Yuchie Language Project (*yUdjEha gO'wAd-AnApA k'ak'ûnEchE*), directed by activist, scholar, and director of the UN Permanent Forum on Indigenous Issues' Indigenous Language Revitalization Project, Richard Grounds, from the mid 2000s, and model for other Indigenous peoples around the Earth Mother struggling to recover and speak ancestral languages, at https://www.yuchilanguage.org/. The UN General Assembly proclaimed 2019 as the Year of Indigenous Languages and 2022–2032 as the International Decade of Indigenous Languages.

62. See Vine Deloria, *The World We Used to Live In*, especially chapter 4 on "Interspecies Relations." The amazing documentation of the special and mutually reciprocal and respectful relationship between the crocodiles and the community is noted and visible at "This Community in Ghana lives with Crocodiles-II Paga Crocodile Pond," October 27, 2020, at https://www.youtube.com/watch?v=4ltGMQRKmVU.

Epilogue

1. "13 Quotes that Remind Us to Protect Mother Earth," *Indian Country Today*, January 27, 2016, updated September 13, 2018.

2. Eduardo Galeano, *Upside Down: A Primer for the Looking Glass World*, 337.

3. Malidoma Somé, *Of Water and the Spirit*, 75.

4. Ibid., 20.

5. Fiona Harvey, "Experts Call for Global Moratorium on Efforts to Geoengineer Climate," *The Guardian*, September 14, 2023.

6. Thinnapat Poorbootdee, "UN Issues Drought Warning," *Russian Times*, May 11, 2022.

7. Oxfam International, "Ten Richest Men Double their Fortunes in Pandemic While Incomes of 99 Percent of Humanity Fall," January 17, 2022. The figure for total billionaire wealth accumulation of $1.8 trillion during the pandemic is taken from Sharon Zang, "Top 1 Percent in U.S. Now Have More Wealth than Entire Middle Class Combined," *Truthout*, October 13, 2021.

8. Oxfam International, "Ten Richest Men Double their Fortunes in Pandemic While Incomes of 99 Percent of Humanity Fall," January 17, 2022.

9. Robert Frank, "Soaring Markets Helped the Richest 1% Gain $6.5 Trillion in Wealth Last Year, According to the Fed," *CNBC*, April 1, 2022.

10. *Associated Press*, "The Latest: UN Chief Decries Pandemic's Harm to the Poor," October 11, 2021.

11. Adam Smith, *The Theory of Moral Sentiments*, and *The Wealth of Nations*.

12. Adam Smith, *The Theory of Moral Sentiments*, 296–297.

13. U.S. Department of the Treasury Press Releases, "Treasury Sanctions Public Ministry of Nicaragua and Nine Government Officials Following Sham November Elections," November 15, 2021. The *New York Times* correspondent charged with reporting on the elections, Natalie Kietroff, reported from Mexico, not Nicaragua, to provide a specious article on corrupt elections that falsely claimed opposition parties and right-wing groups were banned from participation and thus the results were unfair. See, for instance, Ben Norton, "Debunking Myths About Nicaragua's 2021 Election, Under Attack by the U.S./EU/OAS," *The Grayzone*, November 11, 2021. Further, campaign rallies were banned for all political parties due to Covid-19 health restrictions, not part of political repression. See a detailed report by Roger D. Harris, "Nicaragua Has a Public Relations Problem, Not a Democracy Problem," *Black Agenda Report*, November 13, 2021.

14. Austin Ahlman, "Biden's Decision on Frozen Afghanistan Money is Tantamount to Mass Murder," *The Intercept*, February 11, 2022; *UN News*, "Afghanistan's Healthcare System on Brink of Collapse, as Hunger Hits 95% of Families," September 22, 2021.

15. Sharon Zang, "Top 1 Percent in U.S. Have More Wealth than Entire Middle Class Combined," *Truthout*, October 13, 2021.

16. *Oxfam*, "Pfizer, BioNTech and Moderna Making $1,000 Profit Every Second While World's Poorest Countries Remain Largely Unvaccinated," November 16, 2021.

17. *Ibid*.

18. Nancy Turner Banks, *AIDS, Opium, Diamonds, and Empire: The Deadly Virus of International Greed*, 244.

19. *Ibid*.

20. Melchior Ware in *Standing on Sacred Ground: Profit and Loss Part 1*, Sacred Land Film Project, 2014, produced by Christopher Macleod, et al.

21. Yaya Diallo and Mitchell Hall, *The Healing Drum: African Wisdom Teachings*, 156–169.

22. *Ibid*.

23. Molly Gott and Derek Seidman, "Corporate Enablers of Israel's War on Gaza," *Eyes on the Ties*, October 26, 2023.

24. Jesse Eisinger, Jeff Ernsthausen, and Paul Kiel, "The Secret IRS Files: Trove of Never-Before Seen Records Reveal How the Wealthiest Avoid Income Tax," *ProPublica*, June 8, 2021.

25. Alicia Phaneuf, "Top 10 Biggest U.S. Banks by Assets in 2022," *Insider Intelligence*, January 2, 2022.

26. Pam Martens and Russ Martens, "When Repos Blew Up in 2019, Hedge Funds Were $800 Billion Short U.S. Treasury Futures: Then Margins Blew Out," *Wall Street on Parade*, February 3, 2022.

27. Arun Viswanatha, "Banks to Pay $5.6 Billion in Probes," *Wall Street Journal*, May 20, 2015.

28. Pam Martens and Russ Martens, "Fed Chair Powell Telegraphs the Perfect Storm for Wall Street's Megabanks: Rapid Rate Hikes Hitting $234 Trillion in Derivatives," *Wall Street on Parade*, April 25, 2022.

29. *Ibid*.

30. Office of the Comptroller of the Currency, "Quarterly Derivatives Report Fourth Quarter 2021: Quarterly Report on Bank Trading and Derivatives Activities: Fourth Quarter."

31. Nomi Prins, *Collu$ion: How Central Bankers Rigged the World*, 1. This is an excellent book to demonstrate how capitalism's core, the banking industry, is fundamentally corrupt and determines the lives of virtually every person on Earth since it determines the cost of living, whether people can afford housing, schooling, access to food, etc.

32. *The Rational Walk* (author anonymous), "Robert Rubin Exemplifies the Revolving Door Between Washington and Wall Street," *Seeking Alpha*, April 11, 2010.

33. Simon Johnson and James Kwak, *Thirteen Bankers: The Wall Street Takeover and the Next Financial Meltdown*.

34. Michael Lewis, *Flash Boys: A Wall Street Revolt*.

35. Christopher Leonard, *The Lords of Easy Money: How the Federal Reserve Broke the American Economy*, 2022.

36. See, for instance, Fijian scholar Wadan Harsey's *British Imperialism and the Making of Colonial Currency Systems*.

37. Nomi Prins, *Collu$ion: How Central Bankers Rigged the World*, xvii.

38. *Ibid.*, 2.

39. *Ibid*.

40. Aiwa Ong, *Neo-Liberalism as Exception*, 238.

41. Jamie Robinson, "China's Planting a Forest the Size of Ireland," THE SPACES.

42. "As China Continues Planting Trees, 23% of the Country is Now Covered in Forest," *South China Morning Post,* March 26, 2020.

43. Ibid.

44. Ying Wang, Gustaf Arrhenius, and Yongzhan Zhang, "Drought in the Yellow River—An Environmental Threat to the Coastal Zone," *Journal of Coastal Research* (JCR), Special Issue 34. International Coastal Symposium (ICS 2000): *CHALLENGES FOR THE 21ST CENTURY IN COASTAL SCIENCES, ENGINEERING AND ENVIRONMENT* (August 2001), 503–515; and David Stanway, "China Government Sees its Own Reflection in Water Crisis," *Reuters,* September 23, 2013. The article has a pro-western capitalist and even anti-China bias, as is typical of most *Reuters* news reports.

45. Luxembourg Centre for Contemporary and Digital History, "Gorbachev's Reforms in the Soviet Union."

46. Kurt Biray, "Communist Nostalgia in Eastern Europe: Longing for the Past." See also Stephen White, "Communist Nostalgia and its Consequences in Russia, Belarus and Ukraine" in *The Transformation of State Socialism.*

47. John Bellamy Foster, Brett Clark, and Richard York, *The Ecological Rift,* 416, 417, 441. The prominent books by István Mészáros mentioned in these pages are *Beyond Capital,* which Hugo Chávez read and discussed in his meeting with Mészáros in 2001, and *The Structural Crisis of Capital.*

48. María de los Ángeles Ramírez, Joseph Poliszuk, and María Antonieta Segovia, "Indigenous Resistance Organized in the Venezuelan Jungle," *El Pais,* February 12, 2022.

49. István Mészáros, *The Structural Crisis of Capital,* 140. Also cited in John Bellamy Foster, Brett Clark, and Richard York, *The Ecological Rift,* 416.

50. See, for instance, former Indigenous Aymara Bolivian president Evo Morales and the book, *The Earth Does Not Belong to Us, We Belong to the Earth,* 2011.

51. Vine Deloria Jr., "Circling the Same Old Rock" in *Marxism and Native Americans,* edited by Ward Churchill, 135–136.

52. Dina Gilio-Whitaker, *As Long as Grass Grows,* 138.

53. Ibid., 139–140. The reference to Tinker is from his article in Weaver (ed.), *Defending Mother Earth,* 165.

54. Greg Cajete, *Native Science: Natural Laws of Interdependence,* chapters 1 and 14.

55. Lame Deer, John (Fire), and Richard Erdoes. *Lame Deer: Seeker of Visions,* 211.

56. See Dona Richards (Marimba Ani), "The Ideology of European Dominance," *Présence Africaine,* 1979.

57. Cited by John Bellamy Foster, Brett Clark, and Hannah Holleman, "Marx and the Indigenous," *Monthly Review,* endnotes. Titles of books read by Marx and identified by Foster, Clark, and Holleman are found in the bibliography.

58. Irving Howe (ed)., "Introduction," *Essential Works of Socialism,* 5.

59. Sara Flounders and Lee Siu Hin, *Capitalism on a Ventilator: The Impact of COVID-19 in China & the U.S.,* 2020. Patently, it is capitalist greed that has resulted in over a million deaths and millions infected and ill from Covid-19, receiving inadequate medical care and attention in the U.S., compared to China's aggressive, but effective, health intervention system that has kept pandemic deaths to around 5,000, in a nation of 1.4 billion, over four times the size of the U.S.'s 320 million people.

60. Danny Haiphong, *Black Agenda Report,* "The Biden-Harris Administration is a Political Expression of the Empire's Crisis of Legitimacy," January 27, 2022.

61. Ngũgĩ wa Thiong'o, "Introduction," *Decolonizing the Mind,* Kindle version.

62. See Samir Amin, "The Future of Global Polarization" in *Globalization and Social Change,* edited by Johannes Dragsbaeck Schmidt and Jacques Hersh, 41, and Julian E. Kunnie, *The Cost of Globalization,* 12.

63. Becky Oskin, "Reclaiming Languages, Revitalizing Culture: How Ines Hernández-Ávila Redefines Indigenous Scholarship."

64. Yuchie Language Project, yUdjEha gO'-wAdAnA-A k'ak'ûnEchE.

65. Salish School of Spokane, http://www.salishschoolofspokane.org/revitalizingsalish.html

66. Margaret Kovach, *Indigenous Methodologies,* 2021.

67. Linda Tuhiwai Smith, *Decolonization Methodology,* 1999, 2001, 173.

68. W.E.B. Du Bois, *The Souls of Black Folk,* 1903, 3.

69. Linda Tuhiwai Smith, *Decolonization Methodology,* 1999, 2001, 173.

70. Kathleen S. Fine-Dare, *Grave Injustice.*

71. See Dena Tekruri's excellent video production, "Mark Zuckerberg Sued Native Hawaiians for their Own Land" on *YouTube.*

72. Mariflor Parpan-Pagusara, "The Kalinga Ili: Cultural-Ecological Reflections on Indigenous Theoria and Praxis of Man-Nature Relationship" in *Tidagaketbiag: Land is Life: Selected Papers from the Cordillera Multi-Sectoral Land Congresses* (1983,1994, and 2001), 41.

73. Caitlin Looby, Frank Vaisvilas, and Madeline Helm, "Great Lakes Tribes' Knowledge of Nature Could be Key to Climate Change. Will People Listen?," *USA Today,* November 26, 2023.

74. Ibid.

75. Theda New Breast presentation, American Indian Movement West Conference on the 50th Anniversary of the Indigenous Occupation of Alcatraz Island, November 2019. Also cited in Julian Kunnie, "The Cost of Globalization to

Indigenous Peoples: The Need for Decolonization and Constructive Social Work Strategies in Turtle Island (North America)," in *Recognition, Reconciliation and Restoration: Applying a Decolonized Understanding in Social Work and Healing Processes*, edited by Jan Erik Henriksen, Ida Hydie and Britt Kramvig, 43–44, and in Julian Kunnie's chapter, "Indigenous Spirituality, Decolonization, and Restoration of Traditional Elders' Ancestral Knowledge Today: Social Work and Aging," in *Social Aspects of Aging in Indigenous Communities*, edited by Jordan P. Lewis and Tuula Heinonen, 2023, 195–224.

76. N. Scott Momaday, *Earth Keeper: Reflections on the American Land*, 2020.

77. Leon Shenandoah, *To Become a Human Being: The Message of Tadodaho Chief Leon Shenandoah*, edited by Steve Wall, 43–45.

Bibliography

Books (Print and Ebook)

Abimbola, Kola. *Yoruba Culture: A Philosophical Account*. Birmingham, England: Ìrókò Academic Publishers, 2006.

Abu-Jamal, Mumia, and Stephen Vittoria. *Murder Incorporated: Empire, Genocide, Manifest Destiny: America's Favorite Pastime: Book Two*. San Francisco: Prison Radio, 2019.

_____. *Murder Incorporated: Empire, Genocide, Manifest Destiny: Dreaming of Empire: Book One*. San Francisco: Prison Radio, 2018.

_____. *Murder Incorporated: Empire, Genocide, Manifest Destiny: Perfecting Tyranny: Book Three*. San Francisco: Prison Radio, 2020.

Acheson, Dean. *Present at the Creation: My Years in the State Department*. New York; W.W. Norton & Co., 1987.

Aguon, Julian. *What We Bury at Night: Disposable Humanity*. Tokyo: Blue Ocean Press, 2008.

Alfred, Taiaiake. *Wasáse: Indigenous Pathways of Action and Freedom*. Toronto: Broadview Press, 2005.

Allen, Paula Gunn. *Grandmothers of the Light: A Medicine Woman's Sourcebook*. Boston: Beacon Press, 1991.

Amadiume, Ifi. *Male Daughters, Female Husbands: Gender and Sex in an African Society*. London: Zed Books, 2015.

Amin, Samir. "The Future of Global Polarization." *Globalization and Social Change*, edited by Johannes Dragsbaeck Schmidt and Jacques Hersh. London: Routledge, 2000.

Anderson, M. Kat. *Tending the Wild: Native American Knowledge and the Management of California's Natural Resources*. Berkeley and Los Angeles, CA: University of California Press, 2005.

Ani, Marimba. *Yurugu: An African-Centered Critique of European Cultural Thought and Behavior*. Trenton, NJ: Africa World Press, 1994.

Aptheker, Herbert. *American Negro Slave Revolts*. New York: International Publishers, 1983.

Arthur, Charles, and Michael Dash (eds.). *A Haiti Anthology: Libète*. Princeton, NJ: Markus Wiener Publishers, 2009.

Bacon, Francis. *Rerum Organum (New Instrument)*. Whithorn, Scotland: Anodos Books, 2019. Originally published 1620.

Bah, Chernoh Alpha M. *The Ebola Outbreak in West Africa: Corporate Gangsters, Multinationals & Rogue Politicians*. Philadelphia: Africanist Press, 2015.

Banks, Nancy Turner. *AIDS, Opium, Diamonds, and Empire: The Deadly Virus of International Greed*. Bloomington, IN: iUniverse, 2010.

Baptist, Edward. *The Half Has Never Been Told: Slavery and the Making of American Capitalism*. New York: Basic Books, 2014.

Barrett, Ross, Daniel Worden, and Allan Stoeki (eds.). *Oil Culture*. Minneapolis: University of Minnesota Press, 2014.

Bartlett, Robert. *The Making of Europe: Conquest, Colonization, and Cultural Change 950–1350*. Princeton, NJ: Princeton University Press, 1994.

Bauval, Robert, and Thomas Brophy. *Black Genesis: The Prehistoric Origins of Ancient Egypt*. Rochester, VT: Bear & Company, 2011.

Beeching, Jack. *The Chinese Opium Wars*. Orlando, FL: Harcourt Brace Jovanovich, 1975.

Bennett, Lily, Margaret Katherine, Sybil Ranch, and Liz Thompson. *Living with the Land: Bush Tucker and Medicine of the Jawoyn*. Melbourne, Australia: Pearson Library, 2009.

Benton, Ted. "Marxism and Natural Limits: An Ecological Critique and Reconstruction." *New Left Review*. No. 178, Nov/Dec 1989, 51–86.

Berat, Lynn. *Walvis Bay: Decolonization and International Law*. New Haven, CT: Yale University Press, 1990.

Bernal, Martin. *Black Athena: The Afroasiatic Roots of Classical Civilization*. Vol. 2: *The Archaeological and Documentary Evidence*. New Brunswick, NJ: Rutgers University Press, 1991.

_____. *Black Athena: The Afroasiatic Roots of Classical Civilization, Vol. 1*. New Brunswick, NJ: Rutgers University Press, 1987.

_____. *Black Athena: The Afroasiatic Roots of Classical Civilization*. Vol. 3: *The Linguistic Evidence*. New Brunswick, NJ: Rutgers University Press, 2006.

Bevilacqua, Piero. *La Mucca è Savia*. Rome: Donzelli, 2002.

Biko, Steve. *I Write What I Like*. Johannesburg/Gauteng: Picador Africa, 2004.

Black Elk, Frank. "Observations on Marxism and Lakota Tradition." *Marxism and Native Americans*. Edited by Ward Churchill. Boston: South End Press, 1999, 137–157.

Blum, William. *Rogue State: A Guide to the World's Only Superpower.* Monroe, ME: Common Courage Press, 2005.

Bonwick, James. *The Last of the Tasmanians.* n.p.: Low, 1870.

Boyer, Carl, and Ute Merzbach. *A History of Mathematics.* Hoboken, NJ: John Wiley & Sons, 1991.

Bradley, James. *The China Mirage: The Hidden History of American Disaster in Asia.* New York: Little, Brown and Company, 2015.

Brodie, James Michael. *Created Equal: The Lives and Ideas of Black American Innovators.* New York: Bill Adler Books, Inc., William Morrow and Co. Inc., 1993.

Brodie, Nick. *The Vandemonian War: How Britain Annihilated Tasmania's Tribes and Fabricated History.* Richmond, Victoria, Australia: Hardie Grant Books, 2017.

Bullchild, Percy. *The Sun Came Down: The History of the World as My Blackfeet Elders Told It.* San Francisco: Harper & Row, Publishers, 1985.

Burkett, Paul. *Marx and Nature: A Red and Green Perspective.* New York: St. Martin's Press, 1999.

Burnard, Trevor, and John Garrigus. *The Plantation Machine: Atlantic Capitalism in French Saint Domingue and British Jamaica.* Philadelphia: University of Pennsylvania Press, 2016.

Buxton, Thomas Fowell. *The African Slave Trade and the Remedy* (1840). London: Cass, 1967.

Cajete, Greg. *Native Science: Natural Laws of Interdependence.* Santa Fe, NM: Clear Light Publishers, 2000.

Calloway, Colin G. *Crown and Calumet: British-Indian Relations 1783–1815.* Norman, OK: University of Oklahoma Press, 1987.

Castles, Alex C. "An Australian Legal History." *Aboriginal Legal Issues: Commentary and Materials.* Ed. H. McRae et al. Holmes Beach, FL: Wm. W. Gaunt & Sons, 1991.

Césaire, Aimé. *Discourse on Colonialism.* New York: Monthly Review Press, 2001, originally published by Présence Africaine as *Discourse sur la Colonialism*, 1955.

Chambers, Cynthia. "'We Are All Treaty People': The Contemporary Countenance of Canadian Curriculum Studies." *Reconsidering Canadian Curriculum Studies: Provoking Historical, Present, and Future Perspectives.* Edited by Nicholas Ng-A-Fook and Jennifer Rottmann. New York: Palgrave Macmillan, 2012.

Chinweizu. *White Predators, Black Slavers, and the African Elite.* New York: Random House, 1975.

Churchill, Ward. *A Little Matter of Genocide: Holocaust and Denial in the Americas, 1492 to the Present.* San Francisco: City Lights Books, 1997.

_____ (ed.). *Marxism and Native Americans.* Boston: South End Press, 1999.

Clark, Ramsey. *The Fire This Time: U.S. War Crimes in the Gulf.* New York: International Action Center, 2002.

Conrad, Cecilia, John Whitehead, Patrick Mason, and James Stewart (eds.). *African Americans in the U.S. Economy.* Lanham, MD: Rowman & Littlefield, 2005.

Cook, Sherburn F., and Woodrow Borah. *Essays in Population History, Vol. 1.* Berkeley, CA: University of California Press, 1971.

Coulthard, Glen. *Red Skin, White Masks: Rejecting the Colonial Politics of Recognition.* Minneapolis: University of Minnesota Press, 2014.

Dangarembga, Tsitsi. *Nervous Conditions.* Harare: Zimbabwe Publishing House, 1988.

d'Anghiera, Pietro Martire. *De Orbe Novo, the Eight Decades of Peter Martyr d'Anghera.* Translated by Francis August MacNutt. New York: G.P. Putnam, 1912.

Davis, Lance E. *In Pursuit of Leviathan: Technology, Institutions, Productivity, and Profits in American Writing, 1816–1906.* Chicago: University of Chicago, 1997.

Deloria, Vine. "Circling the Same Old Rock." *Marxism and Native Americans.* Edited by Ward Churchill. Boston: South End Press, 1999, 113–136.

_____. *Custer Died for Your Sins: An Indian Manifesto.* Norman, OK: University of Oklahoma Press, in association with Macmillan Publishing, 1988.

_____. *For This Land: Writings on Religion in America.* New York: Routledge, 1999.

_____. *Red Earth, White Lies: Native Americans and the Myth of Scientific Fact.* New York: Scribner's, 1995.

_____. *The World We Used to Live In: Remembering the Powers of the Medicine Men.* Golden, CO: Fulcrum Publishing, 2006.

Deneven, William. *The Native Population of the Americas in 1492.* Madison, WI: University of Wisconsin Press, 1976.

Diallo, Yaya, and Mitchell Hall. *The Healing Drum: African Wisdom Teachings.* Rochester, VT: Destiny Books, 1989.

Diop, Cheikh Anta. *The African Origin of Civilization: Myth or Reality.* Chicago: Lawrence Hill Books, 1974.

Dobyns, Henry. *Their Numbers Become Thinned: Native American Population Dynamics in Eastern North America.* Knoxville: University of Tennessee Press, 1983.

Dorris, Michael. "Why I'm Not Thankful for Thanksgiving." *Rethinking Columbus: Teaching About the 500th Anniversary of Columbus's Arrival in America: A Special Issue of* Rethinking Schools: *Dedicated to the Children of the Americas.* Milwaukee, WI: Rethinking Schools; Washington, D.C: Network of Educators on Central America, November 1991.

Dowie, Mark. *Conservation Refugees: The Hundred-Year Conflict between Global Conservation and Native Peoples.* Cambridge, MA: MIT Press, 2011.

Du Bois, W.E.B. *The Souls of Black Folk.* Chicago: A.C. McClurg & Co., 1903.

Dunbar-Ortiz, Roxanne. *An Indigenous Peoples' History of the United States: Revisioning History.* Boston: Beacon Press, 2014.

_____. *Not a Nation of Immigrants: Settler Colonialism, White Supremacy, and a History of Erasure and Exclusion*. Boston: Beacon, 2022.

Dutta, Soumya. "The Basic Science of Global Warming and Climate Justice." *Critical Issues of Justice, Equity and the Climate Crisis: Justice and Equity Violated: Those Populations that Pollute/Emit the Most Are Not the Ones Most Threatened by Impacts; The Non-Contributing Sufferers Have the Least Financial & Technological Capability to Cope*. By Meher Engineer, Soumya Dutta, and Asit Dash. New Delhi: Vasudhaiva Kutumbakam, South Asian Dialogues on Ecological Democracy (SADED), 2010.

Eisinger, Chester E. "The Puritans' Justification for Taking the Land." *Essex Institute Hist. Collins*, 1948.

El, Cozmo. *Moor Vol I. and II: What They Didn't Teach You in Black History Books*. CreateSpace Independent Publishing Platform, 2016.

Elder, Hinemoa. *Aroha*. Auckland, Aotearoa (New Zealand): Penguin Books, 2000.

Ellison, Graham. *Destined for War: Can America and China Escape Thucydides' Trap*. Boston: Houghton Mifflin, 2017.

Elm, Demus, and Harvey Antone. *The Oneida Creation Story*. Translated and edited by Floyd Lounsbury and Bryan Glick. Lincoln: University of Nebraska Press, 2000.

Eltis, David, and David Richardson. *Atlas of the Transatlantic Slave Trade*. New Haven, CT: Yale University Press, 2010.

Erdoes, Richard, and Alfonso Ortiz (eds.). *American Indian Myths and Legends*. New York: Pantheon Books, 1984.

Fanon, Frantz. *A Dying Colonialism*. New York: Grove Press, 1967.

_____ *Wretched of the Earth*. New York: Grove Press, 1965.

Fine-Dare, Kathleen S. *Grave Injustice: The American Indian Repatriation Movement and NAGPRA*. Lincoln, NE: University of Nebraska Press, 2002.

Flounders, Sara, and Lee Siu Hin, *Capitalism on a Ventilator: The Impact of COVID-19 in China & the U.S.* New York: World View Forum, for the International Action Center and China U.S. Solidarity Network, 2020.

Floyd, Troy. *The Columbus Dynasty in the Caribbean, 1492–1526*. Albuquerque: University of New Mexico, 1973.

Fogel, Robert. *Without Consent or Contract: The Rise and Fall of American Slavery*. New York: W.W. Norton & Co., 1989.

Foner, Philip. "The International Slave Trade." *African Americans in the U.S. Economy*. Edited by Cecilia A. Conrad, John Whitehead, Patrick Mason, and James Stewart. Lanham, MD: Rowman & Littlefield Publishers, Inc., 2005, 9–13.

Foster, John Bellamy, Brett Clark, and Richard York. *The Ecological Rift: Capitalism's War on the Earth*. New York: Monthly Review Press, 2010.

Francis, John. *Planet Walker*. Washington, D.C.: National Geographic Books, 2008.

Franklin, John Hope, and Loren Schweninger. *Runaway Slaves*. New York: Oxford University Press, 1999.

Freire, Paulo. *Pedagogy of the Oppressed*. New York: Bloomsbury Academic, 2000, 30th Anniversary Edition, originally published 1968.

Furtado, Celso. *The Economic Development of Latin America: A Survey from Colonial Times to the Cuban Revolution*. Cambridge, England: Cambridge University Press, 1970.

Galeano, Eduardo. *Open Veins of Latin America: Five Centuries of the Pillage of a Continent*. New York: Monthly Review Press, 1997.

Genovese, Eugene. *Roll, Jordan, Roll: The World the Slaves Made*. New York: Vintage Books, 1976.

Gilio-Whitaker, Dina. *As Long as the Grass Grows: The Indigenous Fight for Environmental Justice, from Colonization to Standing Rock*. Boston: Beacon Press, 2019.

Gills, Barry K. (ed). *Globalization and the Politics of Resistance*. London: Palgrave Macmillan, 2000.

Goodstein, David. *Out of Gas: The End of the Age of Oil*. New York: W.W. Norton & Co., 2004.

Gordon, Willie (with Jody Bennett). *Gurrbi: My Special Place*. Cooktown, Queensland, Australia: Gurrbi Tours, 2014.

Green, Toby, Philip Havik, and F. Ribeiro da Silva (eds.). *African Voices from the Inquisition, Volume 1: The Trial of Crispina Peres of Cacheu, Guinea-Bissau 1646–1668*. Oxford, England: Oxford University Press, 2021.

Gunn, John M. *Schat-Chen: History, Traditions, and Narratives of the Queres Indians of Laguna and Acoma*. Albuquerque: Albright and Anderson, 1917; New York, AMS, reprint, 1977.

Hallward, Peter. *Damming the Flood: Haiti, Aristide, and the Politics of Containment*. New York: Verso, 2007.

Hamilton, Virginia. *In the Beginning: Creation Stories from Around the World*. New York: Harcourt Brace Jovanovich, Publishers, 1988.

Hanke, Lewis. *The Spanish Struggle for Justice in the Conquest of America*. Philadelphia: University of Pennsylvania Press, 1947.

Harjo, Suzan Shown. "We Have No Reason to Celebrate an Invasion." *Rethinking Columbus: Teaching About the 500th Anniversary of Columbus's Arrival in America: A Special Issue of* Rethinking Schools: *Dedicated to the Children of the Americas*. Milwaukee, WI: Rethinking Schools; Washington, D.C: Network of Educators on Central America. November 1991.

Harsey, Wadan. *British Imperialism and the Making of Colonial Currency Systems*. Basingstoke, England: Palgrave Macmillan, 2016.

Hendicott, Stephen, and Edward Hagerman. *The United States and Biological Warfare*. Bloomington, IN: Indiana University Press, 1998.

Henriksen, Jan Erik, Ida Hydie and Britt Kramvig (eds.). *Recognition, Reconciliation and Restoration: Applying a Decolonized Understanding in Social Work and Healing Processes*. Stamsund, Norway: Orkana Akademisk, 2020.

Hersh, Seymour. *Chemical and Biological Warfare*. New York: Bobbs-Merrill, 1968.

Ho, Mae-Wan. *Genetic Engineering: Dream of Nightmare: Turning the Tide on the Brave New World of Bad Science and Big Business*. New York: Continuum International Publishing Group, 2000.

_____. "The New Genetics and Dangers of GMO." *Seed Sovereignty, Food Security: Women in the Vanguard of the Fight Against GMOs and Corporate Agriculture*. Edited by Vandana Shiva. Berkeley, CA: North Atlantic Books, 2014.

_____. *The Rainbow and the Worm*. Hackensack, NJ: World Scientific Publishing Company, 2008.

Hochschild, Adam. *King Leopold's Ghost: A Story of Greed, Terror and Heroism in Colonial Africa*. New York: Houghton Mifflin, 1998.

Hodgkin, John. "Proper Terms II." *Transactions of the Philological Soc*. Vol. 27, No. 1 (1907-10), Suppl., 1–187.

hooks, bell. *Teaching to Transgress: Education as the Practice of Freedom*. New York: Routledge, 1994.

Horne, Gerald. *The Dawning of the Apocalypse: The Roots of Slavery, White Supremacy, Settler Colonialism, and Capitalism in the Long Sixteenth Century*. New York: Monthly Review Press, 2018.

Howe, Irving (ed.). *Essential Works of Socialism*. New Haven, CT: Yale University Press, 1976.

Howitt, William. *Colonization and Christianity: A Popular History of the Treatment of the Natives by the Europeans in all their Colonies*. London: 1838.

Illich, Ivan. *Deschooling Education*. London: Marion Boyars, 2000, originally published 1971.

Im Hof, Ulrich. *The Enlightenment: The Making of Europe*. Hoboken, NJ: John Wiley & Sons, 1997.

Inikori, Joseph. *Africans and the Industrial Revolution in England: A Study in International Trade and Economic Development*. Cambridge, England: Cambridge University Press, 2002.

Isaacson, Walter, and Evan Thomas. *The Wise Men: Six Friends and the World They Made*. New York: Simon & Schuster, 2013.

Ishtar, Zohl dé. "A-Bombs to Star Wars—The Sixty-Six Years War on Marshall Islanders." *Paradigm Wars: Indigenous Peoples' Resistance to Globalization*. Edited by Jerry Mander and Victoria Tauli-Corpuz, International Forum on Globalization. San Francisco: Sierra Club Books, 2006.

Jackson, John. *Introduction to African Civilizations*. New York: University Books, 1970.

Jaimes, M. Annette (ed.). *The State of Native America: Genocide, Colonization and Resistance*. Boston: South End Press, 1992.

James, C.L.R. *Black Jacobins*. New York: Vintage Books, 1989.

James, Portia. *The Real McCoy: African American Invention and Innovation, 1619–1930*. Washington, D.C.: Anacostia Museum of the Smithsonian Institution by the Smithsonian Institution Press, 1989.

Johnson, Simon, and James Kwak. *Thirteen Bankers: The Wall Street Takeover and the Next Financial Meltdown*. New York: Pantheon, 2010.

Johnson, Walter. *River of Dark Dreams: Slavery and Empire in the Cotton Kingdom*. Cambridge, MA: Harvard University Press, 2013.

Jones, James H. *Bad Blood: The Tuskegee Syphilis Experiment*. New York: The Free Press, 1992.

_____. *Bad Blood: The Tuskegee Syphilis Experiment*. New York: The Free Press, 1992.

Kamen, Henry. *The Spanish Inquisition: A Historical Revision*. New Haven, CT: Yale University Press, 1998.

Kamenka, Eugene (ed.). *The Portable Karl Marx*. New York: Viking Press and Penguin Books, 1983.

Kant, Immanuel. *Critique of Judgement*. Translated by J.H. Bernard. New York: Hafner Press, 1951.

_____. *Critique of Pure Reason*. Translated by Werner S. Pluhar. Indianapolis: Hackett Publishing Company, 2002.

Kovach, Margaret. *Indigenous Methodologies: Characteristics, Conversations, and Contexts*. Second Edition. Toronto: University of Toronto Press, 2021.

Krishnamurthi, J. *The Impossible Question*. New York: Perennial Library, Harper & Row Publishers, 1978.

Kunnie, Julian E. "The Cost of Globalization to Indigenous Peoples: The Need for Decolonization and Constructive Social Work Strategies in Turtle Island (North America)." *Recognition, Reconciliation and Restoration: Applying a Decolonized Understanding in Social Work and Healing Processes*. Edited by Jan Erik Hendriksen, Ida Hydie and Britt Kramvig. Stamsund, Norway: Orkana Akademisk, 2020.

_____. *The Cost of Globalization: Dangers to the Earth and Its People*. Jefferson, NC: McFarland & Co., 2015.

_____. "Indigenous Spirituality, Decolonization, and Restoration of Traditional Elders' Ancestral Knowledge Today: Social Work and Aging." *Social Aspects of Aging in Indigenous Communities*. Edited by Jordan P. Lewis and Tuula Heinonen. New York: Oxford University Press, 2023, 195–224.

LaDuke, Winona. *All Our Relations: Native Struggles for Land and Life*. Cambridge, MA: South End Press, 1999.

LaDurie, La Roy. *Montaillou: The Promised Land of Error*. New York: George Braziller, 2008.

Lame Deer, John (Fire), and Richard Erdoes. *Lame Deer: Seeker of Visions*. New York: Pocket Books, 1994, first published 1976.

Lampe, Armando. "Las Casas and African Slavery in the Caribbean: A Third Conversion." *Bartolomé de las Casas, O.P.: History, Philosophy, and Theology in the Age of European Expansion*. Edited by David Thomas Orique, O.P., and Rady Roldán-Figueroa. Leiden, the Netherlands: Brill, 2018.

Lane-Poole, Stanley. *The Story of the Moors After*

Spain. Washington, D.C.: Traffic Output Publication, 2017. Previously titled *The Barbary Corsairs*, published by G.P. Putnam's, New York, 1890.

⸺. *The Story of the Moors in Spain: A History of the Moorish Empire in Europe: Their Conquest, Book of Laws and Code of Rites.* 1886. Baltimore: Black Classic Press, 1990; New York: Pantianos Classics, 2018.

Las Casas, Bartolomé de. *A Short Account of the Destruction of the Indies.* New York: Penguin, 1992.

Laslett, Peter (ed.). *Locke: Two Treatises of Government.* New York: Cambridge University Press, 1988.

Lawson, Tom. *The Last Man: A British Genocide in Tasmania.* London: I.B. Tauris/Bloomsbury Publishing, 2014.

Lea, Charles. *The Inquisition of Spain, Volumes I, II, III, and IV.* New York: Macmillan, 1906–1907.

Leonard, Christopher. *The Lords of Easy Money: How the Federal Reserve Broke the American Economy.* New York: Simon & Schuster, 2022.

Lewis, Michael. *Flash Boys: A Wall Street Revolt.* New York: W.W. Norton & Co., 2015.

Lindqvist, Sven. *Exterminate All the Brutes: One Man's Odyssey into the Heart of Darkness and the Origins of European Genocide.* New York: The New Press, 1997, Kindle, 2021.

Loewen, James. *Lies My Teacher Told Me: Everything Your American History Textbook Got Wrong.* New York: Touchstone, 1996.

Losada, Angel. "Hernán Cortés en la Obra del Cronista Sepúlveda." *Revista de Indias.* Madrid, Vol. 9, 1948. No publisher listed.

Love, Hallie N. *Watakame's Journey.* Santa Fe, NM: Clear Light Books, 1999.

Mabo, Koiko, and Noel Loos. *Edward Koiki Mabo: His Life and Struggle for Land Rights.* La Lucia, Queensland, Australia: University of Queensland Press, 1996.

Madariaga, Salvador de. *The Rise of the Spanish American Empire.* London: Hollis and Carter Publishers, 1947.

Mander, Jerry, and Victoria Tauli-Corpuz. (eds.). *Paradigm Wars: Indigenous Peoples' Resistance to Globalization.* International Forum on Globalization. San Francisco: Sierra Club Books, 2006.

Mann, Barbara Alice. *Spirits of Blood, Spirits of Breath: The Twinned Cosmos of Indigenous America.* New York: Oxford University Press, 2016.

Marchand, Michael E., and Kristiina A. Vogt (eds.). *The River of Life: Sustainable Practices of Native Americans and Indigenous Peoples.* East Lansing: Michigan State University Press, 2016.

Markel, Howard. *An Anatomy of Addiction: Sigmund Freud, William Halsted, and the Miracle Drug, Cocaine.* New York: Vintage, 2012.

Martines, Lauro. *Power and Imagination: City-States in Renaissance Italy.* New York: Knopf, 1979.

Marx, Karl. *Capital: A Critique of Political Economy, Volume One: The Process of Capitalist Production.* New York: International Publishers, 1967.

⸺. *Capital: A Critique of Political Economy, Volume Three.* New York: Vintage Books, 1981.

⸺. *Economic and Political Manuscripts of 1844.* New York: International Publishers, 1964.

⸺. *Grundrisse.* New York: Vintage Books, 1973.

Matthias, Peter. *Science and Society, 1600–1900.* Cambridge, England: Cambridge University Press, 1972.

Mbiti, John S. *Introduction to African Religion, Second Edition.* Long Grove, IL: Waveland Press, Inc., 2015.

McKenna, Eric, and Scott L. Pratt. *American Philosophy: From Wounded Knee to the Present.* New York: Bloomsbury, 2015.

McKinney, Cynthia (ed.). *The Illegal War on Libya: A Dignity Project.* Atlanta, GA: Clarity Press, 2012.

McPhee, Allan. *The Economic Revolution in British West Africa, Second Edition.* London: Frank Cass, 1971.

Memmi, Albert. *The Colonizer and the Colonized.* Boston, MA: Beacon, 1991.

Merivale, Herman. *Lectures on Colonization and Colonies.* London: Longman, Brown, Green, and Longmans, 1841.

Mészáros, István. *Beyond Capital: Toward a Theory of Transition.* New York: Monthly Review Press, 2000.

⸺. *The Structural Crisis of Capital.* New York: Monthly Review Press, 2010.

Miller, Robert J., Jacinta Ruru, Larissa Behrendt, and Tracey Lindberg. *Discovering Indigenous Lands: The Doctrine of Discovery in the English Colonies.* Oxford, England: Oxford University Press, 2012.

Mink, John (ed.). *Teaching Resistance: Radicals, Revolutionaries, and Cultural Subversives in the Classroom.* San Francisco: PM Press, 2019.

Mintz, Sidney W. *Sweetness and Power: The Place of Sugar in Modern History.* New York: Penguin, 1986.

Momaday, N. Scott. *Earth Keeper: Reflections on the American Land.* New York: Harper, 2020.

Moore, MariJo, and Tracey A. Demeyer. *Unraveling the Spreading Cloth of Time: Indigenous Thoughts Concerning the Universe.* Candler, NC: Renegade Planets Publishing, 2013.

Moore, R.I. *The War on Heresy: Faith and Power in Medieval Europe.* London: Profile Books, 2012, 2014.

Morales, Evo. *The Earth Does Not Belong to Us, We Belong to the Earth.* Ministry of Exterior Relations, Plurinational State of Bolivia, 2011.

Morgan, Lewis Henry. *League of the Ho-dé-no-saunee or Iroquois.* New York: M.H. Newman & Co., 1851.

Mumford, Lewis. *Technics and Civilization.* Chicago: University of Chicago Press, 2010.

Nkrumah, Kwame. *Neo-Colonialism: The Last Stage of Imperialism.* New York: International Publishers, 1966.

Nzongola-Ntalaja, Georges. *The Congo from Leopold to Kabila: A Peoples' History*. New York: Zed Books, 2002, second printing, 2003.

Obenga, Théophile. *A Lost Tradition: African Philosophy in World History*. Philadelphia: Source Editions, 1995.

Obi, Cyril I. "Globalization and Local Resistance: The Case of Shell Versus the Ogoni." *Globalization and the Politics of Resistance*. Edited by Barry K. Gills. London: Palgrave Macmillan, 2000, pp. 280–294.

Ong, Aiwa. *Neo-Liberalism as Exception: Mutations in Citizenship and Sovereignty*. Durham, NC: Duke University Press, 2006.

Oyewumi, Oyeronke. *The Invention of Women: Making an African Sense of Western Gender Discourses*. Minneapolis: University of Minnesota Press, 1997.

Pagán, Luis Rivera. *A Violent Evangelism: The Political and Religious Conquest of the Americas*. Louisville, KY: Westminister/John Knox Press, 1992.

Parpan-Pagusara, Mariflor. "The Kalinga *Ili*: Cultural-Ecological Reflections on Indigenous Theoria and Praxis of Man-Nature Relationship." *Tidagaketbiag: Land Is Life: Selected Papers from the Cordillera Multi-Sectoral Land Congresses (1983, 1994, and 2001)*. Baguio City, Cordillera, Philippines, 2009.

Paul, Heike. *The Myths that Made America: An Introduction to American Studies*. Bielefeld, Germany: Transcript Verlag, 2014.

Pearce, Roy Harvey. *Savagism and Civilisation*. Baltimore: Johns Hopkins Press, 1965.

Petrini, Carlo. *Slow Food Nation: Why Our Food Should be Good, Clean, and Fair*. New York: Rizzoli ex libris, Rizzoli International Publications, 2007.

Physicians for Social Responsibility. "Body Count: Casualty Figures of the War on Terror: Iraq, Afghanistan, and Pakistan." Washington, D.C.: Physicians for Social Responsibility, 2015.

Pieth, Reto. "Toxic Military." *The Nation*. June 8, 1992. 773.

Podur, Justin, and Sasha Liley. *Haiti's New Dictatorship: The Coup, the Earthquake, and the UN Occupation*. Toronto: Between the Lines Press, 2012.

Prescott, William. *The History of the Conquest of Mexico*. 1843.

———. *The History of the Conquest of Peru*. 1847.

Prince, Rod. *Family Business*. London: Latin American Bureau, 1985.

Prins, Nomi. *Collu$ion: How Central Bankers Rigged the World*. New York: Bold Type Books, 2018.

Purchas, Samuel. *Hakluytus Posthumus or Purchas his Pilgrims*. Glasgow, Scotland, 1906, no publisher listed.

Rabisha, Will. *The Whole Body of Cookery Dissected*. London: E. Calvert, et al., 1661.

Raboteau, Albert. *Slave Religion*. New York: Oxford University Press, 1978.

Raffles, Thomas Stamford. *The History of Java*. London: John Murray, 1817.

Rawlings, Helen. *The Spanish Inquisition*. Malden, MA: Blackwell Publishing, 2006.

Reséndez, Andrés. *The Other Slavery: The Uncovered Story of Indian Enslavement in America*. New York: Mariner Books, 2016, reprint.

Reynolds, Henry. *The Law of the Land*. New York: Viking, 1987.

Robinson, Claire, Michael Antoniou, and John Fagan. *GMO Myths and Truths: A Citizen's Guide to the Evidence on the Safety and Efficacy of Genetically Modified Crops and Foods, 4th Edition*. London: Earth Open Source, 2012.

Robinson, Randall. *An Unbroken Agony: Haiti, from Revolution to the Kidnapping of a President*. New York: BasicCivitas Books, 2008.

Rodney, Walter. *How Europe Underdeveloped Africa*. Washington, D.C.: Howard University Press, 1980.

Roessler, Shirley Elson, and Reny Miklos. *Europe 1715–1919: From Enlightenment to World War*. Lanham, MD: Rowman & Littlefield, Publishers, 2003.

Sale, Kirkpatrick. *The Conquest of Paradise: Christopher Columbus and the Columbian Legacy*. New York: Plume, Penguin Books, 1991.

Sampson, Anthony. *The Seven Sisters: The Great Oil Companies and the World They Shaped*. New York: Viking, 1980.

Sanders, Barry. *The Green Zone: The Environmental Costs of Militarism*. Oakland, CA: AK Press, 2009.

Sarkin, Jeremy. *Germany's Genocide of the Herero: Wilhelm Kaiser II: His General, His Settlers, His Soldiers*. Cape Town: UCT Press, 2010 and Suffolk, England: James Currey, 2011.

Saro-Wiwa, Ken. *Genocide in Nigeria: The Ogoni Tragedy*. Port Harcourt, Nigeria and London: Saros International Publishers, 1992.

Schwarz, Stuart B. *The Iberian Mediterranean and Atlantic Traditions in the Formation of Columbus as a Colonizer*. Minneapolis: University of Minnesota Press, 1986.

Scott, Heidi. "Whale Oil Culture, Consumerism, and Modern Conservation." *Oil Culture*. Edited by Ross Barrett, Daniel Worden, and Allan Stoeki. Minneapolis: University of Minnesota Press, 2014, pp. 3–18.

Sharpe, Christina. *In the Wake: Blackness and Being*. Durham, NC: Duke University Press, 2016.

Shenandoah, Leon. *To Become a Human Being: The Message of Tadodaho Chief Leon Shenandoah*. Edited by Steve Wall. Charleston, VA: Hampton Roads Publishing Company, Inc., 2001.

Shiva, Vandana (ed.). *Making Peace with the Earth*. London: Pluto Press; Winnipeg, Manitoba: Fernwood Publishing, 2013.

———. *Seed Sovereignty, Food Security: Women in the Vanguard of the Fight Against GMOs and Corporate Agriculture*. Berkeley, CA: North Atlantic Books, 2014.

Shiva, Vandana, with Kartikey Shiva. *Oneness Vs the 1%: Shattering Illusions, Seeding Freedom*. White River Junction, VT: Chelsea Press, 2020.

Smith, Adam. *The Theory of Moral Sentiments*. Cambridge, England and New York: Cambridge University Press, 2002.

_____. *The Wealth of Nations*. New York: Start Publishing, distributed by Simon & Schuster, New York, 2013.

Smith, Linda Tuhiwai. *Decolonization Methodologies: Research and Indigenous Peoples*. London and New York: Zed Books, and Dunedin, Aotearoa/New Zealand: University of Otago Press, 1999, 2001.

Somé, Malidoma. *Of Water and the Spirit: Ritual, Magic, and Initiation in the Life of an African Shaman*. New York: Arkana/Penguin Group, 1995.

Spence, Mark David. *Dispossessing the Wilderness: Indian Removal and the Making of the National Parks*. New York: Oxford University Press, 1999.

Stennard, David. *American Holocaust: The Conquest of the New World*. New York: Oxford University Press, 1992.

Stiffarm, Leonore A., with Phil Lane, Jr. "The Demography of Native North America: A Question of American Indian Survival." *The State of Native America: Genocide, Colonization and Resistance*. Edited by M. Annette Jaimes. Boston: South End Press, 1992, pp. 23–53.

Strachey, Charles (ed.). *The Letters of the Earl of Chesterfield to his Son*. England, 1901.

Stuckey, Sterling. *Slave Culture: Nationalist Theory and the Foundations of Black America*. New York: Oxford University Press, 1987.

Sweet, James H. *Domingos Alvarez: African Healing and the Intellectual History of the Atlantic World*. Durham, NC: University of North Carolina Press, 2011.

Sze, Julie. *Noxious New York: The Racial Politics of Urban Health and Environmental Justice*. Cambridge, MA: MIT Press, 2007.

Tannenbaum, Edward. *European Civilization Since the Middle Ages, Second Edition*. New York: John Wiley and Sons, Inc., 1971.

Thatcher, J.B. *Christopher Columbus, Vol. 2*. New York: Putnam's Sons Publishers, 1903–1904.

Thomas, Keith. *Man and the Natural World: A History of the Modern Sensibility*. New York: Pantheon Books, 1983.

Thornton, Russell. *American Indian Holocaust and Survival: A Population History since 1491*. Norman, OK: University of Oklahoma, 1992.

Tinker, George. "An American Indian Theological Response to Ecojustice." *Defending Mother Earth: Native American Perspectives in Environmental Justice*. Edited by Jace Weaver. New York: Orbis Books, 1996, pp. 153–76.

Tockner, Klement, Urs Uehlinger, and Christopher T. Robinson. *Rivers of Europe*. Amsterdam, The Hague: Elsevier, 2009.

Torres, Pat Mamanyjun. "Indigenous Knowledge Systems in Yawuru Aboriginal Australia." *Indigenous Peoples' Wisdom and Power: Affirming our Knowledge Through Narratives*. Edited by Julian Kunnie and Nomalungelo Goduka. Aldershot, Hampshire, England and Burlington, VT: Ashgate Publishing Company, 2006.

Trevor-Roper, Hugh. *The Rise of Christian Europe*. New York: Harcourt, Brace, and Jovanovich, 1975.

Trotter, Joe William. *The African American Experience*. Boston: Houghton Mifflin, 2001.

Trouillot, Michel-Rolph. *Silencing the Past: Power and the Production of History*. Boston: Beacon, 2015.

Trusler, John. *The Honours of the Table*. 2nd edition. England, 1791.

Turuki Pere, Rangimarie. "A Celebration of Maori Sacred and Spiritual Wisdom." *Indigenous Peoples' Wisdom and Power: Affirming our Knowledge Through Narratives*. Edited by Julian Kunnie and Nomalungelo Goduka. Aldershot, Hampshire, England: Ashgate, 2006.

Uhl, Michael, and Tod Ensign. *GI Guinea Pigs: How the Pentagon Exposed Our Troops to Dangers More Deadly than War: Agent Orange and Atomic Radiation*. New York: Playboy Press, 1980.

Varela, Consuela. "Prólogo," *Textos y Documentos Completes: Relaciones de Viajes, Cartas y Memorials*. Madrid: Allianza Editorial, 1982.

Vaughan, Alden T. *New England Frontier: 1620–1675*. Boston: Little, Brown, 1965.

wa Thiong'o, Ngũgĩ. *Caitaani utharaba-ini*. Nairobi: East African Educational Publishers, 1980.

_____. *Decolonizing the Mind: The Politics of Language in African Literature*. London: James, Currey Ltd./Heinemann, 1986; Kindle version.

Wakefield, Edward Gibbon (ed.). *England and America*. New York: Harper & Brothers, 1834.

Warren, Wendy. *New England Bound: Slavery and Colonization in Early America*. New York: W.W. Norton, 2016.

Washington, Harriet. *Medical Apartheid: The Dark History of Medical Experimentation on Black Americans from Colonial Times to the Present*. New York: Doubleday, 2006.

Weiner, Tim. *Legacy of Ashes: The History of the CIA*. New York: Anchor, 2008.

Williams, Eric C. *Capitalism and Slavery*. Chapel Hill: University of North Carolina Press, 1994.

Williams, Victoria R. *Indigenous Peoples: An Encyclopedia of Culture, History, and Threats to Survival, Volume 4*. Santa Barbara, CA: ABC-CLIO, 2020.

Wilson, James. *The Earth Shall Weep: A History of Native America*. New York: Grove Press, 1998.

Woodson, Carter G. *The Miseducation of the Negro*. First edition, 1933. Trenton, NJ: Africa World Press, 1990.

Woolford, Andrew, Jeff Benvenuto, and Alexander Laban Hinton. *Colonial Genocide in Indigenous North America*. Durham, NC: Duke University Press, 2014.

Woolley, Hannah. *The Queen-Like Closet*. 5th ed. Richard Lowndes, 1684.

Zaloom, Caitlin. *Indebted: How Families Make College Work at Any Cost.* Princeton, NJ: Princeton University Press, 2019.

Newspaper Article (Print)

Arizona Daily Star, "Report: Teachers in Arizona are Less Diverse than Students." *Associated Press,* March 1, 2020.

Online Resources (websites)

Abramson, Jeff. "Proposed U.S. Arms Export Agreements from January 1, 2001 to December 31, 2001," *Facts Sheets & Briefs.* https://www.armscontrol.org/factsheets/armexp2001 (accessed February 18, 2022).

Ahlman, Austin. "Biden's Decision on Frozen Afghanistan Money is Tantamount to Mass Murder." *The Intercept,* February 11, 2022. https://theintercept.com/2022/02/11/-afghanistan-frozen-assets-economy/ (accessed June 5, 2022).

Air Weather Association. *Air Force Weather: Our Heritage 1937–2012: An Illustrated Chronology: Meeting the Challenge for 75 Years.* July 1, 2012. https://www.557weatherwing.af.mil/Portals/62/documents/Air%20Force%20Weather%20-%20Our%20Heritage%202nd%20Ed.pdf?ver=2019-01-09-141532-173 (accessed January 5, 2022).

Al Jazeera. "Canada: At Least 600 Graves Found at Indigenous Boarding School." https://www.aljazeera.com/news/2021/6/24/canada-hundreds-of-graves-found-at-indigenous-boarding-school. June 24, 2021 (accessed November 25, 2021).

———. "Nations Where 3.3bn People Live Spend More on Debt than Health Care, Schools." July 13, 2023. https://www.aljazeera.com/news/2023/7/13/nations-where-3-3bn-live-spend-more-on-debt-than-health-schools (accessed September 27, 2023).

———. "US Tech Giants Sued Over DRC Cobalt Mines Over Child Labor Deaths," *Reuters.* December 17, 2019. https://www.aljazeera.com/economy/2019/12/17/us-tech-giants-sued-over-drc-cobalt-mine-child-labour-deaths (accessed August 1, 2023).

al-Rubaie, Ahmed. "How the US, UAE and Israel Plundered Iraq's Antiquities." *The Cradle.* September 8, 2023. https://new.thecradle.co/articles/how-the-us-uae-and-israel-plundered-iraqs-antiquities (accessed September 8, 2023).

Albright, Madeleine. "Punishing Saddam." Interview with Leslie Stahl clip. *CBS 60 Minutes.* May 12, 1996. https://www.youtube.com/watch?v=FbIX1CP9qr4 (accessed August 24, 2023).

Aldrian, Edvin. "Indonesia's Capital Jakarta is Sinking: Here's How to Stop This." *The Conversation,* November 11, 2021. https://theconversation.com/indonesias-capital-jakarta-is-sinking-heres-how-to-stop-this-170269 (accessed May 22, 2022).

American Friends Service Committee. "As Global Military Expenditures Top $1.9 Trillion, 39 Organizations Call for Cuts to the Pentagon." https://www.afsc.org/newsroom/-global-military-expenditures-top-19-trillion-39-organizations-call-cuts-to-pentagon (accessed on November 19, 2021).

———. "Federal Budget Priorities." https://www.afsc.org/key-issues/issue/federal-budget-priorities; "One Minute for Peace Petition." https://www.afsc.org/action/one-minute-peace-petition (accessed November 19, 2021).

American Petroleum Institute. "How Much Water Does Hydraulic Fracturing Use?" https://www.api.org/oil-and-natural-gas/energy-primers/-hydraulic-fracturing/how-much-water-does-hydraulic-fracturing-use-2 (accessed September 11, 2023).

Amnesty International. "Israel/OPT: Civilian Deaths and Extensive Destruction in Latest Gaza Offensive Highlight Human Toll of Apartheid." 2017. https://www.amnesty.org/en/latest/news/2023/06/israel-opt-civilian-deaths-and-extensive-destruction-in-latest-gaza-offensive-highlight-human-toll-of-apartheid/ (accessed June 20, 2023).

Anderson, Catherine, and Rebecca Engebretsen. "The Impact of Coronavirus (COVID-19) and the Global Oil Price Shock on the Fiscal Position of Oil Exporting Developing Countries." Organization of Economic Co-operation and Development (OECD), September 30, 2020. https://www.oecd.org/coronavirus/policy-responses/-the-impact-of-coronavirus-covid-19-and-the-global-oil-price-shock-on-the-fiscal-position-of-oil-exporting-developing-countries-8bafbd95/ (accessed March 25, 2022).

Anderson, Tim. "Washington Rushes to Hide Its 'Octopus' NED Funding in Ukraine." *Al Mayadeen (English) Net,* March 15, 2022. https://english.almayadeen.net/articles/analysis/-washington-rushes-to-hide-its-octopus-ned-funding-in-ukraine (accessed March 26, 2022).

Animal Matters. "Facts-Wildlife." https://www.animalmatters.org/facts/wildlife/.(accessed November 9, 2021).

"As China Continues Planting Trees, 23% of the Country is Now Covered in Forest." *South China Morning Post,* March 26, 2020. https://www.youtube.com/watch?v=HP-iBKeqcF0 (accessed October 10, 2021).

Asia-Pacific Forum on Sustainable Development 2022. March 31, 2022. Side event organized by the Asia Pacific Network of Environment Defenders and IPMSDL. https://www.ipmsdl.org/campaign/ensuring-the-participation-of-environmental-human-rights-defenders-to-achieve-the-sdgs-towards-development-justice/ (accessed March 11, 2022).

Associated Press. "Federal Probe Blames EPA for Gold King Mine Spill." October 22, 2015.

_____. "The Latest: UN Chief Decries Pandemic's Harm to the Poor." October 11, 2021. https://apnews.com/article/coronavirus-pandemic-lifestyle-business-sri-lanka-health-6c9780d0d29568650437994ea7fdfdc5 (accessed October 17, 2021).

_____. "Over 50 Whales Found Dead in Scottish Beach in Worst Mass-Stranding in Area, Experts Say." *USA Today*, July 17, 2023. https://www.usatoday.com/story/news/world/2023/07/17/whale-stranding-scotland-kills-55-pilot-whales/70421541007/ (accessed September 4, 2023).

Atkin. Paul. "Let's Get Out from the Carbon Military Boot Print." *urbanramblings*, June 21, 2023. https://urbanramblings19687496.city/2023/06/21/lets-get-out-for-under-the-carbon-military-boot-print/ (accessed June 25, 2023).

Atomic Heritage Foundation. "Nevada Test Site." https://www.atomicheritage.org/location/nevada-test-site (accessed June 26, 2022).

Avery, John Scales. "The Illegality of NATO." *Transcend Media Service: Solutions-Oriented Peace Journalism,* January 28, 2022. https://www.transcend.org/tms/2022/01/the-illegality-of-nato-2/ (accessed February 1, 2022).

Aydelotte Foundation podcast. "Land Grab Universities." Tristan Ahtone and Maria Parazo Rose. Swarthmore College, May 18, 2023. https://open.spotify.com/episode/1X1FoNz9k8gf762EyqgphO?si=NmdaJf52Q4-Otv5kG60BSA&nd=1 (accessed August 22, 2023).

Badrakumar, M.K. *Punchline,* "India and the Donbass Republics." February 22, 2022. https://www.indianpunchline.com/india-and-the-donbass-republics/ (accessed on March 1, 2022).

Bailey, Diane. Natural Resources Defense Council. "Breathe Wheezy: Traffic Pollution Not Only Worsens Asthma, but May Cause It." *Scientific American*, January 9, 2013. https://www.scientificamerican.com/article/traffic-pollution-and-asthma/ (accessed March 1, 2022).

Banks, Jasmine. "Elite Schools Must Quit Koch Money." *The Progressive*, December 1, 2021. https://progressive.org/op-eds/elite-schools-quit-koch-money-banks-211201/ (accessed June 15, 2022).

Bardin, Jon. "Fish Getting Skin Cancer from UV Radiation, Scientists Say." *Los Angeles Times,* August 2, 2012. https://www.latimes.com/science/la-xpm-2012-aug-02-la-sci-fish-skin-cancer-20120802-story.html (accessed April 14, 2022).

Barringer, Felicity. "Climate Change Will Disrupt Half of North America's Bird Species, Study Says." *New York Times*, September 8, 2014. https://www.nytimes.com/2014/09/09/us/climate-change-will-disrupt-half-of-north-americas-bird-species-study-says.html (accessed May 2, 2022).

Barry-Shaw, Nikolas, and Dru Oja Jay. "New Documents Detail How Canada Helped Plan 2004 Coup d'État in Haiti." *The Breach*, July 15, 2021. https://breachmedia.ca/new-documents-detail-how-canada-helped-plan-2004-coup-detat-in-haiti/ (accessed October 19, 2021).

Baumeister, Christine, and Lutz Kilian. "Forty Years of Oil Price Fluctuations: Why the Price of Oil May Still Surprise Us." *Journal of Economic Perspectives,* Volume 30, Number 1, Winter, 2016, 139–160.

Beals, Rachel Koning. "Every Whale is Worth $2 Million? Why It's Time to Add the Value of Nature to GDP." *Market Watch,* October 28, 2021. https://www.marketwatch.com/story/every-whale-is-worth-2-million-why-its-time-to-add-the-value-of-nature-to-gdp-11632852074?mod=article_inline&mod=article_inline (accessed March 10, 2022).

Behar, Michael. "Fracking's Latest Scandal: Earthquake Swarms." *Mother Jones*-March/April 2013 Issue. https://www.motherjones.com/environment/2013/03/does-fracking-cause-earthquakes-wastewater-dewatering/ (accessed on February 24, 2022).

Bell, Horace. *Reminiscences of a Ranger: Or, Early Times in Southern California.* Los Angeles: Yarnell, Caystille & Mathes, Printers, 1881. *Internet archive.* https://archive.org/details/reminiscencesofr00bellrich/page/48/mode/2up (accessed September 30, 2021).

Bello, Walden. "Post-9/11 'Nation-Building' in Afghanistan and Iraq Was Nothing But Destruction." *Common Dreams,* September 12, 2021. https://www.commondreams.org/views/2021/09/12/post-911-nation-building-afghanistan-and-iraq-was-nothing-destruction (accessed September 15, 2021)_https://archive.org/details/reminiscencesofr00bellrich/page/48/mode/2up (accessed September 30, 2021).

Benjamin, Medea, and Nicholas J.S. Davies. "The Staggering Death Toll in Iraq." *Salon,* March 19, 2018. https://www.salon.com/2018/03/19/the-staggering-death-toll-in-iraq_partner/ (accessed April 4, 2022).

"Bio-Weapons: Truth or Fiction." *Russian Times Documentary Channel,* March 2022. https://rtd.rt.com/films/bio-weapons-truth-or-fiction/ (accessed March 30, 2022).

Biray, Kurt. "Communist Nostalgia in Eastern Europe: Longing for the Past." *openDemocracy.* November 10, 2015. https://www.opendemocracy.net/en/can-europe-make-it/communist-nostalgia-in-eastern-europe-longing-for-past/ (accessed August 31, 2023).

Bishop, Michelle. "Indigenous Education Sovereignty: Another Way of 'Doing' Education." *Critical Studies in Education,* November 30, 2020. https://doi.org/10.1080/17508487.2020.1848895 (accessed February 10, 2021).

Black Alliance for Peace. "Afghanistan News Update Update #6." https://blackallianceforpeace.com/afghanistannewsupdate/2021/11/23/afghanistan-news-update-6 (accessed November 23, 2021).

_____. "A Brief Guide on the Situation in

Ukraine." https://blackallianceforpeace.com/resourcesonukraine (accessed February 1, 2022).

Bloomberg Government. "The Top 10 Defense Contractors." June 10, 2021. https://about.bgov.com/top-defense-contractors/ (accessed January 10, 2022).

B:M 2022: Dig Deep, Look Closer, Think Bigger. "Taíno: Indigenous Caribbeans." February 12, 2021. https://www.blackhistorymonth.org.uk/article/section/pre-colonial-history/taino-indigenous-caribbeans/ (accessed June 26, 2022).

"Bolivia: Evo Morales Warns of DEA-Driven Destabilization Plan." *Telesur*, February 12, 2022. https://www.telesurenglish.net/news/Bolivia-Evo-Morales-Warns-of-DEA-driven-Destabilisation-Plan-20220212-0008.html?utm_source=planisys&utm_medium=NewsletterIngles&utm_campaign=NewsletterIngles&utm_content=8 (accessed February 13, 2022).

Bretton Woods Project. "What are the Bretton Woods Institutions?" January 14, 2018. https://www.brettonwoodsproject.org/2019/01/art-320747/ (accessed June 26, 2022).

Briggs, Helen. "Biodiversity Loss Risks 'Ecological Meltdown—Scientists." *BBC News,* October 10, 2021. https://www.bbc.com/news/science-environment-58859105 (accessed November 7, 2021).

Budanovic, Nikola. "Liquid Fire—How Napalm Was Used in the Vietnam War." *War History Online,* June 1, 2016. https://www.warhistoryonline.com/vietnam-war/history-napalm-vietnam-war.html?firefox=1 (accessed April 1, 2022).

Burger, Joanna. "Ecological Concerns Following Superstorm Sandy: Stressor Level and Recreational Activity Levels Affect Perceptions of Ecosystem." *Urban Ecosyst,* 18 (2), June 1, 2015, 553–575, PMC PubMed Central, National Institute of Health (NIH): National Library of Medicine. https://www.ncbi.nlm.nih.gov/pmc/articles/PMC4800746/ (accessed April 5, 2022).

Burton, Eric. "Socialisms Between Cooperation and Competition: Ideology, AID and Cold-War Politics in Tanzania's Relation with East Germany." *Socialismes en Afrique: Socialisms in Africa*. Edited by Françoise Blum, Héloïse Kiriakou, Martin Mourre, et al. *Éditions de la Maison des sciences de l'homme*/OpenEdition Books, 2021. https://books.openedition.org/editionsmsh/51520?lang=en (accessed June 25, 2022).

Byrd, Robert C., John B. Smith, Sr., and Steven J. Spease. "The Challenges Facing the Industry in Offshore Facility Decommissioning on the California Coast Offshore, Inc." Technology Conference, April 30-May 3, 2018. (emailed by John B. Smith on November 1, 2023). file:///Users/aswan/Downloads/OTC%20Paper%20CA%20Decom%20Challenges%202018.pdf.

Canadian Museum of History. https://www.historymuseum.ca/cmc/exhibitions/civil/egypt/egcr09e.html (accessed June 20, 2021).

Carrere, Michelle. "To Fight Climate Change, Save the Whales, Some Scientists Say." Translated by Sarah Engel. https://news.mongabay.com/2021/03/to-fight-climate-change-save-the-whales-some-scientists-say/ (accessed March 5, 2023).

Carson, J.H. *Recollections of the California Mines: An Account of the Early Discoveries of Gold, with Anecdotes and Sketches of California and Miners' Life, and a Description of the Great Tulare Valley*. Oakland, CA: Biobooks, 1950. https://babel.hathitrust.org/cgi/pt?id=ucl.32106000661600&view=1up&seq=9 (accessed September 28, 2021).

Chalmin, Anja. "Geo-Engineering Activities on the African Continent." *Geoengineering Monitor,* January 12, 2021. https://www.geoengineeringmonitor.org/2021/01/geoengineering-activities-on-the-african-continent/ (accessed May 1, 2022).

Chawla, Dalmeet Singh. "Millions of Animals May be Missing from Scientific Studies." *Science,* October 14, 2020. https://www.science.org/content/article/millions-animals-may-be-missing-scientific-studies (accessed November 12, 2021).

Cho, Renee. "The Fracking Facts." Columbia Climate School. *Climate, Earth, and Society: State of the Planet,* June 6, 2014. https://news.climate.columbia.edu/2014/06/06/the-fracking-facts/ (accessed February 28, 2022).

The Citizen. "U.S. Admits There Are 'Biological Facilities' in Ukraine," *The Citizen,* March 9, 2022. https://www.thecitizen.co.tz/tanzania/news/international/us-admits-there-are-biological-facilities-in-ukraine-3741878 (accessed March 11, 2022).

Civil Georgia. "Georgian Speaker Refutes Moscow's Bio-Warfare Allegations." March 11, 2022. https://civil.ge/archives/478723 (accessed March 29, 2022).

Climate Home News. "Kiribati President: My People Are the Polar Bears of the Pacific." September 9, 2014. https://www.climatechangenews.com/2014/09/23/kiribati-president-my-people-are-the-polar-bears-of-the-south/ (accessed December 9, 2021).

Cohen, Dan. "Ukraine's Propaganda War: International PR Firms, DC Lobbyists and CIA Cutouts." *Mint Press News,* March 22, 2022. https://www.mintpressnews.com/ukraine-propaganda-war-international-pr-firms-dc-lobbyists-cia-cutouts/280012/ (accessed on March 22, 2022).

―――. "US Admits to Funding Biological Laboratories in Ukraine, with Dilyana Gaytandzhieva." *Mint Press News,* March 10, 2022. https://www.mintpressnews.com/us-admits-funding-biological-laboratories-ukraine-dilyana-gaytandzhieva/279904/ (accessed on March 10, 2022).

Comparative Constitutions Project. *Constitution, United States of America 1789* (rev 1992). https://www.constituteproject.org/constitution/

United_States_of_America_1992 (accessed June 21, 2022).

"Conspiracy Theories Abound as US Military Closes HAARP." *NBC News*, May 22, 2014. https://www.nbcnews.com/science/weird-science/conspiracy-theories-abound-u-s-military-closes-haarp-n112576 (accessed May 21, 2022).

Corbett, Jessica. "'Nothing More Grotesque than a Media Pushing for War,' says Edward Snowden." *Common Dreams*, February 12, 2022. https://www.commondreams.org/news/2022/02/11/-nothing-more-grotesque-media-pushing-war-says-edward-snowden (accessed February 13, 2022).

———. "Reports Expose U.S. Billionaires and Corporate Profiteers Enabling Israel's War on Gaza." *Common Dreams*, October 27, 2023. https://www.commondreams.org/news/us-weapons-to-israel (accessed October 27, 2023).

Cornwall, Warren. "Europe's Deadly Floods Leave Scientists Stunned." *Science*, July 20, 2021. https://www.science.org/content/article/-europe-s-deadly-floods-leave-scientists-stunned (accessed May 22, 2022).

Costa, Daniel, and Philip Martin. "Record-Low Number of Federal Wage and Hour Investigations of Farms in 2022." *Economic Policy Institute*. Aug 22, 2023. https://www.epi.org/publication/record-low-farm-investigations/?mc_cid=2f44c64208&mc_eid=49bb4f1fa1 (accessed August 25, 2023).

Cotton, Simon. "Explainer: What is Napalm?" *The Conversation*, September 5, 2013. https://theconversation.com/explainer-what-is-napalm-17795 (accessed June 10, 2022).

Crawford, Neta C. "Pentagon Fuel Use, Climate Change, and the Costs of War." *Costs of War*. Boston University: Watson Institute of International and Public Affairs, November 13, 2019. https://watson.brown.edu/costsofwar/files/(cow/imce/papers/Pentagon%20Fuel%20Use%2C%20Climate%20Change%20and%20the%20Costs%20of%20War%20Revised%20November%202019%20Crawford.pdf (accessed February 4, 2022).

Cuemath: The History of Algebra. https://www.cuemath.com/learn/mathematics/algebra-history-of-algebra/ (accessed September 2, 2021).

Daurelle, B.C. "Haitian Textile Workers Strike Against U.S. and International Sweatshops." *Left Voice*, February 18, 2022. https://www.leftvoice.org/haitian-textile-workers-strike-against-sweatshops/ (accessed August 5, 2023).

David, Javier E. "Ukraine Gas Producer Appoints Biden's Son to Board." *CNBC, Energy*, May 13, 2014. https://www.cnbc.com/2014/05/13/bidens-son-joins-ukraine-gas-companys-board-of-directors.html (accessed March 26, 2022).

DefenseWorld.Net. "United States' Arms Exports Totaled $175.08 Billion in 2020, Up 2.8% Over 2019." December 5, 2020. https://www.defenseworld.net/news/28470/United_States____Arms_Exports_Totaled__175_08_billion_in_2020__up_2_8__Over_2019#.YhQpnR17kk8 (accessed February 21, 2022).

"Definition, Importance, and History of Organometallics." *LibreTexts Chemistry*. https://chem.libretexts.org/Bookshelves/Inorganic_Chemistry/Supplemental_Modules_and_Websites_(Inorganic_Chemistry)/Advanced_Inorganic_Chemistry_(Wikibook)/01%3A_Chapters/1.18%3A_Definition_Importance_and_History_of_Organometallics (accessed September 18, 2023).

DeLetter, Emily. "Endangered North Atlantic Right Whales Are in 'Crisis': Species Approaching Extinction." *USA Today*, July 19, 2023. https://www.usatoday.com/story/news/nation/2023/07/19/north-atlantic-right-wales-noaa-crisis-approaching-extinction/70429914007/ (accessed September 5, 2023).

Dhillon, Jaskiran, and Nick Estes. "Standing Rock, #NoDAPL, and Mini Wiconi." *Society for Cultural Anthropology*, December 22, 2016. https://culanth.org/fieldsights/series/standing-rock-nodapl-and-mni-wiconi (accessed August 18, 2023).

DiChristopher, Tom. "The Financial Crisis Crushed Record Oil Prices but the Market is Still Gripped." *CNBC, Energy*, September 12, 2018. https://www.cnbc.com/2018/09/12/financial-crisis-crushed-record-oil-prices-but-another-boom-looms.html (accessed March 25, 2022).

Dickinson, Tim. "Inside the Koch Brothers' Toxic Empire." *Rolling Stone*, September 24, 2015. https://www.rollingstone.com/politics/-politics-news/inside-the-koch-brothers-toxic-empire-164403/ (accessed June 21, 2022).

———. "Koch Industries Responds to Rolling Stone—And We Answer Back." *Rolling Stone*, September 29, 2014. https://www.rollingstone.com/politics/politics-news/koch-industries-responds-to-rolling-stone-and-we-answer-back-237090/ (accessed on June 21, 2022).

Digital History, ID 1248. "Columbus' Journal Entries from August to November 1492." https://www.digitalhistory.uh.edu/disp_textbook.cfm?smtID=3&psid=1248 (accessed September 16, 2021).

Ding, Jamie. "California Urban Water Use Rose 19% in March Despite Worsening Drought." *Los Angeles Times*, May 10, 2022. https://www.latimes.com/environment/story/2022-05-10/california-water-use-rose-19-percent-in-march?utm_id=55165&sfmc_id=764628 (accessed May 10, 2022).

Dolgin, Ellie. "The Tangled History of mRNA Vaccines." *Nature*, September 14, 2021. https://www.nature.com/articles/d41586-021-02483-w (accessed August 15, 2021).

Dongping, Han. "No Country Can be the World's Policeman: Debt-Ridden U.S. Should Focus on Itself." *Think China*, September 17, 2020. https://www.thinkchina.sg/no-country-can-be-worlds-policeman-debt-ridden-us-needs-focus-itself (accessed on February 10, 2022).

Earth Justice. "Texas and Fracking." September 29, 2015. https://earthjustice.org/features/texas-and-fracking (accessed April 22, 2022).

Earthworks. "Fracking-Related Earthquakes." https://earthworks.org/issues/fracking_earthquakes/ (accessed February 23, 2022).

Edelstein, Karen. "Impacts of 2020 Colonial Pipeline Rupture Continue to Grow." *FracTracker Alliance*, May 26, 2021. https://www.fractracker.org/2021/05/august-2020-colonial-pipeline-spill-in-north-carolina/

Eisinger, Jesse, Jeff Ernsthausen, and Paul Kiel. "The Secret IRS Files: Trove of Never-Before-Seen Records Reveal How the Wealthiest Avoid Income Tax." *ProPublica*, June 8, 2021. https://www.propublica.org/article/the-secret-irs-files-trove-of-never-before-seen-records-reveal-how-the-wealthiest-avoid-income-tax (accessed June 3, 2022).

Encyclopedia Virginia: Virginia Humanities. https://encyclopediavirginia.org/7001-a62d2a140fd1ff0/ (accessed September 27, 2021).

Environment. July-August 2010. http://www.environmentmagazine.org/Archives/Back%20Issues/July-August%202010/securing-foreign-oil-full.html (accessed February 4, 2022).

Environment Health Trust. "5G: Environmental Effects, Birds, Bees, Trees." YouTube educational research video. https://www.youtube.com/watch?v=V1THhDzqZD0&t=124s (accessed September 22, 2023).

———. "Liability, 5G and Cell Tower Radiation." February 26, 2022. https://ehtrust.org/liability-and-risk-from-5g-and-cell-towers/ (accessed September 22, 2023).

Erickson, P., and P. Achakulwisut. "How Subsidies Aided the U.S. Shale Oil and Gas Boom." Stockholm, Sweden: *Stockholm Environmental Institute*, June 23, 2021. https://www.sei.org/publications/subsidies-shale-oil-and-gas/ (accessed February 5, 2022).

European Centre for Medium-Range Weather Forecasts. "Flood List," May 22, 2022. https://floodlist.com/europe (accessed May 22, 2022).

Ewald, Megan. "The Oil Pollution Act of 1990: 30 Years of Oil Spills." National Oceanic and Atmospheric Administration, U.S. Department of Commerce, August 18, 2020. https://storymaps.arcgis.com/stories/61565da528724ef3a542a766389f3087 (accessed March 21, 2022).

Ewepu, Gabriel. "Imo Explosion: 11 CSOs Accuse Government of Failure to Address Factors Behind Proliferation of Illegal Refineries." *Vanguard News Nigeria*, April 25, 2022. https://www.vanguardngr.com/2022/04/imo-explosion-11-csos-accuse-fg-of-failure-to-address-factors-behind-proliferation-of-illegal-refineries/ (accessed April 26, 2022).

"Federal Probe Blames EPA For Gold King Mine Spill." *Associated Press*, October 22, 2015. https://www.cpr.org/2015/10/22/federal-probe-blames-epa-for-gold-king-mine-spill/ (accessed October 25, 2022).

Fields, Samantha. "Nearly 9 Million Older Americans Still Have Student Loan Debt." *Marketplace, Morning Report.* June 21, 2021. https://www.marketplace.org/2021/06/21/nearly-9-million-older-americans-still-have-student-loan-debt/ (accessed June 1, 2022).

Fihn, Beatrice. "Universities Across America Profit from Developing Weapons. It's Unconscionable." *Newsweek.* November 13, 2019. https://www.newsweek.com/universities-funding-nuclear-weapons-research-1471572 (accessed November 27, 2023).

5GFree.org, Arizona action against 5G. "Josh Del Sol's 5G Solutions Summit with Lena Pu. Watch This." June 5, 2020. https://5gfree.org/groups/arizona/ (accessed September 21, 2023).

"5 Wild Weather Control Ideas." *Live Science,* August 6, 2012. https://www.livescience.com/22131-wild-weather-control-ideas.html (accessed May 21, 2022).

Flood, Alison. "Einstein's Travel Diaries Reveal 'Shocking' Xenophobia." *The Guardian,* June 12, 2018. https://www.theguardian.com/books/2018/jun/12/einsteins-travel-diaries-reveal-shocking-xenophobia (accessed July 28, 2021).

Foster, John Bellamy, Brett Clark, and Hannah Holleman. "Marx and the Indigenous." *Monthly Review,* February 2020. https://monthlyreviewarchives.org/index.php/mr/article/view/MR-071-09-2020-02_1 (accessed August 16, 2023).

Fountain, Henry. "Shift to a Not-So-Frozen North Is Well Underway, Scientists Warn," *New York Times,* December 8, 2020. https://www.nytimes.com/2020/12/08/climate/arctic-climate-change.html (accessed December 8, 2020).

Frank, Robert. "Soaring Markets Helped the Richest 1% Gain $6.5 Trillion in Wealth Last Year, According to the Fed." *CNBC,* April 1, 2022. https://www.cnbc.com/2022/04/01/richest-one-percent-gained-trillions-in-wealth-2021.html (accessed June 1, 2022).

Friedman, Zack. "Student Loan Statistics in 2021: A Record $1.7 Trillion." *Forbes,* February 20, 2021. https://www.forbes.com/sites/zackfriedman/2021/02/20/student-loan-debt-statistics-in-2021-a-record-17-trillion/?sh=96cac5514310 (accessed June 20, 2022).

Fulton, Deirdre. "Climate Change Threatens 'Rampant Bird Extinctions,' Study Finds." *Common Dreams,* September 9, 2014. https://www.commondreams.org/news/2014/09/09/climate-change-threatens-rampant-bird-extinctions-study-finds (accessed May 2, 2022).

Garrison, Ann. "Rwanda and Uganda's M23 Reappears to Slaughter and Plunder in DRC: U.S. backs Rwanda and Uganda." *Black Agenda Report.* September 7, 2022. https://www.blackagendareport.com/rwanda-and-ugandas-m23-militia-reappears-slaughter-and-plunder-drc-us-backs-rwanda-and-uganda (accessed June 30, 2023).

Gaytandhzieva, Dilyana. "Experiments Expose

U.S. Biological Experiments on Allied Soldiers in Ukraine and Georgia." *Arms Watch,* January 25, 2022. https://armswatch.com/documents-expose-us-biological-experiments-on-allied-soldiers-in-ukraine-and-georgia/ (accessed March 12, 2022).

———. "New Data Leak from the Pentagon Biolaboratory in Georgia." *Armswatch.* https://armswatch.com/new-data-leak-from-the-pentagon-biolaboratory-in-georgia/ (accessed September 20, 2020).

———. "Pentagon Unit A1266 Studies Bioterrorism Agents in Kazakhstan." *Arms Watch,* July 21, 2020. https://armswatch.com/pentagon-unit-a1266-studies-bioterrorism-agents-in-kazakhstan/ (accessed September 20, 2020).

———. "Salisbury Attack Reveals $70 Million Pentagon Program at Porton Down." *Armswatch.* http://armswatch.com/salisbury-attack-reveals-70-million-pentagon-program-at-porton-down/ (accessed May 10, 2020).

Generation Zapped. Film written and produced by Sabine El Gemayel, 2017. https://www.imdb.com/title/tt6264018/ (accessed January 10, 2018).

Geneva International Centre for Justice. "Razing the Truth about Sanctions Against Iraq." https://www.gicj.org/positions-opinons/gicj-positions-and-opinions/1188-razing-the-truth-about-sanctions-against-iraq (accessed December 13, 2021).

Genomics Education Program. "Why mRNA Vaccines Aren't Gene Therapies." June 11, 2021. https://www.genomicseducation.hee.nhs.uk/blog/why-mrna-vaccines-arent-gene-therapies/ (accessed September 20, 2023).

"Geoengineering: Parts I, II, and III: Hearing Before the Committee on Science and Technology, House of Representatives, One Hundred and Eleventh Congress First Session and Second Session." November 5, 2009, February 4, 2010, and March 18, 2010. Serial No. 111–62, Serial No. 111–75, and Serial No. 111–88. Washington, D.C.: Printed for the use of the Committee on Science and Technology, 2010. https://www.govinfo.gov/content/pkg/CHRG-111hhrg53007/pdf/CHRG-111hhrg53007.pdf (accessed May 17, 2022).

Geoengineering Watch: Exposing the Climate Engineering. "HAARP—An Assessment, What You Don't Understand CAN Hurt You." October 31, 2016. https://www.geoengineeringwatch.org/?s=HAARP and https://www.geoengineeringwatch.org/haarp-a-short-summary-video/#more-33110 (accessed September 21, 2021).

"Georgian Speaker Refutes Moscow's Bio-Warfare Allegations." *Civil Georgia,* March 11, 2022. https://civil.ge/archives/478723 (accessed March 29, 2022).

Germanos, Andrea. "Biden Urged to 'Be the Hero' to Save American Bumblebee from Extinction." *Common Dreams,* February 1, 2021.

Gohd, Chelsea. "NASA and U.S. Space Force Team Up for Planetary Defense, Moon Trips and More." *Space.com,* part of Future U.S., Inc. https://www.space.com/nasa—space-force-moon-planetary-defense-collaboration.html (accessed November 12, 2021).

Goodman, Peter S. "'He Has an Incredible Knack to Smell the Latest Fad': How Klaus Schwab Built a Billionaire Circus at Davos." *Vanity Fair,* January 18, 2022. https://www.vanityfair.com/news/2022/01/how-klaus-schwab-built-a-billionaire-circus-at-davos (accessed June 16, 2022).

Gott, Molly, and Derek Seidman. "Corporate Enablers of Israel's War on Gaza." *Eyes on the Ties,* October 26, 2023. https://news.littlesis.org/2023/10/26/corporate-enablers-of-israels-war-on-gaza/ https://news.littlesis.org/2023/10/26/corporate-enablers-of-israels-war-on-gaza/ (accessed October 27, 2023).

Greenpeace. "Commercial Whaling." https://www.greenpeace.org/usa/oceans/issues/commercial-whaling/ (accessed September 2, 2023).

Greenwood, Michael. "Chemicals in Fracking Fluid and Wastewater Are Toxic, Study Shows." *YaleNews,* January 6, 2016. https://news.yale.edu/2016/01/06/toxins-found-fracking-fluids-and-wastewater-study-shows (accessed February 28, 2022).

Gross, Daniel A. "Chemical Warfare: From the European Battlefield to the American Laboratory." *Distillation: Using Stories from Science's Past to Understand our World.* Philadelphia: Science History Institute, April 14, 2015. https://www.sciencehistory.org/distillations/chemical-warfare-from-the-european-battlefield-to-the-american-laboratory (accessed January 9, 2022).

Hahn, Alicia, and Jordan Tarver. "2022 Student Loan Statistics: Average Student Loan Debt." *Forbes,* June 9, 2022. https://www.forbes.com/advisor/student-loans/average-student-loan-statistics/ (accessed June 18, 2022).

Haiphong, Danny. "The Biden-Harris Administration is a Political Expression of the Empire's Crisis of Legitimacy." *Black Agenda Report,* January 27, 2022. https://blackagendareport.com/biden-harris-administration-political-expression-empires-crisis-l (accessed February 1, 2022).

Hanink, Johanna. "A New Path for Classics: The Field Is a Product and Accomplice of White Supremacy, Scholars Are Fighting to Change That." *Chronicle of Higher Education,* February 11, 2021. https://www.chronicle.com/article/if-classics-doesnt-change-let-it-burn?cid2=gen_login_refresh&cid=gen_sign_in (accessed February 15, 2021).

Harding, Lee. "Gates, WHO, and Abortion Vaccines." *Frontier Centre for Public Policy,* July 19, 2020. https://fcpp.org/2020/07/19/gates-who-and-abortion-vaccines/ (accessed on August 6, 2021).

Harman, Kristyn. "Explainer: The Evidence for the Tasmanian Genocide." *The Conversation,* January 17, 2018. https://theconversation.com/-explainer-the-evidence-for-the-tasmanian-genocide-86828 (accessed December 15, 2021).

Harris, Roger D. "Nicaragua Has a Public Relations Problem, Not a Democracy Problem." *Black Agenda Report,* November 13, 2021. https://www.blackagendareport.com/nicaragua-has-public-relations-problem-not-democracy-problemhttps://www.blackagendareport.com/nicaragua-has-public-relations-problem-not-democracy-problem (accessed November 15, 2021).

Hartung, William, and Christina Arabia. "Trends in Major U.S. Arms Sales in 2018: The Trump Record-Rhetoric Versus Reality." *Security Assistance Monitor,* Center for International Policy, Washington, D.C, 2019. https://securityassistance.org/publications/trends-in-major-u-s-arms-sales-in-2018-the-trump-record-rhetoric-versus-reality/ (accessed February 21, 2022).

Harvey, Fiona. "Experts Call for Global Moratorium on Efforts to Geoengineer Climate." *The Guardian,* September 14, 2023. https://www.theguardian.com/environment/2023/sep/14/experts-call-for-global-moratorium-on-efforts-to-geoengineer-climate (accessed November 28, 2023).

Healing Foundation. https://healingfoundation.org.au/who-are-the-stolen-generations/ (accessed November 25, 2021).

Hendricks, Mike. "Kansas Has a Lot More Earthquakes. Is that Because of Fracking?" *Kansas City Star,* January 28, 2014. https://www.governing.com/news/headlines/kansas-has-a-lot-more-earthquakes-is-that-because-of-fracking.html (accessed February 28, 2022).

Heubi, Ben. "How a Company Got Away After Committing the Worst Oil Spill in History." *E&T, Engineering and Technology,* February 4, 2021. https://eandt.theiet.org/content/articles/2021/01/lasting-effects-from-the-rainforest-oil-spill-in-ecuador/ (accessed March 7, 2022).

Hill, Ronald, Michelle Scharer, Michael Nemeth, and Andy Bruckner. "Reef Fish Habitat Use as a Measure of Coral Reef Restoration Success at the Fortuna Reefer Grounding Site, Mona Island, Puerto Rico." *Proceedings of the 50th Gulf and Caribbean Fisheries Institute,* November 5–9, 2007, Punta Cana, Dominican Republic. https://nsgl.gso.uri.edu › flsgp › data › papers PDF (accessed March 30, 2022).

Hirschlag, Ally. "How to Fix Our Ocean Noise Problem?" *BBC,* July 12, 2022. https://www.bbc.com/future/article/20220712-how-to-fix-our-ocean-noise-pollution-problem (accessed July 27, 2022).

Hoch, Maureen. "New Estimate Puts Gulf Oil Leak at 205 Million Gallons." *PBS News Hour,* August 2, 2010. https://www.pbs.org/newshour/science/new-estimate-puts-oil-leak-at-49-million-barrels (accessed April 5, 2022).

Hoegh-Guldberg, Ove, Elvira S. Poloczanska, William Skirving, and Sophie Dovie. "Coral Reef Ecosystems under Climate Change and Ocean Acidification." *Frontiers in Marine Science,* May 29, 2017. https://doi.org/10.3389/fmars.2017.00158 (accessed January 10, 2022).

Hoffman, Meredith. "Live Near a Fracking Site in Pennsylvania? You Might Be Going to the Hospital More than Others." *Vice News,* July 21, 2015. https://www.vice.com/en/article/9kjy93/live-near-a-fracking-site-in-pennsylvania-you-might-be-going-to-the-hospital-more-than-others (accessed February 28, 2022).

Hood, Marlowe. "World Seeing 'Catastrophic Collapse of Insects: Study." *Science × Daily, Phys Org,* February 11, 2019. https://phys.org/news/2019-02-world-catastrophic-collapse-insects.html (accessed February 18, 2022).

Human, L. "Farm Workers March Against Pesticides." *GroundUp,* September 9, 2023. https://www.groundup.org.za/article/farm-workers-march-against-pesticides-in-paarl/ (accessed September 10, 2023).

The Humane Society of the United States. "Animals Used in Experiments." https://www.humanesociety.org/resources/animals-used-experiments-faq#many (accessed November 9, 2021).

"Hunter Biden Did Fund Ukraine Biolabs, Emails Published by Media Suggest." *Russian Times,* March 26, 2022. https://www.rt.com/russia/552733-hunter-biden-arranged-ukraine-biolab-financing/ (accessed March 26, 2022).

Hunziker, Robert. "Insect Decimation Upstages Global Warming." *Counterpunch,* March 27, 2018. https://www.counterpunch.org/2018/03/27/insect-decimation-upstages-global-warming/ (accessed April 1, 2018).

Husseini, Sam. "Peter Daszak's EcoHealth Alliance Has Hidden Almost $40 Million in Pentagon Funding and Militarized Pandemic Science." *Independent Science News for Food and Agriculture,* December 16, 2020. https://www.independentsciencenews.org/news/peter-daszaks-ecohealth-alliance-has-hidden-almost-40-million-in-pentagon-funding/ (accessed January 13, 2022).

Hynes, H. Patricia. "Pentagon Pollution, 7: The Military Assault on Global Climate." *Climate & Capitalism,* February 8, 2015. https://climateandcapitalism.com/2015/02/08/pentagon-pollution-7-military-assault-global-climate/ (accessed August 10, 2020).

"Imminent Ozone Layer Collapse: A Dire Warning from a Former NASA Contract Engineer." www.geoengineeringwatch.com, April 13, 2022. https://www.geoengineeringwatch.org/imminent-ozone-layer-collapse-a-dire-warning-from-a-former-nasa-contract-engineer/ (accessed April 14, 2022).

Indigenous Peoples' Movement for Self-Determination and Liberation (IPMSDL). "Stop the Attacks Against Lumad! No to Campus Militarization." IPMSDL website, February 15, 2021. https://www.ipmsdl.org/statement/stop-the-attacks-against-lumad-no-to-campus-militarization/ (accessed July 30, 2021).

International Atomic Agency. "Shaping a Secure and Sustainable Future for All." Video. December 2, 2019. https://www.iea.org/ (accessed February 8, 2022).

International Campaign to Abolish Nuclear Weapons (ICAN). "Schools of Mass Destruction: Universities List." https://universities.icanw.org/universities_list#h_2618224144401573592440433 (accessed November 27, 2023).

International Energy Agency. December 2, 2019. https://www.iea.org/areas-of-work/ensuring-energy-security (accessed February 8, 2022).

International Federation of Red Cross and Red Crescent. "India: Monsoon Rains and Floods Final Report DREF No. MDRIN024," July 23, 2020. Provided by the UN Office for the Coordination of Humanitarian Affairs, at https://reliefweb.int/report/india/india-monsoon-rains-and-floods-final-report-dref-n-mdrin024 (accessed on May 17, 2022).

International Union for Conservation of Nature (IUCN) Red List. "More than 41,000 Species Are Threatened with Extinction: That Is Still 28% of All Assessed Species." https://www.iucnredlist.org/ (accessed January 10, 2022).

Ivanova, Irin. "Who Are the Biggest U.S. Methane Emitters?" *CBS News*, August 30, 2019. https://www.cbsnews.com/news/who-are-the-biggest-us-methane-emitters/ (accessed March 4, 2022).

James, Ian. "As Water Crisis Worsens on Colorado River, an Urgent Call for Western States to 'Act Now,'" *Los Angeles Times* "News Alert," June 20, 2022. https://www.latimes.com/environment/story/2022-06-20/as-colorado-river-reservoirs-drop-states-urged-to-act-now?utm_id=58791&sfmc_id=764628 (accessed June 20, 2022).

———. "Major Water Cutbacks Loom as Shrinking Colorado River Nears 'Moment of Reckoning,'" *LA Times*, June 14, 2022. https://www.latimes.com/environment/story/2022-06-14/-big-water-cutbacks-ordered-amid-colorado-river-shortage (accessed June 14, 2022).

Janson, Jay. "Hillary Laughed, 'We Came, We Saw, He Died.'" *brattleboro: write where you live*. https://www.ibrattleboro.com/opinion/op-ed/2016/10/hillary-laughed-we-came-he-died-while-a-bombed-million-demonstrated-for-their-beloved-hero-gadaffi/ (accessed March 31, 2022).

Jay, Paul. "Peter Kuznick: Why Did Americans Accept Barbaric Slaughter of Japanese Civilians." Podcast and transcript. *Naked Capitalism: Fearless Commentary on Finance, Economics, Politics and Power*, August 6, 2020. https://www.nakedcapitalism.com/2020/08/peter-kuznick-why-did-americans-accept-barbaric-slaughter-of-japanese-civilians.html (accessed April 4, 2022).

Johnson, Jake. "Dozens Arrested as Scientists Worldwide Mobilize to Demand 'Climate Revolution.'" *Common Dreams*, April 7, 2022. https://www.commondreams.org/news/2022/04/07/-dozens-arrested-scientists-worldwide-mobilize-demand-climate-revolution (accessed April 22, 2022).

Joukowsky Institute for Archaeology and the Ancient World. "Thirteen Things: Gunpowder: Origins in the East." Brown University, Providence, Rhode Island. https://www.brown.edu/Departments/Joukowsky_Institute/courses/13things/7687.html, 2009 (accessed on November 5, 2021).

Juhasz, Aubri. "NOLA-PS Know They Need to Close, Consolidate Schools. But Which Ones?" *WWNO-New Orleans Public Radio*, 89.9 FM, May 30, 2022. https://www.wwno.org/education/2022-05-20/enrollment-follow (accessed June 16, 2022).

Kavoussi, Bonnie. "Number of Ph.D. Recipients Using Food Stamps Surged During Recession: Report." *Huffington Post*, May 8, 2012. https://www.huffpost.com/entry/food-stamps-phd-recipients-2007-2010_n_1495353 (accessed June 23, 2022).

Kent State University. "University Budget and Financial Analysis: RCM Operating Manual." https://www.kent.edu/budget/rcm-manual. December 3, 2015 (accessed June 16, 2022).

Kesler, Sam Yellowhorse. "Indian Boarding Schools Traumatic Legacy, and the Fight to Get Native Ancestors Back." *Code Switch: Race in Your Face: National Public Radio*. August 28, 2021. https://www.npr.org/sections/codeswitch/2021/08/28/1031398120/native-boarding-schools-repatriation-remains-carlisle (accessed November 25, 2021).

Klecka, Joey. "Tug Collision Spills Unknown Amount of Diesel Fuel in Waters Near Sitka." *KTUU: Alaska's News Source*, March 22, 2022. https://www.alaskasnewssource.com/2022/03/22/tug-collision-spills-unknown-amount-diesel-fuel-waters-near-sitka/ (accessed April 7, 2022).

Kopec, Kelsey, and Lori Ann Burd. "A Systematic Status Review of North American and Hawaiian Native Bees." *Pollinators in Peril*. Center for Biological Diversity, February 2017. https://www.biologicaldiversity.org/campaigns/native_pollinators/pdfs/Pollinators_in_Peril.pdf (accessed April 9, 2022).

Kotch, Alex. "The Kochs' Mad Science: The Billionaire Barons Have a New Libertarian Laboratory." *Salon*, May 25, 2016. https://www.salon.com/2016/05/25/the_kochs_sinister_science_project_partner/ (accessed June 15, 2022).

Kumar, KP Narayana. "Controversial Vaccine Studies: Why is Bill & Melinda Gates Foundation Under Fire from Critics in India." *Economic Times India-English edition*, August 31, 2014. https://economictimes.indiatimes.com/industry/healthcare/biotech/healthcare/controversial-vaccine-studies-why-is-bill-melinda-gates-foundation-under-fire-from-critics-in-india/articleshow/41280050.cms (accessed December 10, 2021).

Kunz, Kathleen. "A Closer Look at Arizona's Funding

for Higher Education." *Tucson Local Media*, July 31, 2018. https://www.tucsonlocalmedia.com/back_to_school/article_d77578f0-89f8-11e8-a856-f33110f6df79.html (accessed June 10, 2022).

Kuteyi, Eyitope. "Many Burnt Beyond Recognition in Imo Illegal Refinery Explosion." *Channels TV*, Nigeria, April 23, 2022. https://www.channelstv.com/2022/04/23/many-killed-in-illegal-refinery-fire-in-imo/ (accessed April 26, 2022).

Kuzmarov, Jeremy, and Roger Peace. "Was There a Diplomatic Alternative: The Atomic Bombing and Japan's Surrender." *The Asia-Pacific Journal: Japan Focus*, Volume 19, Issue 20, Number 4, October 15, 2021. https://apjjf.org/2021/20/-Kuzmarov-Peace.html (accessed April 4, 2022).

KwaZulu-Natal and Parts of Eastern Cape. Document Reference: CS-CMS-LETT-003. April 12, 2022. www.weathersa.co.za. (accessed April 30, 2022).

Lancaster University. "U.S. Military Consumes More Hydrocarbons than Most Countries—Massive Hidden Impact on Climate." *ScienceDaily*, June 20, 2019. https://www.sciencedaily.com/releases/2019/06/190620100005.htm (accessed February 1, 2022).

Lawler, Dave. "U.S. Dominates Global Arms Sales with Russia Falling Far Behind." *Axios*, March 14, 2023. https://www.axios.com/2023/03/14/-global-arms-sales-us-dominates-russia (accessed June 30, 2023).

Leading Doctors. *Total Health: Written by Leading Doctors*. https://www.totalhealth.co.uk/blog/-are-people-getting-full-facts-covid-vaccine-risks (accessed August 1, 2021).

Lee, Robert, and Tristan Ahtone. "Land Grab Universities: Expropriated Indigenous Land is the Foundation of the Land-Grant University System." *High Country News*, March 30, 2020. https://www.hcn.org/issues/52.4/indigenous-affairs-education-land-grab-universities (accessed August 22, 2023).

Lehner, Peter. "Why Bayer Won't Tell You About Why Bees Are Dying." *National Resources Defense Council* (NRDC), July 11, 2014. https://www.nrdc.org/experts/peter-lehner/what-bayer-wont-tell-you-about-why-bees-are-dying (accessed on December 10, 2021).

Lester, Charles. "California's Latest Offshore Oil Spill Could Fuel Pressure to End Oil Production Statewide." *The Conversation*, October 5, 2021. https://theconversation.com/californias-latest-offshore-oil-spill-could-fuel-pressure-to-end-oil-production-statewide-169215 (accessed April 15, 2022).

Levine, Jon, and Jesse O'Neill. "Hunter Biden Helped Secure Funds for U.S. Biolab Contractor in Ukraine." *New York Post*, March 26, 2022. https://nypost.com/2022/03/26/-hunter-biden-played-role-in-funding-us-biolabs-contractor-in-ukraine-e-mails/ (accessed March 26, 2022).

Levinthal, Dave. "How the Koch Brothers are Influencing U.S. Colleges." *Time*, December 15, 2015. https://time.com/4148838/koch-brothers-colleges-universities/ (accessed June 15, 2022).

Lienhard, John. *Engines of Our Ingenuity*. "No. 127: Black Inventors." University of Houston. https://www.uh.edu/engines/epi127.htm (accessed August 6, 2023).

Light, Malcolm P.R. "Lucy-Alamo Projects-Hydroxyl Generation and Atmospheric Methane Destruction." *American Meteorological Society*, January 13, 2016. https://ams.confex.com/ams/96Annual/webprogram/Paper275345.html (accessed May 21, 2022).

Lima, Christiano, and Aaron Schaffer. "U.S. Quietly Paying Millions to Send Starlink Terminals to Ukraine, Contrary to SpaceX Claims." *Washington Post*, April 8, 2022. https://www.washingtonpost.com/politics/2022/04/08/-us-quietly-paying-millions-send-starlink-terminals-ukraine-contrary-spacexs-claims/ (accessed June 8, 2022).

Lisa, Andrew. "Environment: States with the Worst Droughts." *Stacker*, June 20, 2022. https://stacker.com/stories/3053/states-worst-droughts (accessed June 20, 2022).

Lisbdnet.com. "How Many Offshore Oil Rigs in the Gulf of Mexico." https://lisbdnet.com/-how-many-offshore-oil-rigs-in-the-gulf-of-mexico-2/ (accessed April 7, 2022).

Live Science. "5 Wild Weather Control Ideas." August 6, 2012. https://www.livescience.com/22131-wild-weather-control-ideas.html (accessed May 21, 2022).

Looby, Caitlin, Frank Vaisvilas, and Madeline Helm. "Great Lakes Tribes' Knowledge of Nature Could be Key to Climate Change. Will People Listen?" *USA Today*, November 26, 2023. https://www.usatoday.com/story/news/nation/2023/11/26/great-lakes-native-tribes-knowledge-key-to-climate-change/71602778007/ (accessed November 27, 2023).

Lowkey. *An Indigenous Peoples History of the United States with Roxanne Dunbar-Ortiz*. https://www.youtube.com/watch?v=MJfaZ7wAbv4. May 27, 2022. (accessed May 27, 2022).

Luqman, Jacquelyn. "Pentagon in Hollywood." *Toward Freedom*, December 2, 2021. https://towardfreedom.org/story/archives/americas/jacqueline-luqman-on-the-u-s-state-controlling-hollywood/ (accessed December 3, 2021).

Lustgarten, Abrahm. "Injection Wells: The Poison Beneath Us." *ProPublica*, June 21, 2012. https://www.propublica.org/article/injection-wells-the-poison-beneath-us (accessed February 25, 2022).

Luxembourg Centre for Contemporary and Digital History. "Gorbachev's Reforms in the Soviet Union." https://www.cvce.eu/en/recherche/unit-content/-/unit/02bb76df-d066-4c08-a58a-d4686a3e68ff/cf38d617-0419-4e69-9cd2-37215b9bae6b (accessed August 31, 2023).

Lynch, Matthew. "Which Countries Provide Free Education at A University Level?" *The Advocate*, March 26, 2020. https://www.theedadvocate.org/which-countries-provide-free-education-at-a-university-level/ (accessed June 18, 2022).

Marcus, Steve. "As Lake Mead Water Levels Fall, Government Must Rise to the Occasion." *Las Vegas Sun*, May 3, 2022. https://m.lasvegassun.com/news/2022/may/03/as-lake-mead-water-levels-fall-government-must-ris/ (accessed May 9, 2022).

Marnef, A., and Legube, G. "m6A RNA Modification as a New Player in R-loop Regulation." *Natural Genetics* 52, 27–28 (2020). https://doi.org/10.1038/s41588-019-0563-z (accessed November 5, 2023).

Martens, Pam, and Russ Martens. "Fed Chair Powell Telegraphs the Perfect Storm for Wall Street's Megabanks: Rapid Rate Hikes Hitting $234 Trillion in Derivatives." *Wall Street on Parade*, April 25, 2022. https://wallstreetonparade.com/2022/04/fed-chair-powell-telegraphs-the-perfect-storm-for-wall-streets-megabanks-rapid-rate-hikes-hitting-234-trillion-in-derivatives/ (accessed April 25, 2022).

_____. "The Money Trail to the January 6 Attack on the Capitol is Ignored in Last Night's Public Hearing." *Wall Street on Parade*, June 10, 2022. https://wallstreetonparade.com/2022/06/the-money-trail-to-the-january-6-attack-on-the-capitol-is-ignored-in-last-nights-public-hearing/ (accessed June 10, 2022).

_____. "When Repos Blew Up in 2019, Hedge Funds were $800 billion Short U.S. Treasury Futures: Then Margins Blew Out." *Wall Street on Parade*, February 3, 2022. https://wallstreetonparade.com/2022/02/when-repos-blew-up-in-2019-hedge-funds-were-800-billion-short-u-s-treasury-futures-then-margins-blew-out/ (accessed February 3, 2022).

Massachusetts Peace Action. "Meet the Big 5: How the Military-Industrial Complex Controls Politics." https://www.youtube.com/watch?v=6EhyIwxwQ0 (viewed live on January 24, 2022).

Masters, Jeff. "The Top 10 Global Weather and Climate Change Events of 2021." *Yale Climate Connections*. https://yaleclimateconnections.org/2022/01/the-top-10-global-weather-and-climate-change-events-of-2021/ (accessed May 17, 2022).

Matthews, Jay. "The Philadelphia Experiment: The Story Behind Philadelphia's Edison Contract." *Education Next*, Vol. 3, No. 1, July 14, 2006 (updated). https://www.educationnext.org/thephiladelphiaexperiment/ (accessed June 18, 2022).

Mayer, Emma. "More Teachers Are Facing Penalties for Quitting During Pandemic." *Newsweek*, March 25, 2022. https://www.newsweek.com/teachers-quitting-during-pandemic-face-penalties-1691945 (accessed June 20, 2022).

McConnaughey, Janet, and Matthew Brown. "Pipeline Spills Over 300,000 Gallons of Diesel Near New Orleans." *US News and World Report*, January 12, 2022. https://www.usnews.com/news/politics/articles/2022-01-12/pipeline-spills-300-000-gallons-of-diesel-near-new-orleans (accessed April 17, 2022).

McKey, Maureen. "$4 Billion and Rising: BP's Cost to Date." *Fiscal Times*, July 20, 2010. https://www.thefiscaltimes.com/Articles/2010/07/20/-Oil-Spill-Bold-Look-Bare-Economic-Numbers (accessed April 6, 2022).

McLeod, Alan. "Musk is Not a Renegade Outsider: He is a Massive Pentagon Contractor." *Mint Press News*, May 31, 2022. https://www.mintpressnews.com/elon-musk-not-renegade-outsider-cia-pentagon-contractor/280972/ (accessed May 31, 2022).

_____. "New Report Links Mass Extinction of Animals to Human Activity." *Mint Press News*, September 23, 2020. https://www.mintpressnews.com/new-report-links-mass-extinction-of-animals-to-human-activity/271437/. This article extracts from the 2020 *World Wildlife Fund Report*. https://c402277.ssl.cf1.rackcdn.com/publications/1371/files/original/ENGLISH-FULL.pdf?1599693362https://c402277.ssl.cf1.rackcdn.com/publications/1371/files/original/-ENGLISH-FULL.pdf?1599693362 (accessed October 4, 2020).

_____. "The Spin War: Revealed: Documents Show Bill Gates Has Given $319 Million to Media Outlets." *Mint Press News*, November 15, 2021. https://www.mintpressnews.com/documents-show-bill-gates-has-given-319-million-to-media-outlets/278943/ (accessed December 1, 2021).

Meding, Jason von, and Hang Thai T.M. "Agent Orange, Exposed: How U.S. Chemical Warfare in Vietnam Unleashed a Slow-Moving Disaster." *The Conversation*, October 3, 2017. https://theconversation.com/agent-orange-exposed-how-u-s-chemical-warfare-in-vietnam-unleashed-a-slow-moving-disaster-84572 (accessed April 1, 2022).

Mercola, Joseph. "Inventor of mRNA Interviewed About Injection Dangers." *Global Research*, June 22, 2021. https://www.globalresearch.ca/-inventor-mrna-interviewed-about-injection-dangers/5748269 (accessed on July 28, 2021).

_____. "mRNA Vaccine Inventor Erased from History Books." *Global Research*, July 6, 2021. https://www.globalresearch.ca/mrna-vaccine-inventor-erased-history-books/5749413 (accessed July 28, 2021).

Metz, Sam. "Multibillion Dollar Uinta Basin Railway Chugs Along Amid Environment and Derailment Concerns." *CPR (Colorado Public Radio) News*, August 10, 2023. https://www.cpr.org/2023/08/10/uinta-basin-railway-chugs-along-amid-environment-and-derailment-concerns/ (accessed September 15, 2023).

Michulich, Alex. "Indigenous Scholars Invite Decolonization of the Anthropocene." *National Catholic Reporter*, October 8, 2019. https://

www.ncronline.org/earthbeat/decolonizing-faith-and-society/indigenous-scholars-invite-decolonization-anthropocene (accessed September 12, 2023).

Miguel, Edward, and Gérard Roland. "The Long-Run Impact of Bombing Vietnam." *Journal of Development Economics*. Vol. 96, Issue 1, 2011. https://doi.org/10.1016/j.jdeveco.2010.07.004 (accessed September 9, 2023).

"Military Greenhouse Gas Emissions: EPA Should Recognize Environmental Impact of Protecting Foreign Oil, Researchers Urge." University of Nebraska–Lincoln." *ScienceDaily*, July 22, 2010. https://www.sciencedaily.com/releases/2010/07/100721121657.htm (accessed February 4, 2022).

Miller, David. Interview. *Politics Today*, November 8, 2021. https://politicstoday.org/i-was-fired-from-the-university-of-bristol-despite-being-cleared-of-anti-semitism-an-interview-with-david-miller/ (accessed January 15, 2022).

Min, Sarah. "US Oil and Gas Boom Powers Nation's Fastest-Growing Industry." *CBS News*, February 7, 2019. https://www.cbsnews.com/news/us-oil-and-gas-boom-powers-nations-fastest-growing-industry/ (accessed March 4, 2022).

Mitchell, Michael, Michael Leachman, and Matt Saenz. "State Higher Education Funding Cuts Have Pushed Costs to Students, Worsened Inequality." *Center on Budget and Policy Priorities*, October 24, 2019. https://www.cbpp.org/research/state-budget-and-tax/state-higher-education-funding-cuts-have-pushed-costs-to-students (accessed June 15, 2022).

Mohanty, Abihijit. "Extracting a Radioactive Disaster in Niger." *DownToEarth*, March 5, 2019. https://www.downtoearth.org.in/blog/health-in-africa/extracting-a-radioactive-disaster-in-niger-63451 (accessed October 17, 2021).

Montoya Bryan, Susan. "New Mexico Reaches $32 Million Settlement Over 2015 Gold King Mine Spill that Polluted Colorado Water." *CPR (Colorado Public Radio) News*, June 17, 2022. https://www.cpr.org/2022/06/17/new-mexico-settlement-over-2015-gold-king-mine-spill/ (accessed September 15, 2023).

Morton, Adam. "Football Pitch-Sized Area of Tropical Rainforest Lost Every Six Seconds." *The Guardian*, June 2, 2020. https://www.theguardian.com/environment/2020/jun/02/football-pitch-area-tropical-rainforest-lost-every-six-seconds (accessed January 14, 2021).

Moscow Times. "Nearly 100 Fires Blazing Across Russia's Republic of Sakha." August 7, 2023. https://www.themoscowtimes.com/2023/08/07/nearly-100-wildfires-blazing-across-russias-republic-of-sakha-a82080 (accessed September 23, 2023).

Mulvihill, Conor. "How Nature Shaped Celtic Culture in Ireland." *Green News*, March 29, 2016. https://greennews.ie/how-nature-shaped-celtic-culture-in-ireland/ (accessed August 29, 2023).

Munguia, Hayley. "The 2,128 Native American Mascots People Aren't Talking About." *FiveThirtyEight*, September 5, 2014. https://fivethirtyeight.com/features/the-2128-native-american-mascots-people-arent-talking-about/ (accessed November 9, 2021).

Musiyiwa, Mickias. "The Significance of Myths and Legends in Children's Literature in Contemporary Zimbabwe." Presented at the 31st *International Board on Books for Young Children Congress*, Copenhagen, 2008. https://www.ibby.org/index.php?id=913 (accessed August 27, 2021).

Mwai, Peter. "Durban Floods: Is It a Consequence of Climate Change." *BBC News*, April 19, 2022. https://www.bbc.com/news/61107685 (accessed April 20, 2022).

Nairn, Carly. "Where Does the Money Go in Environmental Grantmaking?" *Nonprofit Quarterly*, August 23, 2023. https://nonprofitquarterly.org/-where-does-the-money-go-in-environmental-grantmaking/ (accessed September 9, 2023).

"Nanodiamonds Made Under Ambient Conditions." *Science Daily*, reporting on researchers from Case Western University in Cleveland, Ohio, October 21, 2013. https://www.sciencedaily.com/releases/2013/10/131021095030.htm (accessed May 21, 2022).

National Academy of Sciences, National Research Council (U.S.) and Institute of Medicine (U.S.), and Institute of Medicine (U.S.) Committee on the Use of Laboratory Animals in Biomedical and Behavioral Research. "Use of Laboratory Animals in Biomedical and Behavioral Research." Washington, D.C.: National Academies Press, 1988. National Center for Biotechnology Information publication at https://www.ncbi.nlm.nih.gov/books/NBK218261/ (accessed November 10, 2021).

National Aeronautics and Space Administration (NASA). "NASA Announces Summer 2023 Hottest on Record." September 14, 2023. https://www.nasa.gov/press-release/nasa-announces-summer-2023-hottest-on-record (accessed September 21, 2023).

National Association for Biomedical Research. "The Importance of Animal Research." https://www.nabr.org/biomedical-research/-importance-biomedical-research (accessed November 12, 2021).

National Collaborating Center for Environmental Health. *Neonicotinoid Pesticides*, Vancouver, Canada, January 21, 2020. https://ncceh.ca/environmental-health-in-canada/health-agency-projects/neonicotinoid-pesticides (accessed January 10, 2022).

"National Endowment for Destabilization? CIA Funds for Latin America." *Telesur*, April 4, 2019. https://www.telesurenglish.net/analysis/National-Endowment-for-Destabilization-CIA-Funds-for-Latin-America-in-2018-20190403-0042.html (accessed March 31, 2022).

National History Day, National Archives &

Records Administration, and U.S. Freedom Corps. "President Dwight Eisenhower's Farewell Address 1961." https://www.ourdocuments.gov/doc.php?flash=true&doc=90 (accessed January 24, 2022).

National Humanities Center Resource Toolbox. "American Beginnings: The European Presence in North America, 1492–1690." *Requerimiento 1510: Requirement: PRONOUNCEMENT TO BE READ BY SPANISH CONQUERORS TO DEFEATED INDIANS.* https://nationalhumanitiescenter.org/pds/amerbegin/contact/text7/requirement.pdf (accessed September 26, 2021).

National Oceanic and Atmospheric Administration (NOAA), National Integrated Drought System (NIDS). *Office of Response and Restoration.* "Incident Responses for February 2022." Posted on March 11, 2022. https://blog.response.restoration.noaa.gov/incident-responses-february-20 (accessed April 17, 2022).

———. "Special Edition Drought Status Update for the Western United States," May 11, 2023. https://www.drought.gov/drought-status-updates/special-edition-drought-status-update-western-united-states-2023-05-11#:~:text=As%20such%2C%20all%2011%20western%20states,still%20report%20some%20degree%20of%20drought (accessed September 29, 2023).

———. U.S. Department of Commerce. "Deepwater Horizon Oil Spill: Final Programmatic Damage Assessment and Restoration Plan and Final Programmatic Environmental Impact Statement," February 11, 2016. https://repository.library.noaa.gov/view/noaa/18084 (accessed April 5, 2022).

National Wildlife Federation. "Deepwater Horizon's Impact on Wildlife." https://www.nwf.org/oilspill (accessed April 5, 2022).

Native American Rights Fund. "Keystone KX Pipeline: Case Updates." https://www.narf.org/cases/keystone/ (accessed November 12, 2021).

Native Americans in Philanthropy. "Thanksgiving Massacres." https://nativephilanthropy.candid.org/events/thanksgiving-day-massacres/ (accessed September 4, 2023).

NBC News. "Conspiracy Theories Abound as U.S. Military Closes HAARP." May 22, 2014. https://www.nbcnews.com/science/weird-science/conspiracy-theories-abound-u-s-military-closes-haarp-n112576. (accessed May 21, 2022).

NEA (National Education Association) Research Land Grant University Brief No. 1. "Land Grant Institutions: An Overview." March 2022. https://www.nea.org/sites/default/files/2022-03/Land%20Grant%20Institutions%20-%20An%20Overview.pdf (accessed November 27, 2023).

New Bedford Whaling Museum. https://www.whalingmuseum.org/learn/research-topics/-whaling-history/yankee-whaling/ (accessed September 2, 2023).

New World Coming. "Race in Cuba: Everything Within the Revolution." Interview with Cuban intellectual and academic Estaban Morales Dominguez, April 22, 2022. https://www.youtube.com/watch?v=mQn_y5GjjG4 (accessed May 5, 2022).

Nguyen, David H. "Toxic Chemicals in Solar Panels." *Sciencing,* April 30, 2018. https://sciencing.com/toxic-chemicals-solar-panels-18393.html (accessed October 19, 2023).

Nguyen, Patricia K., and Joseph C. Wu. "Radiation Exposure from Imaging Tests: Is There an Increased Cancer Risk?" *Expert Review of Cardiovascular Therapy,* Volume 9, 2011, Issue 2. https://www.tandfonline.com/doi/full/10.1586/erc.10.184 and https://www.ncbi.nlm.nih.gov/pmc/articles/PMC3102578/ (accessed January 27, 2022).

Nikiforuk, Andrew. "The Oil Spill Cleanup Illusion: Why Do We Pretend to Clean Up Oil Spills in the Ocean?" *Hakai Magazine,* July 12, 2016. https://hakaimagazine.com/features/oil-spill-cleanup-illusion/ (accessed April 21, 2022).

Nilsen, Ella. "Exclusive: Experts Say the Term 'Drought' May be Insufficient to Describe What's Happening in the West." *CNN,* February 16, 2022. https://www.cnn.com/2022/02/16/politics/west-drought-water-shortage-reclamation-touton-climate/index.html (accessed May 9, 2022).

Nodrada. "Karl Marx and Radical Indigenous Critiques of Capitalism." October 10, 2022. https://nodrivers.medium.com/karl-marx-and-radical-indigenous-critiques-of-capitalism-fd27169c357 (accessed August 5, 2023).

NoGeoengineering: Portal Against Climatic and Environmental Manipulation. "1962: 'HE WHO CONTROLS THE WEATHER, WILL CONTROL THE WORLD' (LBJ)." August 2, 2018. https://www.nogeoingegneria.com/news-eng/-1962-he-who-controls-the-weather-will-control-the-world-lbj/ (accessed December 9, 2021).

Norton, Ben. "Debunking Myths About Nicaragua's 2021 Election, Under Attack by the US/EU/OAS." *The Grayzone,* November 11, 2021. https://thegrayzone.com/2021/11/11/nicaragua-2021-elections/ (accessed November 5, 2021).

Nuclear Information and Resource Center. https://www.nirs.org/radiation/ (accessed September 10, 2023).

Nuclear Monitor. "U.S. Radiation Panel Recognizes No Safe Radiation Dose." Washington, D.C.: Nuclear Information & Resource Service World Information Service on Energy, 2005. https://www.nirs.org/wp-content/uploads/radiation/radtech/nosafedose072005.pdf (accessed January 27, 2022).

Office of the Comptroller of the Currency. "Quarterly Derivatives Report Fourth Quarter 2021: Quarterly Report on Bank Trading and Derivatives Activities: Fourth Quarter." Report on Bank Trading and Derivative Activities, PDF https://www.occ.gov/publications-and-resources/publications/quarterly-report-on-bank-trading-

and-derivatives-activities/files/pub-derivatives-quarterly-qtr4-2021.pdf (accessed June 7, 2022).

Officer of the Watch. "The Collision of U.S. Submarine Greeneville and the Fishing Vessel Ehime Maru—Investigative Report." June 10, 2013. https://officerofthewatch.com/2013/06/10/the-collision-of-us-submarine-greeneville-and-the-fishing-vessel-ehime-maru-investigation-report/ (accessed March 24, 2022).

Oller Jr., John W., Christopher A. Shaw, Lucija Tomljenovic, Stephen K. Karanja, Wahome Ngare, Felicia M. Clement, and Jamie Ryan Pillette. "HCG Found in WHO Tetanus Vaccine in Kenya Raises Concern in the Developing World." *Open Access Library Journal*, 04 (10): 1–30, October 2017. DOI:10.4236/oalib.1103937 at https://www.researchgate.net/publication/320641479_HCG_Found_in_WHO_Tetanus_Vaccine_in_Kenya_Raises_Concern_in_the_Developing_World (accessed May 1, 2021).

Olliaro, Piero, Els Torreele, and Michel Vaillant. "Covid-19 Vaccine Efficacy and Effectiveness—The Elephant (Not) in the Room." *The Lancet,* April 20, 2021. https://www.thelancet.com/journals/lanmic/article/PIIS2666-5247(21)00069-0/fulltext (accessed June 21, 2021).

Oneida Indian Nation. *The Haudenosaunee Creation Story.* https://www.oneidaindiannation.com/the-haudenosaunee-creation-story/ (accessed June 27, 2022).

Oskin, Betty. "Reclaiming Languages, Revitalizing Culture: How Ines Hernández-Ávila Redefines Indigenous Scholarship." https://publicengagement.ucdavis.edu/stories/ines-hernandez-avila. (accessed August 17, 2023).

Oxfam International. "Carbon Emissions of Richest 1 Percent More than Double the Emissions of the Poorest Half of Humanity." September 21, 2020. https://www.oxfam.org/en/press-releases/carbon-emissions-richest-1-percent-more-double-emissions-poorest-half-humanity (January 10, 2021).

_____. "Pfizer, BioNTech and Moderna Making $1,000 Profit Every Second While World's Poorest Countries Remain Largely Unvaccinated." November 16, 2021. https://www.oxfam.org/en/press-releases/pfizer-biontech-and-moderna-making-1000-profit-every-second-while-worlds-poorest (accessed June 6, 2022).

_____. "Ten Richest Men Double Their Fortunes in Pandemic While Incomes of 99 Percent of Humanity Fall." January 17, 2022. https://www.oxfam.org/en/press-releases/ten-richest-men-double-their-fortunes-pandemic-while-incomes-99-percent-humanity (accessed January 22, 2022).

Palcy, Euzhan (Producer). *Sugar Cane Alley.* 1983. https://www.imdb.com/title/tt0086213/ (viewed in September 2000).

Parker, Sam. "UA Students, Faculty Outraged Over Financial Crisis at Board Meeting." *The Daily Wildcat,* November 17, 2023. https://wildcat.arizona.edu/151595/news/n-abor-financial-crisis/ (accessed November 18, 2023).

Patton, Stacey. "The Ph.D. Now Comes with Food Stamps." *Chronicle of Higher Education,* May 6, 2012. https://www.chronicle.com/article/the-ph-d-now-comes-with-food-stamps/?cid2=gen_login_refresh&cid=gen_sign_in (accessed June 23, 2022).

Paul, Jean Eddy Saint. "Assassinations and Invasions: How the U.S. and France Shaped the Long History of Haitian Turmoil." *The Conversation,* August 27, 2021. https://theconversation.com/assassinations-and-invasions-how-the-us-and-france-shaped-haitis-long-history-of-political-turmoil-164269 (accessed October 19, 2021).

Payne, Julia. "Israel Accepts First Delivery of Disputed Kurdish Oil." *Reuters,* June 20, 2014. https://www.reuters.com/article/us-israel-iraq-idUSKBN0EV0X620140620 (accessed September 8, 2023).

People for the Ethical Treatment of Animals. "Experiments on Animals: Overview." https://www.peta.org/issues/animals-used-for-experimentation/-animals-used-experimentation-factsheets/-animal-experiments-overview/ (accessed June 20, 2020).

Peoples' Advancement Center. https://peoplescentre.org/about-us-2/ (accessed September 10, 2021).

_____. "Celestine AkpoBari with other participants..." October 29, 2017. https://peoplescentre.org/photos-celestine-akpobari-with-other-participants-at-the-united-nations-environment-consultation-meeting-in-nairobi-kenya/ (accessed June 2021).

Peterson-Smith, Khury. "Twenty-First Century Imperialism in the Pacific." *International Socialist Review,* Issue 82, March 2012. https://isreview.org/issue/82/twenty-first-century-colonialism-pacific/index.html (accessed September 19, 2023).

Phaneuf, Alicia. "Top 10 Biggest U.S. Banks by Assets in 2022." *Insider Intelligence,* January 2, 2022. https://www.insiderintelligence.com/insights/largest-banks-us-list/ (accessed June 6, 2022).

Physicians Committee for Responsible Medicine. "Last Remaining Medical School to Use Live Animals for Training Makes Switch to Human-Relevant Methods." June 30, 2016. https://www.pcrm.org/news/news-releases/last-remaining-medical-school-use-live-animals-training-makes-switch-human (accessed November 12, 2021).

Pilkington, Ed. "U.S. College Rejects Jewish Professor Over Anti–Israel Stance." *The Guardian,* May 11, 2007. https://www.theguardian.com/world/2007/jun/11/internationaleducationnews.usa (accessed July 15, 2008).

Pinto, Black. "Real Problems, Real Solutions: Examining Pollution in Los

Angeles County." *Voices*, Pacific Oak College, Pasadena, CA. https://www.pacificoaks.edu/voices/social-justice/real-problems-real-solutions-examining-pollution-in-los-angeles-county/ (accessed March 1, 2022).

Piper, Julia, and Brian O'Leary. "Executive Compensation at Public and Private Colleges." *The Chronicle of Higher Education*, February 15, 2022. https://www.chronicle.com/article/executive-compensation-at-public-and-private-colleges/?cid=gen_sign_in#id=table_public_2020 (accessed June 16, 2022).

Platt, John R. "Why Don't We Hear about More Species Going Extinct?" *Scientific American*, June 16, 2019. https://blogs.scientificamerican.com/extinction-countdown/why-dont-we-hear-about-more-species-going-extinct/ and https://www.iucnredlist.org/ (accessed January 10, 2022).

Poorbootdee, Thinnapat. "UN Issues Drought Warning." *Russian Times*, May 11, 2022. https://www.rt.com/news/555334-un-droughts-climate-change/ (accessed May 30, 2022).

Pope Alexander VI. *Inter caetera* (May 3, 1493). In *European Treaties Bearing on the History of the United States and its Dependencies to 1648*. Ed. Frances Gardiner Davenport. Washington, D.C.: Carnegie Institution of Washington, 1917. *Encyclopedia Virginia: Virginia Humanities*. https://encyclopediavirginia.org/entries/inter-caetera-by-pope-alexander-vi-may-4-1493/ (accessed September 16, 2021).

Prashad, Vijay. "Our Time is Now #3." COP 21-Coalition's People's Summit for Climate Justice. Scotland, November 10, 2021. https://www.youtube.com/watch?v=Bho6xY-jSu (accessed December 10, 2022).

Pratt, Emma. "Reclaiming Thanksgiving: Including Indigenous History & Perspectives." *NonviolenceNY Network*, August 12, 2022. https://www.nonviolenceny.org/post/reclaiming-thanksgiving-including-indigenous-history-perspectives (accessed September 5, 2023).

Press TV, Iran. "Millions Stranded, Dozens Dead as Deadly Floods Hit Bangladesh, India." May 21, 2022. https://www.presstv.ir/Detail/2022/05/21/682503/Bangladesh-floods-India-Assam (accessed May 22, 2022).

Price, Michael. "Study Reveals Culprit Behind Piltdown Man, One of Science's Most Famous Hoaxes." *Science Magazine*, August 9, 2016. https://www.sciencemag.org/news/2016/08/study-reveals-culprit-behind-piltdown-man-one-science-s-most-famous-hoaxes (accessed July 28, 2021).

Profit and Loss: Standing on Sacred Ground, Episode 2 (2013): Sacred Land Film Project. Produced by Christopher McLeod. https://sacredland.org/profit-and-loss/ (accessed July 14, 2015).

Pryor, Steve. "A Partial List of the 6,000 Products Made from One Barrel of Oil (After Creating 19 Gallons of Gasoline)." July 26, 2016. *Linkedin*. https://www.linkedin.com/pulse/partial-list-over-6000-products-made-from-one-barrel-oil-steve-pryor/ (accessed March 14, 2022).

Queally, Jon. "'Indefensible': U.S. Billionaires Became $2.1 Trillion Richer in 19 Months of Pandemic." *Common Dreams*, October 18, 2021. https://www.commondreams.org/news/2021/10/18/indefensible-us-billionaires-became-21-trillion-richer-19-months-pandemic?utm_term=AO&utm_campaign=Daily%20Newsletter&utm_content=email&utm_source=Daily%20Newsletter&utm_medium=Email (accessed October 18, 2021).

RabbitAir. Southern California Environmental Health Sciences Center. "Pollution in Los Angeles County." https://www.rabbitair.com/pages/pollution-in-los-angeles-county (accessed March 1, 2022).

Raczek, Tracy. "Geoengineering: Reining in the Weather Warriors." Royal Institute of International Affairs, Chatham House. February 15, 2022. https://www.chathamhouse.org/2022/02/geoengineering-reining-weather-warriors (accessed November 24, 2023).

Railroad Commission of Texas. "Texas Oil and Gas Production Statistics for December 2021." https://www.rrc.texas.gov/news/031122-december-production-statistics/ (accessed April 22, 2022).

Ramírez, María de los Ángeles, Joseph Poliszuk, and María Antonieta Segovia. "Indigenous Resistance Organized in the Venezuelan Jungle." *El Pais*, February 12, 2022. https://pulitzercenter.org/stories/indigenous-resistance-organized-venezuelan-jungle-spanish (accessed August 31, 2023).

The Rational Walk (author anonymous). "Robert Rubin Exemplifies the Revolving Door between Washington and Wall Street." *Seeking Alpha*, April 11, 2010. https://seekingalpha.com/article/198065-robert-rubin-exemplifies-the-revolving-door-between-washington-and-wall-street (accessed June 9, 2022).

Readfearn, Graham. "Antarctica Warming Much Faster than Models Predicted in 'Deeply Concerning' Sign for Sea Levels." *The Guardian*, September 7, 2023. https://www.theguardian.com/world/2023/sep/08/antarctica-warming-much-faster-than-models-predicted-in-deeply-concerning-sign-for-sea-levels (accessed September 10, 2023).

Reardon, Marguerite. "How the FAA Went to War Against 5G." *CNET*, January 28, 2022. https://www.cnet.com/tech/mobile/how-the-faa-went-to-war-against-5g/ (accessed January 31, 2022).

Renda, Matthew. "Feds to Take Water from Wyoming to Bolster Lake Powell." *Courthouse New Service*, May 3, 2022. https://www.courthousenews.com/feds-to-take-water-from-wyoming-to-bolster-lake-powell/ (accessed May 9, 2022).

"Residents Urged to Shelter After Chemical Leak in Louisiana." *telesur*, April 19, 2022. https://

www.telesurenglish.net/news/Residents-Urged-To-Shelter-After-Chemical-Leak-In-Louisiana-20220419-0003.html (accessed April 20, 2022).

Resnick, Brian. "An Australian Ecologist Explains Just How Bad the Fires Are for Wildlife." *Vox*, January 10, 2020. https://www.vox.com/energy-and-environment/2020/1/9/21057375/australia-fire-wildlife-extinctions-ecology (accessed April 28, 2022).

Reuters. "Record-Breaking May Rain in Valencia Triggers Floods." yahoo!, May 4, 2022. https://www.yahoo.com/video/record-breaking-may-rain-valencia-113943070.html (accessed May 10, 2022).

Reyes, Theresa. "What Ever Happened to Imelda Marcos' 3,000 Pairs of Shoes." *Vice*, November 13, 2019. https://www.vice.com/en/article/59n8ab/what-ever-happened-imelda-marcos-3000-pairs-shoes-philippines (accessed August 8, 2023).

Richards, Dona. (Marimba Ani). "The Ideology of European Dominance." *Présence Africaine*. Nouvelle série, No. 111 (3e TRIMESTRE 1979), 3–18. 1979. https://www.jstor.org/stable/24350048 (accessed August 26, 2023).

Ridlington, Elizabeth, and John Rumpler. "Fracking by the Numbers: Key Impacts of Dirty Drilling at the State and National Level." Washington, D.C.: Environment America Research and Policy Center, 2013. https://environmentamerica.org (accessed October 10, 2021).

Ridlington, Elizabeth, Kim Norman, and Rachel Richardson. "Fracking by the Numbers: The Damage to Our Water, Land and Climate from a Decade of Dirty Drilling." Denver, CO: Environment America Research & Policy Center, April 14, 2016. https://environmentamerica.org/reports/ame/fracking-numbers-0 (accessed July 21, 2021).

Robinson, Jamie. "China's Planting a Forest the Size of Ireland." *THE SPACES*. https://thespaces.com/chinas-planting-forest-size-ireland/ (accessed April 12, 2022).

Rogers, Nala. "2019: The Year Fracking Earthquakes Turned Deadly." *Inside Science*, February 21, 2020. https://www.insidescience.org/news/2019-year-fracking-earthquakes-turned-deadly (accessed February 22, 2022).

Ross, Sean. "Top 10 Wealthiest Families in the World." *Investopedia*, updated May 9, 2022. https://www.investopedia.com/articles/insights/052416/top-10-wealthiest-families-world.asp (accessed June 21, 2022).

Rott, Nathan. "The World Lost Two-Thirds of Its Wildlife In 50 Years: We Are to Blame." *National Public Radio (NPR)*, September 10, 2020. https://www.npr.org/2020/09/10/911500907/the-world-lost-two-thirds-of-its-wildlife-in-50-years-we-are-to-blame (accessed September 12, 2023).

Rozsa, Matthew. "Elon Musk Becomes Twitter Laughingstock After Bolivian Socialist Movement Returns to Power." *Salon*, October 20, 2020. https://www.salon.com/2020/10/20/elon-musk-becomes-twitter-laughingstock-after-bolivian-socialist-movement-returns-to-power/ (accessed June 20, 2022).

Ruiz, Arabella. "Latest Global E-Waste Statistics and What They Tell Us." *TheRoundup.org*. https://theroundup.org/global-e-waste-statistics/#:~:text=The%20total%20is%20growing%20by%20an%20average%20of,and%20Iceland%20have%20the%20highest%20e-waste%20recycling%20rates (accessed August 28, 2023).

Ruocco, Nadia, Mario Constantini, and Luigia Santella. "New Insights Into Negative Effects of Lithium on Sea Urchin *Paracentrotus lividus* Embryos." *Science Reports*, August 26, 2016, National Library of Medicine. 10.1038/srep32157 (accessed August 7, 2022).

Russian Times. "BRICS Diplomat Comments on What is Drawing Countries to the Bloc." June 30, 2023. https://www.rt.com/africa/578972-brics-attracts-developing-nations/ (accessed June 30, 2023).

———. Documentary Channel. "Bio-Weapons: Truth or Fiction." March 2022. https://rtd.rt.com/films/bio-weapons-truth-or-fiction/ (accessed March 30, 2022).

———. "Hunter Biden Did Fund Ukraine Biolabs, Emails Published by Media Suggest." March 26, 2022. https://www.rt.com/russia/-552733-hunter-biden-arranged-ukraine-biolab-financing/ (accessed March 26, 2022).

———. "Siberian Wildfires Cause Devastation and Deaths (VIDEO)." *Agence France-Presse*, May 7, 2022. https://www.rt.com/russia/555134-siberia-wildires-hundreds-houses-burn/ (accessed May 8, 2022).

———. "Top German Professor Raises Covid-19 Vaccines Alarm." May 4, 2022. https://www.rt.com/news/554976-vaccine-side-effects-pfizer/ (accessed May 4, 2022).

Sadegh, Mojtaba, John Abatzoglou, and Mohammad Reza Alizadeh. "Heat Waves Hit the Poor Hardest—Calculating the Rising Impact on Those Least Able to Adapt to the Warming Climate." *The Conversation*, February 10, 2022. https://theconversation.com/heat-waves-hit-the-poor-hardest-calculating-the-rising-impact-on-those-least-able-to-adapt-to-the-warming-climate-175224 (accessed March 10, 2024).

Salazar, Luis, and Fabian Salazar. "Mother Earth." *Wuauquikuna*. https://www.google.com/search?q=Indigenous+Songs+on+the+beautiful++Earth&sca_esv=561484899&ei=cO7vZM-1LpDbkPIPvKKysAY&ved=0ahUKEwjPrcCA4YWBAxWQLUQIHTyRDGYQ4dUDCA8&uact=5&oq=Indigenous+Songs+on+the+beautiful++Earth&gs_lp=Egxnd3Mtd2l6LXNlcnAiKEluZGlnZW5vdXMgU29uZ3Mgb24gdGhlIGJlYXV0aWZ1bCAgRWFydGGyCBAKGKABGMMESM1wULyFWLNqcAF4AJABAJgBkgGgAfgLqgEEMi4xMbgBA8gBAPgBAcICxAAGIoFGIYDGLADwgIKECECGBiwAZwYQYCsICBBAhGArCAggQABiJBRiiBMICBRAAGKIEsICBgQAYgAQYsAPCAgYQIRgKGKABiBIBiBmCsBggAQGCxEIIGAAGEESGKADmAMAogYCaAGoCQGCAQGSBwIyLQGkoAGIgu4B&sclient=gws-wiz-serp

GAZAGAg&sclient=gws-wiz-serp#fpstate=ive&vld=cid:d5a5f9e9,vid:cPh3Rm9IaJM (accessed August 30, 2023).

Salish School of Spokane. http://www.salishschoolofspokane.org/revitalizingsalish.html (accessed July 21, 2023).

Sánchez-Bayo, Francisco, et al. "Worldwide Decline of the Entomofauna: A Review of Its Drivers." *Biological Conservation* (2019). DOI: 10.1016/j.biocon.2019.01.020. www.sciencedirect.com/science/ … /S0006320718313636#! (accessed February 18, 2022).

Saunders, Tom. "2023 on Track to be World's Hottest Year on Record, Temperatures Exceed 1.5C Above Pre-Industrial Levels for First Time." *ABC Australia*, September 10, 2023. https://www.abc.net.au/news/2023-09-11/global-temperatures-pass-1-5c-above-pre-industrial-levels/102836304 (accessed September 17, 2023).

Schellenberger, Michael. "If Solar Panels Are So Clean, Why Do They Produce So Much Toxic Waste?" *Forbes*, May 23, 2018. https://www.forbes.com/sites/michaelshellenberger/2018/05/23/if-solar-panels-are-so-clean-why-do-they-produce-so-much-toxic-waste/?sh=6af3f82d121c (accessed November 29, 2023).

Schmidt, Anne. "Here's How Much Money These Environmental Organizations Make." *FOXBusiness*, April 22, 2020. https://www.foxbusiness.com/money/how-much-environmental-organizations-make (accessed September 9, 2023).

Schuessler, Jennifer. "Lessons from Ants to Grasp Humanity." *New York Times*, April 9, 2012. https://www.nytimes.com/2012/04/09/books/-edward-o-wilsons-new-book-social-conquest-of-earth.html (accessed June 1, 2013).

Schulz, Matt. "2023 Student Debt Loan Statistics." *LendingTree*. Edited by Dan Shepard and Xiomara Martinez-White. August 10, 2023. https://www.lendingtree.com/student/student-loan-debt-statistics/ (accessed November 27, 2023).

Schwab, Klaus. "Now Is the Time for a 'Great Reset.'" *World Economic Forum*, January 3, 2020. https://www.weforum.org/agenda/2020/06/-now-is-the-time-for-a-great-reset/ (accessed June 16, 2022).

Schwartzstein, Peter, and Rojita Adhikari. "Merchants of Thirst." *New York Times*, January 12, 2020. https://www.nytimes.com/2020/01/11/business/drought-increasing-worldwide.html (accessed January 12, 2020).

ScienceDaily. "Air Pollution Linked to Premature Birth In Pregnant Women." University Of California, Los Angeles. August 27, 2007. www.sciencedaily.com/releases/2007/08/070823150343.htm (accessed March 1, 2022).

_____. "Nanodiamonds Made Under Ambient Conditions." Case Western University. https://www.sciencedaily.com/releases/2013/10/131021095030.htm (accessed May 21, 2022).

Seismological Society of America. "Studies Link Earthquakes to Fracking in the Central and Eastern U.S." *ScienceDaily*, April 26, 2019. https://www.sciencedaily.com/releases/2019/04/190426110601.htm (accessed February 28, 2022).

Semuels, Alana. "Easing Student Debt Makes Economic Sense: So Why Is It So Hard To Do?" *Time* magazine, April 1, 2021. https://time.com/5951766/student-loan-forgiveness-challenges/ (accessed April 15, 2021).

Shelton, Jim. "Wiped Out Forever—the Ecological Impact of Australia's Wildfires." *YaleNews*, January 16, 2020. https://news.yale.edu/2020/01/16/wiped-out-forever-ecological-impact-australias-wildfires (accessed April 28, 2021).

Shvangiradze, Tsira. "The U.S.' Secret War in Laos: The Most Heavily Bombed Country in History." *The Collector*, Jan 22, 2022. https://www.thecollector.com/war-in-laos-most-heavily-bombed-country-in-history/ (accessed September 10, 2023).

Sinclair, Ian. "The U.S. Invasion and Occupation of Iraq that Never Happened." *People's World*, May 26, 2023. https://peoplesworld.org/article/the-u-s-invasion-and-occupation-of-syria-that-never-happened/ (accessed September 8, 2023).

Sinelschikova, Yekaterina. "What Free Benefits Are Available to Russians?" *Russia Beyond*, March 20, 2020. https://www.rbth.com/lifestyle/331861-free-benefits-russians (accessed June 26, 2022).

Smith, Ashley. "Millions of Americans Over 60 Are Lugging Student Loans Into Retirement." *13 News Now*, April 4, 2022. https://www.13newsnow.com/article/money/millions-americans-over-60-student-loans-retirement/-291-3f7d596d-b6ff-4821-adbf-6370029a30bd (accessed June 19, 2022).

Smith, David Michael. "Counting the Dead: Estimating the Loss of the Indigenous Holocaust, 1492-Present." https://www.se.edu/native-american/wp-content/uploads/sites/49/2019/09/-A-NAS-2017-Proceedings-Smith.pdf (accessed January 15, 2019).

Smith, Hayley, and Ian James, "To Survive Drought, Parts of SoCal Must Cut Water Use by 35%. The New Limit is 80 Gallons a Day." *Los Angeles Times*, April 30, 2022. https://www.latimes.com/california/story/2022-04-30/can-you-get-by-on-just-80-gallons-of-water-a-day (accessed April 30, 2022).

Smith, Matt. "Fracking Seems to be Causing Hundreds of Earthquakes in the Country Each Year." *Vice News*, January 8, 2015. https://www.vice.com/en/article/9kvpk5/fracking-seems-to-be-causing-hundreds-of-earthquakes-across-the-country-each-year (accessed February 28, 2022).

Smith, Oliver, and Avi Grisman. "Plastic Waste and the Environmental Crisis Industry." *Critical Criminology*, 2021 (29: 289–309). Springer, March 16, 2021. https://doi.org/10.1007/s10612-021-09562-4 (accessed August 24, 2023).

Smithberger, Mandy, and William Hartung.

"Making Sense of the $1.25 Trillion National Security State Budget: A Dollar-by-Dollar Tour of the National Security State." Center for Defense Information. The Project on Government Oversight. May 7, 2019. https://www.pogo.org/analysis/2019/05/making-sense-of-the-1-25-trillion-national-security-state-budget/ (accessed November 19, 2021).

Soloway, Rose Ann Gould. "BPA and the Controversy about Plastic Food Containers." *Poison Control, National Capital Poison Center.* https://www.poison.org/articles/plastic-containers-are-they-harmful (accessed August 24, 2023).

Sonne, Paul. "How a Secretive Pentagon Agency Seeded the Ground for a Rapid Coronavirus Cure." *Washington Post,* July 30, 2020. https://www.washingtonpost.com/national-security/how-a-secretive-pentagon-agency-seeded-the-ground-for-a-rapid-coronavirus-cure/2020/07/30/ad1853c4-c778-11ea-a9d3-74640f25b953_story.html (accessed August 5, 2021).

Sotero, Eduardo. "Worst Drought in Decades Devastates Ethiopia's Nomads." *Agence France-Presse,* May 4, 2022. https://www.france24.com/en/live-news/20220504-worst-drought-in-decades-devastates-ethiopia-s-nomads (accessed May 5, 2022).

Sousounis, Peter. "The 2011 Thai Floods: Changing the Perception of Risk in Thailand." Edited by Meagan Phelan. *Verisk,* April 19, 2012. https://www.air-worldwide.com/publications/air-currents/2012/The-2011-Thai-Floods—-Changing-the-Perception-of-Risk-in-Thailand/ (accessed May 17, 2022).

South African Weather Service, Department of Forestry, Fisheries & the Environment, Media Release. "Extreme Rainfall and Widespread Flooding Overnight: KwaZulu-Natal and parts of Eastern Cape." April 12, 2022. https://web.archive.org/web/20220425050249/https://www.weathersa.co.za/Documents/Corporate/Medrel12April2022_12042022142120.pdf

Stanway, David. "China Government Sees Its Own Reflection in Water Crisis." *Reuters,* September 23, 2013. https://www.reuters.com/article/us-climate-ipcc-china/chinese-government-sees-its-own-reflection-in-water-crisis-idINBRE98M0BP20130923 (accessed June 26, 2022).

Statista. "Monthly Number of Started Fracking Operations in the United States from January 2019 to May 2020." https://www.statista.com/statistics/615531/us-monthly-fracking-operations-started/ (accessed April 22, 2022).

Stocker, Marc, John Baffes, and Dana Vorisek. "What Triggered the Oil Price Plunge of 2014–2016 and Why It Failed to Deliver an Economic Impetus in Eight Charts." *World Bank, Blogs,* January 18, 2018. https://blogs.worldbank.org/developmenttalk/what-triggered-oil-price-plunge-2014-2016-and-why-it-failed-deliver-economic-impetus-eight-charts (accessed March 25, 2022).

Stone, Mike. "U.S. Weapons Exports Decreased 21% to $138.2 Billion in Fiscal 2021." *Reuters,* December 22, 2021. https://www.reuters.com/world/us/us-weapons-exports-decreased-21-1382-billion-fiscal-2021-2021-12-22/ (accessed February 18, 2022).

Stone, Oliver. *Ukraine on Fire.* Film Documentary, 2016. https://www.youtube.com/watch?v=IwZApPCFXIc (accessed June 25, 2022).

Tailor, Neelam, and Nikhita Chulani. "Indigenous Activists on Tackling the Climate Crisis: 'We Have Done More than Any Government.'" *The Guardian,* November 4, 2021. https://www.theguardian.com/environment/video/2021/nov/04/indigenous-activists-tackling-climate-crisis-done-more-than-any-government-cop26-video (accessed August 18, 2023).

Tanzi, Alexander, and Madison Paglia. "Older Americans Are on the Front Line of the Student Debt Crisis." *Bloomberg,* June 21, 2021. https://www.bloomberg.com/news/articles/2021-06-17/student-loan-growing-share-of-1-7-trillion-debt-pile-held-by-older-americans?srnd=premium&sref=yaJhKSOh#xj4y7vzkg (accessed June 20, 2022).

Taylor, Brett. "Firewatch: California Experiencing Second Worst Fire Season in State's History." *KDRV-TV, ABC News, Newswatch.* September 5, 2021. Updated January 12, 2022. https://www.kdrv.com/community/firewatch-california-experiencing-second-worst-fire-season-in-states-history/article_47c86f1d-f47d-5e05-9082-b7b78b9acccf.html (accessed April 28, 2022).

Tekruri, Dena. "Mark Zuckerberg Sued Native Hawaiians for Their Own Land." 2017. https://www.youtube.com/watch?v=W6_RyE6XZiw (accessed September 29, 2023).

Texas Oil & Gas Association (TOGA). "Shale Story: Barnett." https://www.txoga.org/shale-story-barnett/ (accessed February 16, 2022).

"13 Quotes that Remind Us to Protect Mother Earth." *Indian Country Today,* January 27, 2016. Updated September 13, 2018. https://indiancountrytoday.com/archive/13-quotes-that-remind-us-to-protect-mother-earth (accessed May 28, 2022).

"This Community in Ghana lives with Crocodiles—II Paga Crocodile Pond." October 27, 2020. https://www.youtube.com/watch?v=4ltGMQRKmVU (accessed December 15, 2020).

Thompson, Derek. "The Spectacular Rise and Fall of U.S. Whaling: An Innovation Story." *The Atlantic,* February 22, 2012. https://www.theatlantic.com/business/archive/2012/02/the-spectacular-rise-and-fall-of-us-whaling-an-innovation-story/253355/ (accessed September 2, 2023).

Thrall, Trevor, Jordan Cohen, and Caroline Dorminey. "Power, Profit, or Prudence? U.S. Arms Sales Since 9/11." *Strategic Studies Quarterly,* Summer 2020, Perspective. https://www.airuniversity.af.edu/Portals/10/SSQ/

documents/Volume-14_Issue-2/Thrall.pdf (accessed February 18, 2022).

Timke, Scott. "West Africa is the Latest Testing Ground for US Military Artificial Intelligence." *Mint Press News.* April 22, 2021. https://www.mintpressnews.com/west-africa-latest-testing-ground-us-military-artificial-intelligence/276900/ (accessed November 10, 2021).

Todd, Zoe. "on time." Blog. https://zoestodd.com/2018/11/07/on-time/ (accessed September 12, 2023).

Tohono O'odham Nation, Division of Behavioral Health, Health & Human Services Department, The Respect Life Strategy Project Brand Strategy Report, prepared by Westwordvision. https://www.westwordvision.com/wp-content/uploads/2020/01/TON_BrandStrategyReport_LoRez_FINAL-1.pdf (accessed August 25, 2021).

Towe, Jazmin. "These 4 Afrocentric Schools Teach Kids Their Culture as a Matter of Approach." *Parents,* August 29, 2022. https://www.parents.com/kindred/these-4-afrocentric-schools-teach-kids-their-culture-as-a-matter-of-approach/ (accessed September 27, 2023).

Treuer, David. "Return the National Parks to the Tribes: The Jewels of America's Landscape Should Belong to America's Original Peoples." *The Atlantic,* May 2021. https://www.theatlantic.com/magazine/archive/2021/05/return-the-national-parks-to-the-tribes/618395/ (accessed August 4, 2023).

Tuholske, Cascade, Kelly Caylor, Chris Funk, and Tom Evans. "Global Urban Population Exposure to Extreme Heat." *Proceedings of the National Academy of Sciences* (PNAS), October 4, 2021. https://www.pnas.org/doi/suppl/10.1073/pnas.2024792118 (accessed April 30, 2022).

Turse, Nick. "Exclusive: The U.S. Military's Plan to Cement Its Bases in Africa." *Mail & Guardian,* South Africa, May 1, 2020. https://mg.co.za/article/2020-05-01-exclusive-the-us-militarys-plans-to-cement-its-network-of-african-bases/ (accessed June 17, 2022).

Tweedy, Ann C. "From Beads to Bounty: How Wampum Became America's First Currency—and Lost Its Power." *Indian Country Today,* October 5, 2017. https://indiancountrytoday.com/archive/from-beads-to-bounty-how-wampum-became-americas-first-currencyand-lost-its-power (accessed October 28, 2021).

Twitter, *TeleSur English Post.* "United States Nuclear Tests Are Far More Radioactive than Chernobyl and Fukushima." July 17, 2019. https://twitter.com/telesurenglish/status/1151552144786333696?ref_src=twsrc%5Etfw%7Ctwcamp%5Etweetembed%7Ctwterm%5E1151552144786333696%7Ctwgr%5E6dab84183d1bfc752cf49c147c4a01afd61fed4b%7Ctwcon%5Es1_&ref_url=https%3A%2F%2Fwww.telesurenglish.net%2Fnews%2FFijian-Organizations-Oppose-Japans-Wastewater-Discharge-Plan-20230817-0008.html (accessed August 17, 2023).

Uguen-Csenge, Eva, and Bethany Lindsay. "For 3rd Straight Day, B.C. Village Smashes Record for Highest Canadian Temperature at 49.6C." *CBC News,* June 29, 2021. https://www.cbc.ca/news/canada/british-columbia/bc-alberta-heat-wave-heat-dome-temperature-records-1.6084203 (accessed May 17, 2022).

United American Indians of New England (UAINE). http://www.uaine.org/2022_ndom/2022%20NDOM%20flyer.pdf.

United Farm Workers. "The Story of Cesar Chavez." https://ufw.org/research/history/-story-cesar-chavez (accessed August 2, 2023).

United Nations (UN). *Global Perspective Human Stories.* "UN Chief Calls for Bold Action to End 'Suicidal War with Nature.'" October 11, 2021. https://news.un.org/en/story/2021/10/1102672 (accessed February 22, 2022).

———. *UN News.* "Afghanistan's Healthcare System on Brink of Collapse, as Hunger Hits 95% of Families." September 22, 2021. https://theintercept.com/2022/02/11/afghanistan-frozen-assets-economy/ (accessed September 30, 2021).

United Nations (UN). Department of Economic and Social Affairs. "Impact of the 'Doctrine of Discovery' on Indigenous Peoples." June 1, 2012. https://www.un.org/en/development/desa/newsletter/desanews/dialogue/2012/06/3801.html (accessed July 20, 2015).

United Nations (UN) General Assembly. International Decade of Indigenous Languages. https://www.un.org/development/desa/indigenouspeoples/indigenous-languages.html (accessed June 20, 2021).

United Nations (UN) Office for the Coordination of Humanitarian Affairs (OCHA). Relief Web: Bangladesh: Floods and Landslides. https://reliefweb.int/disaster/fl-2021-000097-bgd (accessed May 22, 2022).

United Nations Conference on Trade and Development (UNCTAD). "Africa: A Sequence of Shocks Beyond Its Borders Diminished Africa's Ability to Develop and Led to Fast Increasing Debt Levels." 2022 Report. https://unctad.org/publication/world-of-debt/regional-stories (accessed September 27, 2023).

University of Alaska. "HAARP: High Frequency Active Auroral Research Program." https://haarp.gi.alaska.edu/faq (accessed September 25, 2023).

U.S. Department of Agriculture (USDA), Forest Service Southwestern Region. Report. *Environmental Assessment for Integrated Treatment of Noxious or Invasive Plants.* Phoenix, AZ: Tonto National Forest and Washington, D.C: USDA Forest Service, August 2012.

———. TP-R16–25, *Review and Assessment of Programs for Invasive Species Management in the Southwestern Region, 2012,* 14. https://www.fs.usda.gov/Internet/FSE_DOCUMENTS/stelprdb5414000.pdf (accessed January 26, 2022).

U.S. Department of Education, Federal Student Aid. "Federal Student Loan Portfolio: Portfolio by Age and Debt Size." https://studentaid.gov/data-center/student/portfolio (accessed June 19, 2022).

U.S. Department of the Interior. "Tesoro Hawaii Single Point Mooring Hose Fuel Oil Spill." https://www.cerc.usgs.gov/orda_docs/CaseDetails?ID=911 (accessed March 24, 2022).

U.S. Department of the Treasury Press Releases. "Treasury Sanctions Public Ministry of Nicaragua and Nine Government Officials Following Sham November Elections." November 15, 2021. https://home.treasury.gov/news/press-releases/jy0481 (accessed June 5, 2022).

U.S. Energy Information Administration. "Frequently Asked Questions: How Much Oil Is Consumed in the United States?" https://www.eia.gov/tools/faqs/faq.php?id=33&t=6 (accessed March 14, 2022).

———. "How Much Oil Does the U.S. Import and Export?" https://www.eia.gov/tools/faqs/faq.php?id=727&t=6 (accessed September 6, 2023).

———. "Today in Energy: Domestic Oil Production Reversed Decades-Long Decline in 2009 and 2010." April 27, 2011. https://www.eia.gov/todayinenergy/detail.php?id=1130 (accessed March 14, 2022).

———. "Today in Energy: EIA Forecasts Oil Prices Will Fall in 2022 and 2023." January 12, 2022. https://www.eia.gov/todayinenergy/detail.php?id=50858 (accessed March 26, 2022).

———. "Today in Energy. U.S. Government Energy Consumption Continues to Decline: U.S. Government Energy Consumption (Fiscal Years 2003–2017) Quadrillion British Thermal Units." July 25, 2019.https://www.eia.gov/todayinenergy/detail.php?id=40192 (accessed March 19, 2024).

———. "What Is OPEC+ and How Is It Different from OPEC?" May 9, 2023. https://www.eia.gov/todayinenergy/detail.php?id=56420# (accessed September 7, 2023).

U.S. Environmental Protection Agency (EPA). "Hydraulic Fracturing for Oil and Gas: Impacts from the Hydraulic Fracturing Water Cycle on Drinking Water Resources in the United States." December 2016. www.epa.gov/hfstudy and https://www.epa.gov/hfstudy/executive-summary-hydraulic-fracturing-study-final-assessment-2016 (accessed November 2, 2023).

———. "Overview of Greenhouse Gases." https://www.epa.gov/ghgemissions/overview-greenhouse-gases (accessed March 22, 2022).

———. "Summary of the Oil Pollution Act, 33 U.S.C. § 2701 et seq. (1990)." https://www.epa.gov/laws-regulations/summary-oil-pollution-act (accessed March 21, 2022).

———. "Underground Injection Control: Class II Oil and Gas Related Injection Wells." September 12, 2021. https://www.epa.gov/uic/class-ii-oil-and-gas-related-injection-wells (accessed February 25, 2022).

U.S. Food and Drug Administration. https://www.fda.gov/ (accessed May 17, 2022).

"U.S. Military Consumes More Hydrocarbons than Most Countries—Massive Hidden Impact on Climate." Lancaster University. *ScienceDaily,* June 20, 2019. https://www.sciencedaily.com/releases/2019/06/190620100005.htm (accessed February 1, 2022).

"U.S. Sends B-52 Nuclear Bombers to U.K. Amid Ukraine Tensions." *TeleSur English.* February 12, 2022. https://www.telesurenglish.net/news/US-Sends-B-52-Nuclear-Bombers-to-UK-Amid-Ukraine-Tensions-20220212-0006.html?utm_source=planisys&utm_medium=NewsletterIngles&utm_campaign=NewsletterIngles&utm_content=14 (accessed February 13, 2022).

Vanamali, Krishna Veera. "Saudi Aramco Becomes World's Most Valuable Company at $1.9 Trillion." *inshorts,* December 11, 2019. https://inshorts.com/en/news/saudi-aramco-becomes-worlds-most-valuable-company-at-$-19-trillion-1576051344426 (accessed March 14, 2022).

Vespucci, Amerigo. *Mundus Novus*. Translated by George Tyler Northrup. Princeton, NJ: Princeton University Press, 1916.

Vetter, David. "Solar Geoengineering: Why Bill Gates Wants It, but these Experts Want to Stop It." *Forbes,* January 20, 2022. https://www.forbes.com/sites/davidrvetter/2022/01/20/solar-geoengineering-why-bill-gates-wants-it-but-these-experts-want-to-stop-it/?sh=1986d7871842 (accessed March 30, 2022).

Viswanatha, Arun. "Banks to Pay $5.6 Billion in Probes." *Wall Street Journal,* May 20, 2015. https://www.wsj.com/articles/global-banks-to-pay-5-6-billion-in-penalties-in-fx-libor-probe-1432130400 (accessed June 8, 2022).

Voices. Pacific Oak College. Pasadena, CA. February 27, 2019. https://www.pacificoaks.edu/voices/social-justice/real-problems-real-solutions-examining-pollution-in-los-angeles-county/ (accessed February 28, 2022).

Voosen, Paul. "Atmospheric Science is Overwhelmingly White: Black Scientists Have Ignited a Change." *Science Magazine,* June 24, 2021. https://www.sciencemag.org/news/2021/06/-atmospheric-science-overwhelmingly-white-black-scientists-have-ignited-change (accessed July 15, 2021).

Vyawahare, Malavika. "South Africa Declares National Emergency as Flood Toll Crosses 440." *Mongabay,* April 19, 2022. https://news.mongabay.com/2022/04/south-africa-declares-national-emergency-as-flood-toll-crosses-440/ (accessed April 30, 2022).

Walker, Jennifer S., Robert E. Kopp, Christopher M. Little, and Benjamin P. Horton. "Timing of Emergence of Modern Sea Level Rise by 1863." *Nature Communications,* 13, Article number 966 (February 18, 2022). https://www.nature.com/articles/s41467-022-28564-6 (accessed May 2, 2022).

Walker, Tim. "Average Teacher Salary Lower Today than Ten Years Ago, NEA Report Finds." *NEA: News Today,* April 26, 2022. https://www.nea.org/advocating-for-change/new-from-nea/-average-teacher-salary-lower-today-ten-years-ago-nea-report-finds (accessed June 18, 2022).

Walton, Kate. "In Java, Water is Running Out." *The Interpreter,* September 4, 2019. Sydney, Australia: Lowy Institute, 2019. https://www.lowyinstitute.org/the-interpreter/in-java-water-is-running-out (accessed May 9, 2022).

Wang, Ying, Gustaf Arrhenius, and Yongzhan Zhang. "Drought in the Yellow River—An Environmental Threat to the Coastal Zone." *Journal of Coastal Research* (JCR), Special Issue 34. International Coastal Symposium (ICS 2000): *CHALLENGES FOR THE 21ST CENTURY IN COASTAL SCIENCES, ENGINEERING AND ENVIRONMENT* (August 2001). Published by Coastal Education & Research Foundation, Inc. https://www.jstor.org/stable/25736316 (accessed June 26, 2022).

Ware, Melchior. *Standing on Sacred Ground: Profit and Loss Part 1.* Sacred Land Film Project, 2014. Produced by Christopher Macleod et al. Distributed by Bullfrog Films. https://standingonsacredground.org/ (accessed on December 5, 2014).

Webb, Whitney. "On Earth Day, Remembering the U.S. Military's Toxic Legacy." *Mint Press News,* April 22, 2019. https://www.mintpressnews.com/on-earth-day-remembering-the-us-militarys-toxic-legacy/227776/ (accessed on April 23, 2019).

Wellerstein, Alex. "Counting the Dead at Hiroshima and Nagasaki." *Bulletin of the Atomic Scientists: 75 Years and Counting,* August 4, 2020. https://thebulletin.org/2020/08/counting-the-dead-at-hiroshima-and-nagasaki/ (accessed November 7, 2021).

White, King. "How Big is the U.S. Call Center Market Compared to India, Latin America, and the Philippines?" *Site Selection Group Blog,* Site Selection Group. July 23, 2018. https://info.siteselectiongroup.com/blog/how-big-is-the-u.s.-call-center-market-compared-to-india-latin-america-and-the-philippines-2 (accessed October 31, 2021).

White, Stephen. "Communist Nostalgia and Its Consequences in Russia, Belarus, and Ukraine." *The Transformation of State Socialism.* Edited by D. Lane. London: Palgrave Macmillan, 2007. https://doi.org/10.1057/9780230591028_2 (accessed August 31, 2023).

Whitney, Jr., W.T. "Ukraine War Reveals Possible US Preparations for Biological Warfare." *Covert Action Magazine,* April 14, 2022. https://covertactionmagazine.com/2022/04/14/ukraine-war-reveals-possible-u-s-preparations-for-biological-warfare/ (accessed April 14, 2022).

Wichner, David. "Tucson-Based Raytheon Unit Nets $400M in Year-End Pentagon Contracts." December 20, 2021. https://tucson.com/business/tucson-based-raytheon-unit-nets-400m-in-year-end-pentagon-contracts/-article_520dd9f2-5f95-11ec-bcba-c7d69a5d266e.html#tracking-source=article-related-bottom (accessed January 10, 2022).

Wigington, Dane. *Geoengineering Watch Global Alert News.* April 30, 2022, #351. https://www.geoengineeringwatch.org/geoengineering-watch-global-alert-news-april-30-2022-351/ (accessed May 1, 2022).

———. "Microplastics in Clouds Could Be 'Contaminating Nearly Everything We Eat and Drink': Study." *Common Dreams.* September 29, 2023. https://www.commondreams.org/news/microplastics (accessed September 29, 2023).

———. "Wildfires Serve Geoengineering Agenda." https://www.youtube.com/watch?v=-nmL0aTXXoM&list=PLwfFtDFZDpwulG0PJ9IID0iypsRXDSa1E www.geoengineeringwatch.org/ (accessed December 31, 2021).

———. Wilkins, Britt. "Big Pharma Monopolies Make Cost of Global Vaccination Against Covid-19 5 Times Costlier Than Needed: Report." *Common Dreams,* July 29, 2021. https://www.commondreams.org/news/2021/07/29/big-pharma-monopolies-make-cost-global-vaccination-against-covid-19-.-times-costlier?utm_term=AO&utm_campaign=Daily%20Newsletter&utm_content=email&utm_source=Daily%20Newsletter&utm_medium=Email (accessed July 30, 2021).

Williams, Eric. *Documents of West Indian History, Vol.1,1492–1655: From the Spanish Discovery to the British Conquest of Jamaica.* Port-of-Spain, Trinidad: PNM Publishing Co., Ltd., 1963. University of Florida Libraries. https://ufdc.ufl.edu/AA00012808/00001/3j (accessed September 20, 2021).

Williams, Kathleen B. "The Military's Role in Stimulating Science and Technology: The Turning Point." *Foreign Policy Research Institute,* May 28, 2010. https://www.fpri.org/article/2010/05/the-militarys-role-in-stimulating-science-and-technology-the-turning-point/#:~:text=Nevertheless%2C%20the%20interconnectedness%20and%20mutual%20dependence%20of%20science, NSF%20is%20responsible%20for%20less%20than%205%20percent (accessed October 30, 2021).

Wilson, Janet. "Solar Surges in the California Desert. So Why Are Environmentalists Upset?" *Desert Sun,* January 3, 2020. https://www.desertsun.com/story/news/environment/2020/01/03/solar-surges-california-desert-environment-trump/2665799001/ (accessed November 29, 2023).

Win, Thin Lei. "Fighting Global Warming, One Belch at a Time." *Thomson Reuters Foundation,* July 19, 2018. https://www.reuters.com/article/us-global-livestock-emissions-fighting-global-warming-one-cow-belch-at-a-time-idUSKBN1K91CU (accessed on November 18, 2021).

World Meteorological Organization (WMO).

"Weather-Related Disasters Increase Over Past 50 Years, Causing More Damage, but Fewer Deaths." August 31, 2021. https://public.wmo.int/en/media/press-release/weather-related-disasters-increase-over-past-50-years-causing-more-damage-fewer (accessed April 28, 2022).

World Wildlife Fund. *Living Planet Report 2020: Bending the Curve of Biodiversity Loss.* https://www.worldwildlife.org/publications/living-planet-report-2020 (accessed September 15, 2020).

World Wildlife Fund Australia. "New WWF Report: 3 Billion Animals Impacted by Australia's Bushfire Crisis." July 28, 2020. https://www.wwf.org.au/news/news/2020/3-billion-animals-impacted-by-australia-bushfire-crisis (accessed September 30, 2020).

Xian, Lu, and Ruo Yan. "After Workers Flee China's Largest iPhone Factory, Activists Demand Accountability from Apple." *Labornotes,* November 10, 2022. https://www.labornotes.org/blogs/2022/11/after-workers-flee-chinas-largest-iphone-factory-activists-demand-accountability-apple (accessed October 10, 2023).

Yancy, Dorothy. "Four Black Inventors with Patents." *Negro History Bulletin* 39 [1976]. http://www.aaregistry.org/historic_events/view/cotton-gin-patented (accessed October 31, 2015).

Yuchie Language Project (*yUdjEha gO'wAdAnApA k'ak'ûnEchE*). Indigenous Language Revitalization Project. https://www.yuchilanguage.org/ (accessed June 1, 2021).

Zabarte, Ian. "A Message from the Most Bombed Nation on Earth." *Al Jazeera,* August 29, 2020. https://www.aljazeera.com/opinions/2020/8/29/a-message-from-the-most-bombed-nation-on-earth/ (accessed September 5, 2020).

Zaloom, Caitlin. "How the Student Debt Complex is Crushing the Next Generation of Americans." *Time,* October 29, 2019. https://time.com/5712504/student-debt-complex-harms-america/ (accessed June 18, 2022).

Zamora-Nipper, Brianna. "Timeline: Inside the 2021 Winter Storm, Power Crisis." Click2Houston.com, February 15, 2022. https://www.click2houston.com/features/2022/02/15/timeline-inside-the-2021-winter-storm-power-crisis/ (accessed May 14, 2022).

Zampa, Matt. "How Many Animals are Killed for Food Every Day?" *Sentient Media,* September 16, 2018. https://sentientmedia.org/how-many-animals-are-killed-for-food-every-day/ (accessed June 20, 2020).

Zang, Sharon. "Top 1 Percent in U.S. Now Have More Wealth than Entire Middle Class Combined." *Truthout,* October 13, 2021. https://truthout.org/articles/top-1-percent-in-us-now-have-more-wealth-than-entire-middle-class-combined/ (accessed June 4, 2022).

Zhouxiang, Zhang. "The Mystery of 336 U.S. Bio-Labs Worldwide." *China Daily,* March 10, 2022. https://global.chinadaily.com.cn/a/202203/10/WS622942dfa310cdd39bc8b92f.html (accessed May 21, 2022).

Interviews and Lectures

Benally, Hataali Jones. Interview with world-renowned Diné healer, world champion hoop dancer, horse rider, and gentle Indigenous teacher, October 7, 2021, Northern Arizona.

Cone, James. Class on Martin and Malcolm. Pacific Lutheran Theological Seminary. Berkeley, CA, Fall 1981.

Index

Numbers in ***bold italics*** indicate pages with illustrations

Abu-Jamal, Mumia 159, 195
Achuar Indians **93**; *see also* Amazon rainforest; genocide
Acoma 17, 177, 197
Afghanistan 47, 99, 102–103, 107, 158, 191, 200, 202–203
Africa 3, 5, 11, 14, 17, 19, 24–25, 29–30, 32, 34–38, 40–42, 44, 45–50, 52–53, 58, 62, 65, 67–68, 70–72, 87, 95, 102, ***104***, 107, 114–117, 123, 134, 142–143, 150–152, 158–159, 161, 164–167, 170, 172–173, 177–182, 187–190, 195–201, 204, 212, 216, 218–221; *see also* AFRICOM; Association of African Women Research and Development (AAWORD); Azania/South Africa; Biko, Steve; Black Consciousness Movement; Burkina Faso; *Chama Cha Mapinduzi* (CCM); Chikoko Movement; Congo; Dagara; Diallo, Yaya; Ebola pandemic; Egypt; Elmina; Ethiopia; Gaddafi, Muammar; Herero; Libya; Lumumba, Patrice; Mengistu; Minianka; Movement for the Survival of the Ogoni People (MOSOP); Nama; Niger Delta; Nigeria; Nkrumah, Kwame; Sankara, Thomas; Somalia; Somé, Malidoma; Tanzania; Zimbabwe
AFRICOM (US Africa Command) 32, 102, 158; *see also* Africa, America in Africa
Agent Orange 46, 108, 186, 201; *see also* Chi Endé (Apache); genocide; Vietnam
"AI" 32, 71, 127
AIDS (Acquired Immunodeficiency Syndrome) 70, 159, 191, 195

Ainu (Indigenous people) 170; *see also* genocide; Japan
Air Force Weather Agency (AFWA) 105
airborne microplastics (AMP) 67
Al-Amin, Jamil Abdullah 159
Algonquian language 49
al-Idrisi 155
Amazon (rainforest) 31, 40, 93, 160; *see also* Achuar Indians; Brazil; ecocide; genocide; Guahibo Indians; Kichwa Indians; Makiritare Indians; Pemon Indians; Tagaeri Indians; Waica Indians; Waorani Indians; Warao Indians; Wayuu Indians; Yanomami Indians; Yasuní preserve
Amazon corporation 53
America in Africa ***104***
Anangu 40; *see also* Australia; genocide
Anishinaabeg 3; *see also* genocide
Aotearoa 43, 126, 134
Apple (corporation) 53, ***60***, 182, 222; *see also* FoxConn
Arawaks 33; *see also* genocide; sex trafficking
Aristide, Jean-Bertrand xii, 36; *see also* Caribbean; Haiti
Armswatch 207; *see also* bio-weapons labs; Gaytandhzieva, Dilyana
Association of African Women Research and Development (AAWORD) 19; *see also* Africa
asthma 83, 185, 203
Atomic Energy Commission 99; *see also* atomic tests; Bikini Atoll; Enewetak Atoll
atomic tests 99, ***105***; *see also* Bikini Atoll; Enewetak Atoll; Hiroshima; Manhattan

Project; "Mike"; Nevada Test Site; Operation Fishbowl; Operation Sandstone
Audubon Report (2014) 71
Australia 15, 40, 44, 51, 69, 71, 110–***111***, ***120–121***, 126, 130, 134, 149, 161, 172, 186–187, 195–196, 199, 201, 216, 221–222, 217; *see also* Anangu; *Buwurr*; ecocide; genocide; Gordon, Willie; Jawoyn; Mabo, Edward Koiki; police brutality; *Uluru Kata-Tjuta*; Yawuru; Yimithirr, Guugu
Aymara 128, 168, 177, 192; *see also* Morales, Evo; *Movimiento al Socialism* (MAS)
Aztec 30, 38, 67; *see also* genocide; Spain

Babad Dawag (Frog Mountain). 128; *see also* O'odham, Tohono
Bank of America 162–163, 165; *see also* military industrial complex; Wall Street
Bank of Japan 165
Barclays 163; *see also* financial crimes
Bayer 68, 71, 210; *see also* ecocide; ecological collapse
Benally, Hataali Jones 2, 18, 22, 23, 127, 128, 150, 171, 222; *see also* Diné
Beothuk Indians 33; *see also* genocide
Bezos, Jeff 159, 162
Biden, Hunter 207–208, 210, 216
Biden, Joe 184, 191–192, 202, 205
Big Bang theory 7
Big Pharma 159, 221; *see also* BioNTech; Moderna; Pfizer
Bikini Atoll ***104–105***; *see also* Atomic Energy Commission;

atomic tests; Enewetak Atoll; genocide
Biko, Steve 48, 151, 190, 195; *see also* Africa; Azania/South Africa; Black Consciousness Movement
BioNTech 158; *see also* Big Pharma;Covid-19
bio-weapons labs (U.S.) 123, 203, 216; *see also* Armswatch; Gaytandhzieva, Dilyana
Black Agenda Report 170, 191–192, 207
Black Alliance for Peace 203
Black Consciousness Movement 151; *see also* Africa; Azania/South Africa; Biko, Steve
Blackfeet 12, 40, 174, 196
BlackRock (asset management conglomerate) 62, 162; *see also* military industrial complex
Boeing corporation 103, 162; *see also* military industrial complex
Bolivia 34, 40, 128, 167–168, 192, 199, 204, 216; *see also* General Motors; lithium; Lithium Americas Corp; lithium mining desecration; Morales, Evo; *Movimiento al Socialismo (MAS)*; Musk, Elon; *PeeMuh'huh* (Thacker Pass)
Brazil 35, 37, 40, 41–42, 86, 138, 150, 179; *see also* Amazon rainforest; BRICS; police brutality
BRICS (Brazil, Russia, India, China, and South Africa) 216; *see also* Brazil; China; India; Russia; South Africa
Britain 35, 42, 36–37, 43–45, 47, 50, 70, **104**, 196; *see also* genocide; Intra-European War; NATO
British Balfour Declaration 45; *see also* Britain; *Nakba*; Palestine
Brooklyn Navy Yard Incinerator 83
Buffett, Warren 162
Bureau of Labor Statistics Data on Monthly Job Quits **146-147**
Burkina Faso 36, 47–48; *see also* Africa
Buwurr (Dreaming or Dreamtime) 15; *see also* Australia; genocide; Jawoyn

California Virtual Academies (CAVA) 143
call centers **50**
Cambodia 47, 63, 99, 103, 186; *see also* genocide
Canadian Natural Resources Limited (CNRL) 85
"Canfield Ocean" **122**
Capital Group (investment company) 162; *see also* military industrial complex
Caribbean 30, 32–33, 34–36, 38, 41, 43, 46–47, 52, 68, 70, 107, **121**, 164, 178, 180–181, 187, 198, 204, 208; *see also* Arawak; Aristide, Jean-Bertrand; Caribs; Castro; Césaire, Aimé; Columbian; Columbus; Cuba; de las Casas, Bartolomé; genocide; Guantanamo Bay; Haiti; Marley, Bob; sex trafficking; Taíno
Caribs **33**; *see also* Caribbean; genocide; sex trafficking
Castro, Fidel 44; *see also* Caribbean; Cuba
Catholic 28, 31–33, 184, 212; *see also* Doctrine of Discovery; genocide
CBPP (Center on Budget and Policy Priorities) 139–140, 212
CBS 60 Minutes 183, 202; *see also* Iraq
Cenovus Energy 85
CEOs 48, 80, 133, 139, 143, 162
Césaire, Aimé 151; *see also* Caribbean
Chama Cha Mapinduzi (CCM) 150; *see also* Africa; Tanzania
Chavez, Cesar 53, 180, 219; *see also* farm workers; United Farm Workers
Chavez, Hugo 67, 168; *see also* Venezuela
Chavista 167; *see also* Chavez, Hugo; Venezuela
Chemical Warfare Service 99; *see also* Atomic Energy Agency; atomic tests; Bikini Atoll; Enewetak Atoll; genocide; Hiroshima; Manhattan Project; Marshall Islands; "Mike"; Nagasaki; Nevada Test Site
chemical weapons 98; *see also* Alfred Nobel
Cherokee 18; *see also* genocide
Chi Chil Bildagoteel (Oak Flat) 160, 166; *see also* Chi Endé (Apache)
Chi Endé (Apache) 46, 160; *see also* genocide
Chiapas 166–167; *see also* Mexico
Chico River Hydro 172; *see also* IPMSDL; Kalinga; the Philippines
China 34, 41–43, 45, 47, 53, 56, 66, 70, 86; 102, **104**, 107, 110, 112–**114**, 138, 150, 158, 165–166, 180–182, 186–187; 192, 196–197, 202, 205, 216, 218, 222; *see also* BRICS; Chinese Communist Party; opium trade; Yangtze River; Yellow River; Yùlóng Xuěshān Mountain
Chinese Communist Party 165; *see also* China
Chitwan National Park 40; *see also* Sherpa
chlorofluorocarbon 121
Citigroup 163–164; *see also* Clinton, Bill; financial crimes; Rubin, Robert; Wall Street
Civil Society Organizations (CSOs) **94**; *see also* Nigeria
Clark Naval Base 42; *see also* the Philippines
Clean Air Act 85
Clean Water Act 77
Clinton, Bill 164–165; *see also* Citigroup; Rubin, Robert
CO_2 emissions 108
Cold War 98, 102, 165, 190
college debt 134, **141**-143
Colorado River 23, 68, 79, 209
Columbian 3, 33–34, 37, 178, 200; *see also* Caribbean; genocide; sex trafficking
Columbus 26, 28, 32–**33**, 34–35, 38, 41, 178, 180, 196–197, 200–201, 205; *see also* Caribbean; genocide; sex trafficking
Colville Indians, Confederation of 1; *see also* genocide
Compagnie des Indes Orientales 29
Congo 44, 46, 48, **61**, 95, 107, 182, 200; *see also* Africa; genocide
COP 26 summit, 2021 107
Corporatization of K-12 Education 143
Correa, Rafael 168; *see also* Cortez 34; Ecuador; genocide
Covid-19 3, 20, 44, 53, 70, 75, 112, 126, 145, 148, 157, 158, 170, 191–192, 197, 202, 214, 216, 221; *see also* Big Pharma; BioNTech; Malone, Robert; Moderna; Pfizer
Cree **3**, 155, 171–172; *see also* genocide
Cuba 44–45, 114, 138, 158, 166, 182, 213; *see also* Caribbean; Guantanamo Bay; Maceo, Antonio

Index

Dagara 24, 152, 155; *see also* Africa; Somé, Malidoma
Dangarembga, Tsitsi 151; *see also* Africa; Zimbabwe
Dann, Mary **138**; *see also* decolonization; genocide; Shoshone; Treaty of Ruby Valley of 1823
"Davos Man" 149; *see also* Schwab, Klaus; World Economic Forum
Decolonization 19, 45, 48, 129, **138**, 151, 153, 162, 171–172, 179, 184, 192–193, 195, 198, 201, 212
Deepwater Horizon oil spill 88, 91–92
Deer, Lame 24, 169, 192, 198; *see also* Lakota
de las Casas, Bartolomé 26, 34, 178, 180, 198; *see also* Caribbean
Deloria, Vine 16, 20–22, 153, 168, 177–179, 190, 192, 196
derivatives 163, 165, 191, 211, 213–214; *see also* Bank of America; Citigroup; Goldman Sachs; Great Recession; JP Morgan Chase; Morgan Stanley; US Bancorp; US Federal Reserve; Wall Street; Wells Fargo
Dessalines 36; *see also* Haiti
Diallo, Yaya 161–162, 191; *see also* Africa; Minianka
Diné 8, 18, 19, 77, 78. 171, 177, 222
DNA 20, 178; *see also* Ho, Mae-Wan
Doctrine of Discovery 32, 179, 199; *see also* genocide
Dogon 12; *see also* Africa; Diallo, Yaya
Dook'oos'liid (San Francisco Peaks) 160
Dow (corporation) 71
Drake, Edward L. 56
Dupont (corporation) 71
Dutch East India Company 29, 42–43; *see also* Africa; Azania/South Africa
Duterte, Rodrigo 42

earthquakes 75, 80, **81–83**, **104**, 185, 203, 206, 208, 216–218; *see also* global weather mega-disasters
Ebola pandemic 159, 195; *see also* Africa
ecocide 4, 20, 21, 29, 5, 55, 57, 67–70, 72, 93, **111**, 126–127, 129; *see also* Amazon rainforest; Australia; wildfires
ecological collapse 96, 108, 128

Ecuador **61**, **93**, 168, 208; *see also* Correa, Rafael; OPEC; Yasuni preserve
EdisonLearning 143
Egypt 44, 103, 138, 195, 204; *see also* Africa
electromagnetic radiation (EMR) 125
Elmina 41; *see also* Africa
Enbridge Pipeline **3**; *see also* Standing Rock
Encomienda system 38
Energy Information Administration (US) 59, 65, 74, 183, 220
"energy security" **107**; *see also* International Energy Agency
Enewetak Atoll **104**; *see also* atomic tests; Bikini Atoll; genocide; Hiroshima; Manhattan Project; "Mike"; Nagasaki; Nevada Test Site; Operation Fishbowl; Operation Sandstone
English East India Trading Company 29
Enslavement of Indigenous and African people 33–41
Environmental Defense Fund 66
environmental racism 77–**78**, 83; ERCOT (Electric Reliability Council of Texas) 118
Ethiopia 44, 68, 167, 184, 218; *see also* Africa
Euro-Christianity 54
European Central Bank (ECB) 165
Everglades National Park 40; *see also* genocide
exemptions (oil and gas) 7; U.S. EPA (Environmental Protection Agency)
Extermination 4, 40–41, 44, 47, 57, 63, 70, 180; *see also* ecocide; ecological collapse
extreme heat waves 110–**111**, 216
Exxon Mobil 62
Exxon Valdes oil spill 85

Fanon, Frantz 151, 190, 197
FAO (Food and Agricultural Organization) 31
farm workers 39, 52–53, 182–183, 208, 219; *see also* Chavez, Cesar; United Farm Workers
Federated States of Micronesia (FSM) 102
Fidelity (investment company) 162; *see also* military industrial complex; Wall Street

financial crimes 163; *see also* Barclays; Citigroup; J.P.Morgan Chase; Royal Bank of Scotland; Union Bank of Switzerland (UBS)
Fisher River Cree Nation 171; *see also* genocide
5G 125, 188, 206, 216
Floyd, George 159
Foundation to Promote Open Society 132; *see also* Soros, George
Foxconn (company) 53; *see also* Apple corporation
"Fraccidents" **76**
fracking 5, 55, 68, **73–74**, **75–77**, 79–**80**, **81–82**, **84**, 85, **93**, 127, 132, 134, 184–185, 203–204, 206–208, 216–218
France 27, 30, 32, 35, 44, 45, 47, 56, 117, 179; 181; 184; *see also* Intra-European war; Haiti
Freire, Paulo 150, 190, 197
Fukushima Daiichi nuclear plant **104**, **112**, 220

Gaddafi, Muammar 48; *see also* Libya; Africa
Gadsen Purchase 9
Garner, Eric 159
Gates, Bill 67, 125–126, 132, 159, 162, 188, 207, 209, 211, 220
Gates Foundation 126
Gaytandhzieva, Dilyana 184, 204, 206; *see also* Armswatch; bio-weapons labs (U.S.)
Gaza 46, 64, 182–183, 191, 202, 205, 207, 217; *see also* British Balfour Declaration; genocide; *Nakba*; Palestine
General Dynamics 162; *see also* military industrial complex
General Motors 40, 128; *see also* Lithium Americas Corp.; lithium mining desecration
Generation Zapped 125, 188, 207
genocide 4, 25, 33, 39, 45–46, 70, 134, 179–180, 195–196, 198–201, 207, 222; *see also* Achuar Indians; Ainu; Aotearoa/New Zealand; Arawaks; Atomic Energy Agency; atomic tests; Australia; Aztec; Bikini Atoll; Britain; Cambodia; Caribs; Chemical Warfare Service; Cherokee; *Chi Chil Bildagoteel* (Oak Flat); Chi Endé; Columbian; Columbus; Congo; Cortez; Diné; Enewetak Atoll; Fish River Cree Nation; Fisher River Cree Nation; Gaza; Guahibo Indians;

Index

Guugu Yimithirr people; Haudenosaunee; Havasupai; Hawaii; Herero; Hiroshima; Hopi; Iraq; Iroquois; Jawoyn; Kichwa Indians; Kiowa; Korea; Lakota; Laos; Maidu Indians; Manhattan Project; Marshall Islands; "Mike"; Miwok Indians; Mohawk; Mole Lake Ojibwe Nation; Muskogee; Nagasaki; Nama; Néhiyaw First Nation; Nevada Test Site; Nez Perce/ Niimiipuu; Nukini Indians; Ohlone; Oneida; Operation Crossroads; Operation Fishbowl; Operation Sandstone; *PeeMuh'huh* (Thacker Pass); Pemon Indians; Pizarro; Potawatomi; San *see* Everglades National Park; Seminole Nation; Shoshone; Spain; Tacana Indians; Tadodaho; Tagaeri Indians; Taíno; Taromenane Indians; "Teapot" 99; *terra nullius*; Tohono O'odham; Tule River Indian Nation; US peace treaties; Vietnam; Waica Indians; Wampanoag; Waorani Indians; Warao Indians; Wayuu Indians; Wixarika: Makiritare Indians; Yanomami Indians; Yellowstone National Park; Yosemite National Park; Yuchie
Geode Capital (asset management and retirement investment company) 162; *see also* military industrial complex; Wall Street
German Democratic Republic (167); *see also* Germany
Germany 32, 45, 56, 71, 99, 98, **104**, 109, **114**, 117, **138**, 190, 200; *see also* NATO
Glacier National Park 40
glasnost 167; *see also* Gorbachev, Mikhail; *perestroika*; Russia
Glen Canyon Dam **2**
global weather mega-disasters **111–119**; *see also* earthquakes; ecocide; wildfires
GMO (genetically modified organism) 20, 67, **100**, 127, 178, 184, 186, 198, 200
Gold King Mine spill 77–**78**, 185, 202, 206, 212
"gold rush" 39
Goldman Sachs 165; *see also* Wall Street
Gorbachev, Mikhail 167; 192; 211; *see also glasnost*; *perestroika*; Russia
Gordon, Willie 15, 110; *see also* Australia; genocide; Guugu Yimithirr people
Grand Canyon 19; National Park 40
Great Depression 64, 165
Great Recession 64, 139, 145, 163; *see also* US Federal Reserve; Wall Street
greenhouse gases **58**, 108–109, 124, 188, 220; *see also* US Department of Defense
Green Movement 66
Grist 134
Guahibo Indians 167; *see also* Amazon rainforest; genocide; Venezuela
Guantanamo Bay 44, 160
Guanxci 165; *see also* China
Gulf of Mexico **58**, **62**, **88**, **92**
Guugu Yimithirr people 15, 110; *see also* Australia; genocide; Gordon, Willie

HAARP (High Frequency Active Auroral Research Program) 123, 188, 205, 207, 213, 219
Haiti 32, 34–36, 43, 45, 107, 117, 180, 195, 197, 200, 203, 214; *see also* Aristide, Jean-Bertrand; France; US CIA
Haudenosaunee 1, 8, 214; *see also* genocide
Havasupai 40; *see also* genocide
Hawaii 46; *see also* Facebook; genocide; Liliuokalani; Zuckerberg, Mark
Herero 46, 200; *see also* genocide
Hiroshima 63, 99, 105, 183, 221; *see also* genocide
Ho, Mae-Wan 20; *see also* DNA
Hokkaido Island people 170; *see also* Japan
Hopi 9, 18; *see also* genocide
Huichol 13–14; *see also* genocide
Hurricane Katrina 113, 160

Imperial Oil Limited 85
Incan empire 38
India 29, 44, 47, 50, 66–67, 86, 107, **118**–119, 183, 188, 209, 216, 221; *see also* BRICS; Dutch East India Company
Indonesia 42, **61**, 69, 71, 107, **113**, **115**, 187, 202
Inquisition 28, 32, 42, 179, 197–200
insect collapse 55, 68, 70, 110, 183, 208

International Energy Agency (IEA) 107, 187, 209
International Indigenous Peoples' Movement for Self Determination and Liberation (IPMSDL) 95, 181, 186, 202, 208
International Union for Conservation of Nature and Natural Resources Red List (IUCN) 70
International Whaling Commission 56
Intra-European war 45, 63, **65**, 99, **101**–103, 107, 164, 167
Iran 53, 56, 58–59, 61–62, 64, 107, 158, 188, 205, 215; *see also* OPEC
Iraq 45, 47, 56–58, 62–63, 69, 86, 103; *see also* CBS 60 Minutes; genocide; OPEC
Iroquois 160, 169; 200; *see also* genocide
ISIS 64
Islamic Revolution 60, 64; *see also* Iran

Japan 44–45, 56, 66–67, **86**, 102, 107, **112**; 164–165, 170, 210, 219; *see also* Ainu people; Fukushima Daiichi nuclear plant; Hiroshima; Nagasaki
Jawoyn 15, 195; *see also* Australia; genocide
Jefferson, Thomas 38; university 133
Johnson, Lyndon B. 97, 103
JP Morgan Chase 162; *see also* financial crimes; military industrial complex; Wall Street

Kalinga 172, 192, 200; *see also* Chico River Hydro; International Indigenous Peoples' Movement for Self-Determination and Liberation (IPMSDL); the Philippines
Kanesatake Resistance 172; *see also* Kanyen'kehà:ka (Mohawk); police brutality
Kanyen'kehà:ka (Mohawk) 172; *see also* Kanesatake Resistance
Kenya 44, 214; *see also* Ngugi wa Thiong'o; police brutality
Keres 18; *see also* genocide
Khaldun, Ibn 27
Kichwa Indians **93**; *see also* Amazon rainforest; genocide
Kiowa 134, 175; *see also* genocide

Kiribati 55, 182, 204
Koch Industries 131–132, 188–189, 205
Korea 45, 47, 63, 86, 107, 114, 138, 150; *see also* genocide
Kosovo 47, 99, 107; *see also* NATO
Kuwait 45, 56, 61–62; *see also* OPEC
Kyoto Convention on Climate Change 106–107

Laguna 17–18, 177, 197
Lake Powell **2**, 68, 184, 216
Lakota 3, 23–24, 31, 40, 156, 179, 19; *see also* genocide
"Land Back" 172
Land Grab Universities 134; *see also* Morrill Act
Land Grant Universities 135–**136**; *see also* Morrill Act
Land, Soil, and Water Degradation 71
Laos 47, 63, 99, 103, 183, 186, 217
Lenin, Vladimir 165; *see also* Russia; Russian Bolshevik Revolution
Libya 44–45, 47, 61, 95, 107, 117, 199; *see also* Gaddafi
Liliuokalani 46, 102; *see also* Hawaii
lithium 23, 40, **72**, 128, 168, 216; *see also* Bolivia; General Motors; Lithium Americas Corp.; lithium mining desecration; *Movimiento al Socialismo* (MAS)
Lithium Americas Corp. 40; *see also* General Motors; lithium mining desecration
Lockheed Martin 103, 162; *see also* military industrial complex
Longview Asset Management 162; *see also* military industrial complex
L'Ouverture, Toussaint 36; *see also* Haiti
Lumumba, Patrice 48; *see also* Africa; Congo

Mabo, Edward Koiki 149; *see also* Australia; genocide
Maceo, Antonio 44; *see also* Cuba
Machu Picchu 40
Maidu Indians 10; *see also* genocide
Makiritare Indians 167; *see also* Amazon rainforest; genocide; Venezuela
Malone, Robert 20; *see also* Covid-19

mana (Maori) 172; *see also* Maori
Manhattan Project 99; *see also* atomic tests; Bikini Atoll; Enewetak Atoll; genocide; Hiroshima; Nagasaki; Nevada Test Site
manoomin 173; *see also* Mole Lake Ojibwe Nation
Maori 16, 172, 177, 201; *see also* genocide; *mana*; *rangatiratanga*
Marcos, Bongbong 42
Marley, Bob 1; *see also* Caribbean
Marshall Islands 12, 47, 102; *see also* genocide
Marx, Karl 5, 31, 41, 48, 49, 50–51, 54, 143, 156–157, 165, 168–169, 179, 181–182, 191–192, 195–196, 198–199, 206, 213
Massachusetts Financial Services Company 162; *see also* military industrial complex
Mayan 18, 23, 67; *see also* genocide
Mengistu (Haile Mariam) 167; *see also* Ethiopia
Meta 162; 172; *see also* Hawaii; Zuckerberg, Mark
methane emissions **31**, **81**, **84**–85, 123–**124**, 185, 188
Mexico 10, 13, 34, 38–39, 53, 61, 64, 67, **114**, 117, 160, 166–167; *see also* Chiapas
MI6 **60**
"Mike" **105**; *see also* Bikini Atoll; Enewetak Atoll; genocide; Hiroshima; Marshall Islands; Nagasaki; Nevada Test Site; Operation Crossroads; Operation Fishbowl; Operation Sandstone; "Teapot"
military industrial complex (U.S.) 58, 98, 99; *see also* Bank of America; BlackRock; Boeing; Capital Group; Fidelity; General Dynamics; Geode Capital; JP Morgan Chase; Lockheed Martin; Longview Asset Management; Massachusetts Financial Services; Morgan Stanley; Newport Trust Company; Northrop Grumman; RTX (formerly Raytheon); State Street; Vanguard
Mindanao 42; *see also* the Philippines
Minianka 161; *see also* Africa; Diallo, Yaya

misogyny 52
Miwok Indians 39; *see also* genocide
Moderna corporation 191, 214; *see also* Big Pharma; Covid-19
Mohawk 8, 30, 172; *see also* genocide; Kanesatake Resistance; *Kanyen'kehà:ka*
Mole Lake Ojibwe Nation 173; *see also* genocide; *manoomin*
Monroe Doctrine 45–46
Moorish 27, 32, 178, 199; *see also* Africa
Morales, Evo 128, 168, 192, 199, 204; *see also* Bolivia; General Motors; lithium; Lithium Americas Corp.; *Movimiento al Socialismo* (MAS)
Morgan Stanley 162; *see also* military industrial complex; Wall Street
Morrill Act 135, 138; *see also* *Land Grab Universities*; Land Grant universities
Mossadegh, Mohammed **60**; *see also* Iran
Movement for the Survival of the Ogoni People (MOSOP) 95; *see also* Africa; Chikoko Movement; Niger Delta; Nigeria
Movimiento al Socialismo (MAS) 168; *see also* Bolivia; General Motors; lithium; Lithium Americas Corp.; Morales, Evo; Musk, Elon
mRNA 20, 205, 207, 210–211
Muscovy Company 29
Musk, Elon 159, 162, 211, 216
Muskogee 22; *see also* genocide

Nagasaki 63, 99, 221; *see also* Atomic Energy Agency; atomic tests; Bikini Atoll; Enewetak Atoll; genocide; Hiroshima; Manhattan Project; Marshall Islands; "Mike"; Nagasaki; Nevada Test Site; Operation Crossroads; Operation Fishbowl; Operation Sandstone
Nairobi National Park 40
Nakba 45; *see also* Balfour Declaration; Gaza; genocide; Palestine
Nama 46; *see also* Africa; genocide
Napalm **105**, 204–205; *see also* Vietnam
Narragansett 49, 57; *see also* genocide
Natchez Nation 33; *see also* genocide

National Bureau of Economic Research **146**
National Education Association (NEA) 145, 147–**148**, 185, 197, 212–213, 221
national security state 108, 132, 186, 218; *see also* police brutality; US CIA
NATO (North American Atlantic Organization) 45, 47, 58, 66, 108, 121, 167, 182, 203; *see also* Kosovo; Serbia; Ukraine
Native American Graves Protection and Repatriation Act (NAGPRA) 137, 172, 197
Native American Team Names 30–31
Nature Conservancy 66
Néhiyaw First Nation 171; *see also* genocide
Nevada Test Site 99, 203; *see also* Atomic Energy Agency; atomic tests; Bikini Atoll; Enewetak Atoll; genocide; Hiroshima; Manhattan Project; Marshall Islands; "Mike"; Nagasaki; Operation Crossroads; Operation Fishbowl; Operation Sandstone; Teapot
New Mexico 10, 38, **75**, 77–**78**, **100**, 137, **141**, 145, **147**, 185, 197, 212
Newport Trust Company (asset management and retirement investment) 162; *see also* military industrial complex
Nez Perce/Niimiipuu 40, 171; *see also* genocide
Nguni **156**, 165, 170; 175, 177, *see also* South Africa/Azania
Nicaragua 128, 158, 167–168, 191, 208, 213
Niger Delta 62, **94**; *see also* Chikoko Movement; Nigeria; Ogoni people
Nigeria 18, 41, 44, **58**, **61**, **65**, **94**–95, 186, 200, 206, 210; *see also* Chikoko Movement; Civil Society Organizations (CSOs); Niger Delta; *Vanguard News Nigeria*
Nkrumah, Kwame 47, 151, 182, 190, 199; *see also* Africa
Nnobi 18
Nobel, Alfred 98; *see also* chemical weapons
Northrup Grumman 162; *see also* military industrial complex
Nukini Indians 40; *see also* genocide

off-shore oil drilling 92
Ogoni people **94**–95, 186, 200; *see also* Niger Delta
Ohlone 23; *see also* genocide
Ojibwe 3, 40, 173; *see also* genocide
Okanagan 14
Oneida 8, 177, 197, 214; *see also* genocide
Onkwehonwe 21
OPEC+ 60–**61**, 220
Opium 34; trade 43; trafficking 34; wars 43, 180; *see also* China
Organization of the Petroleum Exporting Countries (OPEC) 60; *see also* Iran; Iraq; Kuwait; Saudi Arabia; Venezuela
"Operation Crossroads" 104; *see also* Atomic Energy Agency; atomic tests; Bikini Atoll; Enewetak Atoll; genocide; Hiroshima; Manhattan Project; Marshall Islands; "Mike"; Nagasaki; Nevada Test Site; Operation Fishbowl; Operation Sandstone; Teapot
Operation Fishbowl 105; *see also* Atomic Energy Agency; atomic tests; Bikini Atoll; Enewetak Atoll; genocide; Hiroshima; Manhattan Project; Marshall Islands; "Mike"; Nagasaki; Operation Crossroads; Operation Sandstone; Teapot
Operation Sandstone 104; *see also* Atomic Energy Agency; atomic tests; Bikini Atoll; Enewetak Atoll; genocide; Hiroshima; Manhattan Project; Marshall Islands; "Mike"; Nagasaki; Operation Crossroads; Operation Fishbowl; Teapot
Orinoco Basin 168
Otto, Nikolaus 56
Ottoman empire 44–45
outsourcing **50**
"ozone smog" 83

Paga 153
Paha Sapa (the Black Hills) 40; *see also* Lakota
Palestine 45–46, 68; *see also* Gaza; genocide; *Nakba*; police brutality
Papua New Guinea 160, 166
Paris Agreement (COP 21) 68
Pasqua Salteaux First Nation 171; *see also* genocide

Paulson, Henry (Hank) 164; *see also* Goldman Sachs
Peace Treaties (US government) 160; *see also* genocide
PeeMuh'huh (Thacker Pass) 40, 128; *see also* genocide
Pemon Indians 167; *see also* Amazon rainforest; genocide; Venezuela
Pentagon **58**, 102, 103–**104**, 106, 108, **141**, 183, 186–187, 189, 201–202, 205, 207–208, 210–211, 218, 221; *see also* military industrial complex
Pequot 33, 57; *see also* genocide
perestroika 167
Pfizer corporation 158, 191, 214, 216; *see also* Big Pharma; Covid-19
Philadelphia Federation of Teachers Union 144
Philippines 41–42, 69, 95, 103, 172, 181–182; *see also* Clark Naval Base; International Indigenous Peoples' Movement for Self-Determination and Liberation (IPMSDL); Kalinga; Subic Bay
Phoenix program 63; *see also* CIA
Pilgrims 33, 42, 178, 200
Pizarro 34, 38; *see also* genocide
police brutality 4, 35, 58, 151, 159–160, 172–173, 187, 206; *see also* Australia; Brazil; Floyd, George; *Kanesatake* Resistance; Kenya; Palestine; South Africa/Azania; Standing Rock
Pope Alexander VI 32, 179, 21
Porton Down 70, 207; *see also* bio-weapons labs; Gaytandhzieva, Dilyana
Potawatomi 171; *see also* genocide
Potosi 42
pyrocumulonimbus cloud **120-121**

QE (Quantitative Easing) 165

rangatiratanga 172: *see also mana*; Maori
Renaissance (European) 5, 28, 98, 178, 199
Resources Legacy Fund 66
Rio Puerco River spill 77
Rockefeller 62, 159, 183; *see also* Standard Oil
Rodney, Walter 42, 151, 177, 181, 190, 200

Royal African Company 29
Royal Bank of Scotland 163; *see also* financial crimes
RTX Corporation (formerly Raytheon) 106, 125, 162, 221; *see also* military industrial complex
Rubin, Robert 164; *see also* Citigroup; Clinton, Bill
Russia 29, 45, 56, 60–*61*, 66, 85–*86*, *104*, *119*; *see also* BRICS; Russian Bolshevik Revolution
Russia Bolshevik Revolution 165; *see also* Russia

Sale, Kirkpatrick 28–29, 30–31, 178–179, 200
Salish School of Spokane 153, 171, 192, 217
San, Indigenous 40; *see also* genocide
San Quentin prison 160
Sankara, Thomas 48
Saudi Arabia 45, 58, *61–62*, 66, 104, 107, 158; *see also* OPEC; Saudi Aramco
Saudi Aramco 60, 62, 220; *see also* Saudi Arabia
Schwab, Klaus 149; *see also* "Davos Man;" World Economic Forum
Serbia 47, 99; *see also* NATO
Seven Sisters *60*–62, 183, 200
sex trafficking 34, 38; *see also* Arawaks; Caribbean; Caribs; Columbian; Columbus; Taíno
shale gas 58, *61*, *65*, *73–74*, 79–*80*, 82, 184–185, 206, 218
Shell corporation 62, 65, *80*, 82, 186, 200
Shenandoah, Leon 175; *see also* Tadodaho
sherpa 40; *see also* Chitwan National Park
Shona 16, 17; *see also* Tsitsi Dangeremgba; Zimbabwe
Shoshone 40, 46, 138; *see also* Dann, Mary; genocide; Treaty of Ruby Valley of 1823
Silicon Valley 39
Siy'aka 20
Sky Woman 8
Smith, Adam 157, 165, 191
solar panels toxic waste 109–110, 217
Somalia 68, 102, 107, 158; *see also* Africa; AFRICOM
Somé, Malidoma 24, 152, 178, 190, 201; *see also* Africa; Dagara
Soros, George 132; *see also* Foundation to Promote Open Society

South Africa (Azania) 40, 42, 44, 48, 62, 68, 104, ***114–116***, 134, 151, 172–173, 187, 218–220; *see also* BRICS; police brutality
Soviet Union 45, 98, 167; 192, 211; *see also* glasnost; Gorbachev, Mikhail; Lenin; perestroika; Russia; Russian Bolshevik Revolution
"Space Command" 108
Spain 27–28, 30, 32–*33*, 35, 38, 41–42, 87, *91*, 117, 178–179, 199, 221; *see also* Moorish
Spirit of Life 156
Standard Oil *61*–62; *see also* Rockefeller
Standing Rock 3, 169, 197, 205; *see also* Enbridge Pipeline; police brutality
State Street 162; *see also* military industrial complex; Wall Street
Subic Bay 42; *see also* the Philippines
Suncorp Energy 85
superfund toxic sites 106, 108

Tacana Indians 40; *see also* genocide
Tadodaho 175; *see also* genocide; Shenandoah, Leon
Tagaeri Indians *93*; *see also* Amazon rainforest; genocide
Taíno *33*, 204; *see also* Caribbean; genocide; sex trafficking
Tanzania 44, 150, 190, 204; *see also* Africa; *Chama Cha Mapinduzi* (CCM)
Taromenane Indians *93*; *see also* Amazon rainforest; genocide
Tasmania 43, 46, 181, 196, 199, 207
Taylor, Breonna 159
"Teapot" 99; *see also* Atomic Energy Agency; atomic tests; Bikini Atoll; Enewetak Atoll; genocide; Manhattan Project; Marshall Islands; "Mike"; Nagasaki; Nevada Test Site; Operation Crossroads; Operation Fishbowl; Operation Hiroshima; Operation Sandstone
Terra Nullius 32; *see also* genocide
Texaco corporation 60, 93
Thiong'o, Ngugi wa 151; *see also* Africa; Kenya
Three Mile Island 77

TINA syndrome (there is no alternative) *92*, 170
TNT (Trinitrotoluene) 62–63
Tohono O'odham Nation 9–10, 128, 219; *see also* genocide
Treaty of Ruby Valley of 1823 138; *see also* Dann, Mary; Shoshone
Tse-Tung, Mao 165
Tule River Indian Nation 39; *see also* genocide

Ujamaa (cooperative economics) 150; *see also* Africa; Tanzania
Ukraine 66, 99, 103, 121, 169, 182–184, 192, 202, 204–205, 207–208, 210, 216, 218, 220–221
Ukuhlamba-Drakensberg Park 40; *see also* Africa; Azania/South Africa
ultraviolet radiation 109, *122*, 123
Uluru-Kata Tjuta 40; *see also* Australia; genocide
UN Biodiversity Conference (2021) 70
UN Convention Convention on the Prohibition of Military or Any Other Hostile Use of Environmental Modification Techniques (ENMOD) *104*
UN Declaration on the Rights of Indigenous People 160
UNFCCC (United Nations Framework Convention on Climate Change) 69
Union Bank of Switzerland (UBS) 163; *see also* financial crimes
United Farm Workers (UFW) 53, 182, 219; *see also* Chavez, Cesar; farm workers
U.S. Bancorp. 163; *see also* Wall Street
U.S. Bureau of Mines 99
U.S. CIA (Central Intelligence Agency) *60*, 63, 103, 132, 189, 86, 102, 211–213
U.S. Defense Threat Reduction Agency (DTRA) 70; *see also* bio-weapons labs
U.S. Department of Defense (DOD) *58*, 101, 193
U.S. Department of the Treasury 191, 220
U.S. Environmental Protection Agency (EPA) *73*, *77–78*, *85*, *124*, 132, 185, 220; *see also* exemptions
U.S. Federal Communications Commission (FCC) 125
U.S. Federal Reserve 64, 163–

164, 192, 199; *see also* Great Recession; Wall Street
U.S. Food and Drug Administration (FDA) *122*, 188, 220
U.S. National Academy of Science 111; *see also* extreme heat waves
U.S. National Aeronautics and Space Administration (NASA) 41, *93–94*, *120*, *122*, *156*, 187, 189, 207–209, 212
U.S. National Oceanic and Atmospheric Administration (NOAA) 85–*87*, 88–*89*, *101*, 113, *118–119*, 183, 187, 213
U.S. National Security Agency 103; *see also* national security state
U.S. Occupational Safety and Health Administration (OSHA) 52
U.S. Office of the Comptroller of the Currency Report on Bank Trading and Derivative Activities 163
U.S. Oil Pollution Act 5, 55, 85, 86, 185, 186, 206, 220; *see also* Exemptions
UPS (United Parcel Service) 53

Vanguard (asset management and retirement investment company) 62, 165; *see also* military industrial complex; Wall Street
Vanguard News Nigeria *94*, 186
Venezuela 34, 38, 45, *61*, 107, 128, 158, 166–168, 192, 215; *see also* Brazil; genocide; Guahibo Indians; Makiritare Indians; Pemon Indians; Waica Indians; Warao Indians; Wayuu Indians; Yanomami Indians
Vietnam 46, 63, 99, 104–*105*, 183, 186, 204, 211–212; *see also* genocide; 2,4-D (Agent Orange)

Waica Indians 167; *see also* Amazon rainforest; genocide; Venezuela
Wall Street 157, 164–165, 188, 189, 191, 198–199, 211, 215, 220; *see also* Bank of America; BlackRock; Boeing; Capital Group; Citigroup; derivatives; Fidelity; General Dynamics; Geode Capital; Goldman Sachs; JP Morgan Chase; Lockheed Martin; Longview Asset Management; Massachusetts Financial Services; Morgan Stanley; Newport Trust Company; Northrop Grumman; RTX (formerly Raytheon); State Street; US Bancorp.; US Federal Reserve; Vanguard; Wellington; Wells Fargo
Wampanoag 57; *see also* genocide
Wampum 49, 182, 219
Waorani Indians *93*; *see also* Amazon rainforest; genocide
Warao Indians 167; *see also* Amazon rainforest; genocide; Venezuela
Wayuu Indians 167; *see also* genocide; Venezuela
Wellington (investment company) 162; *see also* military industrial complex; Wall Street
Wells Fargo 163, 165; *see also* Wall Street
whales, in crisis 56–57, 182–183, 203–205
WHO (World Health Organization) 70, 126, 207, 213, 217
wildfires 111, *114*, *120*, 187–188, 212, 216–217, 221; *see also* ecocide
Wixárika 13, 14; *see also* genocide
WMO (World Meteorological Organization) *118*, 222

Woodson, Carter G. 149
World Bank xvii, 37, 47, 65, 183, 218
World Economic Forum 149; *see also* Schwab, Klaus
World Trade Organization (WTO) 37
World War III *3*
Wuauquikuna 1, 216
WWF (World Wildlife Fund) 66, 184, 187, 211, 222

Xhosa 170; 177; *see also* Nguni; South Africa/Azania

Yahi 135; *see also* genocide
Yangtze River 166; *see also* China
Yanomami Indians 167; *see also* Amazon rainforest; genocide; Venezuela
Yasuní preserve *93*; *see also* Amazon rainforest
Yawuru 201; *see also* Australia; genocide
Yellow River 266; *see also* China
Yellowstone National Park 40; *see also* genocide
Yemen 66, 68, 158
Yoruba 11, 17, 177, 195; *see also* Africa; Nigeria
Yosemite National Park 39–40; *see also* genocide
Yuchie Language Project 153, *156*, 171, 175, 190, 192, 222
Yugoslavia 47, 99; *see also* NATO
Yùlóng Xuěshān Mountain 166; *see also* China

Zimbabwe 16–17, 177, 196, 212; *see also* Shona; Tsitsi Dangeremgba
Zuckerberg, Mark 163, 172, 192, 218; *see also* Facebook; Hawaii
Zuni 18; *see also* genocide

www.ingramcontent.com/pod-product-compliance
Ingram Content Group UK Ltd.
Pitfield, Milton Keynes, MK11 3LW, UK
UKHW050702160426
5217IPUK00038B/1927